Valentin Heller (Ed.)

Tsunami Science and Engineering

MDPI

This book is a reprint of the Special Issue that appeared in the online, open access journal, *Journal of Marine Science and Engineering* (ISSN 2077-1312) from 2014–2016 (available at: http://www.mdpi.com/journal/jmse/special_issues/ocean-tsunami).

Guest Editor
Valentin Heller
University of Nottingham
UK

Editorial Office
MDPI AG
Klybeckstrasse 64
Basel, Switzerland

Publisher
Shu-Kun Lin

Assistant Editor
Yueyue Zhang

1. Edition 2016

MDPI • Basel • Beijing • Wuhan • Barcelona

ISBN 978-3-03842-233-4 (Hbk)
ISBN 978-3-03842-218-1 (PDF)

Table of Contents

IV

List of Contributors

José M. Abril Departamento de Física Aplicada I, ETSIA, Universidad de Sevilla, Carretera de Utrera km 1, D.P. 41013 Seville, Spain.

Ines Alberico IAMC—Istituto per l'Ambiente Marino Costiero, CNR, Napoli 80133, Italy.

Grezio Anita Istituto Nazionale di Geofisica e Vulcanologia, sezione di Bologna, Via Franceschini 31, 40128 Bologna, Italy.

Aggeliki Barberopoulou National Observatory of Athens, Institute of Geodynamics, Lofos Nymphon, 11810 Athens, Greece; AIR Worldwide, 131 Dartmouth Street, Boston, MA 02116, USA.

Jörn Behrens CEN—Center for Earth System Research and Sustainability, Universität Hamburg, Grindelberg 5, Hamburg, Germany; Department of Mathematics, Universität Hamburg, Bundesstraße 55, Hamburg 20146, Germany.

Nicole Beisiegel CEN—Center for Earth System Research and Sustainability, Universität Hamburg, Grindelberg 5, Hamburg, Germany.

Thomas A. Bianchette Department of Oceanography and Coastal Sciences, School of the Coast and Environment, Louisiana State University, 1002Y Energy Coast and Environment Building, Baton Rouge, LA 70803, USA.

Gian C. Bremm Coastal Research Station, Lower Saxony Water Management, Coastal Defense and Nature Conservation Agency, An der Mühle 5, Norderney D-26548, Germany.

Vincenzo Di Fiore IAMC—Istituto per l'Ambiente Marino Costiero, CNR, Napoli 80133, Italy.

Ulrike Drähne Department of Mechanical Engineering, Faculty of Engineering Science, University College London, Gower Street, London WC1E 6BT, UK.

Frederic M. Evers Laboratory of Hydraulics, Hydrology and Glaciology (VAW), ETH Zurich, CH-8093 Zürich, Switzerland.

Edison Gica Pacific Marine Environmental Laboratory, Center for Tsunami Research, National Oceanic and Atmospheric Administration, 7600 Sand Point Way NE, Seattle, WA 98115, USA; Joint Institute for the Study of the Atmosphere and Ocean, University of Washington, 3737 Brooklyn Ave NE, Box 355672, Seattle, WA 98195-5672, USA.

Nils Goseberg Franzius-Institute for Hydraulic, Estuarine and Coastal Engineering, Leibniz University Hannover, Nienburger Straße 4, Hannover D-30167, Germany; Department of Civil Engineering, Faculty of Engineering, University of Ottawa, 161, Louis Pasteur St., A609, Ottawa, ON K1N 6N5, Canada.

Willi H. Hager Laboratory of Hydraulics, Hydrology and Glaciology (VAW), Swiss Federal Institute of Technology (ETH) Zurich, CH-8093 Zürich, Switzerland.

Valentin Heller Department of Civil and Environmental Engineering, Imperial College London, London SW7 2AZ, UK.

Michael P. Hickey Department of Physical Sciences, Embry-Riddle Aeronautical University, Daytona Beach, FL 32114, USA.

Roberta Iavarone Dipartimento di Architettura, Università degli Studi di Napoli Federico II, Napoli 80134, Italy.

Attila Komjathy NASA Jet Propulsion Laboratory, California Institute of Technology, Pasadena, CA 91109, USA.

Junji Koyama Division of Natural History Sciences, Graduate School of Science, Hokkaido University, N10 W8, Kita-ku, Sapporo, Hokkaido 0650810, Japan; Hyotanjima Scholorship, Sapporo, Hokkaido 151-1854630, Japan.

Mark R. Legg Legg Geophysical, 16541 Gothard St # 107, Huntington Beach, CA 92647, USA.

Kam-biu Liu Department of Oceanography and Coastal Sciences, School of the Coast and Environment, Louisiana State University, 1002Y Energy Coast and Environment Building, Baton Rouge, LA 70803, USA.

John Z. G. Ma Department of Physical Sciences, Embry-Riddle Aeronautical University, Daytona Beach, FL 32114, USA.

Ennio Marsella IAMC—Istituto per l'Ambiente Marino Costiero, CNR, Napoli 80133, Italy.

Terrence A. McCloskey Department of Oceanography and Coastal Sciences, School of the Coast and Environment, Louisiana State University, 1002Y Energy Coast and Environment Building, Baton Rouge, LA 70803, USA.

Ioan Nistor Department of Civil Engineering, University of Ottawa, 161 Louis Pasteur, Ottawa, ON K1N 6N5, Canada.

Raúl Periáñez Departamento de Física Aplicada I, ETSIA, Universidad de Sevilla, Carretera de Utrera km 1, D.P. 41013 Seville, Spain.

Paola Petrosino Dipartimento di Scienze della Terra, dell'Ambiente e delle Risorse Università degli Studi di Napoli Federico II, Napoli 80138, Italy.

Luigi Piemontese Dipartimento di Architettura, Università degli Studi di Napoli Federico II, Napoli 80134, Italy.

Simona Pierdominici Helmholtz Centre Potsdam, GFZ German Research Centre for Geosciences, Potsdam 14473, Germany.

Michele Punzo IAMC—Istituto per l'Ambiente Marino Costiero, CNR, Napoli 80133, Italy.

Laura Sandri Istituto Nazionale di Geofisica e Vulcanologia, sezione di Bologna, Via Franceschini 31, 40128 Bologna, Italy.

Torsten Schlurmann Franzius-Institute for Hydraulic, Estuarine and Coastal Engineering, Leibniz University Hannover, Nienburger Straße 4, Hannover D-30167, Germany.

Jacopo Selva Istituto Nazionale di Geofisica e Vulcanologia, sezione di Bologna, Via Franceschini 31, 40128 Bologna, Italy.

Daniela Tarallo IAMC—Istituto per l'Ambiente Marino Costiero, CNR, Napoli 80133, Italy.

Roberto Tonini Istituto Nazionale di Geofisica e Vulcanologia, Via di Vigna Murata 605, 00143 Roma, Italy.

Motohiro Tsuzuki Division of Natural History Sciences, Graduate School of Science, Hokkaido University, N10 W8, Kita-ku, Sapporo, Hokkaido 0650810, Japan.

Stefan Vater CEN—Center for Earth System Research and Sustainability, Universität Hamburg, Grindelberg 5, Hamburg, Germany.

Kiyoshi Yomogida Division of Natural History Sciences, Graduate School of Science, Hokkaido University, N10 W8, Kita-ku, Sapporo, Hokkaido 0650810, Japan.

About the Guest Editor

Valentin Heller is an Assistant Professor in Hydraulics in the Department of Civil Engineering at the University of Nottingham and a member of the Geohazards and Earth Processes Research Group. His research mainly concerns experimental and computational fluid dynamics into landslide-tsunamis (impulse waves), marine renewables and scale effects. Before moving to the University of Nottingham, Dr Heller held one of the prestigious Junior Research Fellowships at Imperial College London (2011-2014), a Research Fellowship at the University of Southampton (2008–2011) and a postdoctoral position at ETH Zurich (2008). He obtained his PhD at ETH Zurich for his work on landslide generated impulse waves in 2007. Dr Heller was honoured with the Harold Jan Schoemaker Award from IAHR in 2013 (for a review on scale effects published in the Journal of Hydraulic Research) and the Maggia Price 2004 for his diploma thesis on ski jump hydraulics.

Preface to "Tsunami Science and Engineering"

In 2013, I was approached by the editorial team of the *Journal of Marine Science and Engineering* to act as guest editor of a Special Issue on tsunamis. I was very keen to take on this project given that recent tragic events such as the 2004 Indian Ocean Tsunami and the 2011 Tōhoku Tsunami had triggered an increase in research activity in this field. The high interest in this Special Issue *Tsunami Science and Engineering* was reflected by the submission of 21 full length articles. A total of 12 articles were published from 2014 to early 2016 after the rigorous peer-review.

This book comprises all 12 contributions to this Special Issue. The overall aim of this collection is to mitigate the destruction of tsunamis and the negative effects they have on us and our environment. The articles cover a wide range of topics around tsunamis and reflect scientific efforts and engineering approaches in this challenging and exciting research field.

The first three articles address the generation and propagation phases of tsunamis. Heller and Hager (2014) review the significance of the so-called impulse product parameter P, which is believed to be the most universal parameter for subaerial landslide-tsunami (impulse wave) prediction, and they show how this semi-empirical parameter may be instrumental for preliminary hazard assessment. Evers and Hager (2015) show with subaerial landslides impacting a water body that mesh-packed slides result in similar large impulse waves as free granular material. This is very beneficial for a number of reasons including simpler experimental handling. Koyama *et al.* (2015) describe the difference of the earthquake activity in megathrust subduction zones around Japan, depending on whether an along-dip double segmentation or an along-strike single segmentation is encountered. They further show that earthquakes generated by the latter type result on average in twice as large tsunamis as earthquakes generated by an along-dip double segmentation, after identical seismic moments.

The interactions of tsunamis with the shore region and structures are addressed in the following three articles. Barberopoulou *et al.* (2015) show with historical nautical charts and digital elevation models how dredging, expansion of land and the creation of jetties modified the San Diego harbor over the last 150 years and show how these modifications may change the tsunami amplitudes and current speeds based on different tsunami scenarios. Drähne *et al.* (2016) address long wave dynamics on coasts and during on-land propagation, based on physical model tests in a novel wave facility with numerical simulations. The authors further successfully compare some of the findings with established analytical expressions. Bremm *et al.* (2015) investigate the time-history of forces on vertical free-surface-piercing structures exposed to long leading depression waves. The authors characterize and describe in detail the flow patterns during both wave run-up and draw-down and compare the measured total base force with analytically calculated values.

Two interesting tsunami hazard assessment approaches are then presented. Anita *et al.* (2015) propose a comprehensive and total probabilistic approach, in which many different potential source types concur (seismic events, slides, volcanic eruptions, asteroids, *etc.*), to define a total tsunami hazard. The authors apply this approach to two target sites namely the city of Naples and the island of Ischia in Italy. Alberico *et al.* (2015) present a GIS-aided procedure using the Papathoma Tsunami Vulnerability Assessment model of urban environments and use Naples as the target site to illustrate the method. A map shows the vulnerability status of Naples and reveals that approximately 21% of the area shows a very high tsunami vulnerability.

Two potential past tsunamis are addressed next. McCloskey *et al.* (2015) present evidence for the occurrence of a large localized tsunami in the Bay of La Paz, approximately 1100 years ago, based on a field study and core data. The authors suggest that the potential tsunami was triggered by the slumping of an island at the eastern edge of the bay and that the tsunami reached up to 3.6 m above mean water. Abril and Periáñez (2015) aimed at quantitatively assessing the potential role of tsunamis in the parting of the Mediterranean Sea in the context of the narrative of the Biblical Exodus, as previously suggested in a number of studies. The authors numerically model several "best case" scenarios and conclude that it is very unlikely that the investigated tsunamis were the potential cause of the parting of the sea.

The book finishes with two interesting articles showing that seismic tsunamis may affect the Earth's thermosphere hundreds of kilometers above sea level. Ma *et al.* (2015) solve generalized ion momentum and continuity equations and investigate how seismic tsunamis create an ionospheric dynamo electric field on the electron density and total electron content perturbations at an altitude of 150 - 600 km. The authors then apply the solution to two arbitrarily selected locations. Ma (2016) conducts similar theoretical work and describes how seismic tsunami-excited gravity waves may modulate the atmospheric non-isothermality and wind shears.

These brief summaries show the wide range of fascinating topics which are covered in this book. I hope that these articles contribute to the mitigation of the negative effects of tsunamis and inspire many future research activities in this important research field.

Valentin Heller
Guest Editor

A Universal Parameter to Predict Subaerial Landslide Tsunamis?

Valentin Heller and Willi H. Hager

Abstract: The significance of the impulse product parameter P is reviewed, which is believed to be the most universal parameter for subaerial landslide tsunami (impulse wave) prediction. This semi-empirical parameter is based on the streamwise slide momentum flux component and it was refined with a multiple regression laboratory data analysis. Empirical equations based on P allow for a simple prediction of wave features under diverse conditions (landslides and ice masses, granular and block slides, *etc.*). Analytical evidence reveals that a mass sliding down a hill slope of angle 51.6° results in the highest waves. The wave height "observed" in the 1958 Lituya Bay case was well predicted using P. Other real-world case studies illustrate how efficient empirical equations based on P deliver wave estimates which support hazard assessment. Future applications are hoped to further confirm the applicability of P to cases with more complex water body geometries and bathymetries.

Reprinted from *J. Mar. Sci. Eng.* Cite as: Heller, V.; Hager, W.H. A Universal Parameter to Predict Subaerial Landslide Tsunamis? *J. Mar. Sci. Eng.* **2014**, 2, 400–412.

1. Introduction

An important class of tsunamis is caused by mass movements including landslides, rock falls, underwater slumps, glacier calving, debris avalanches or snow avalanches [1–5]; these are commonly referred to as landslide tsunamis or landslide generated impulse waves. Landslide tsunamis (impulse waves) typically occur in reservoirs and lakes, fjords or in the sea at volcanic islands or continental shelves [2,6–9]. The term landslide tsunami is sometimes also applied to waves generated in reservoirs and lakes [10] even though the term impulse waves would then be more correct.

Irrespective of where these waves are caused, they are a considerable hazard and the total cumulative death toll of Unzen (1792), Ritter Island (1888), Vajont (1963) and Papua New Guinea (1998) alone is likely to exceed 22,100 [2,4,11]. Fortunately, landslide tsunamis (impulse waves) may nowadays be predicted with relatively high confidence (much better than an order of magnitude) by given slide parameters using both numerical [3,7,12–19] or physical [14,20–29] model studies.

Subaerial landslide tsunamis are particularly challenging to predict because the mass, initially located above the water surface, impacts the water body and may entrain a large amount of air. Slide velocities of up to 128 m/s were estimated based on slide deposits [30], generating highly turbulent landslide tsunamis (impulse waves) if they interact with a water body. Fortunately, susceptible areas are often monitored, giving prior warning of a potential subaerial landslide tsunami (impulse wave). An active prevention of the wave generation is difficult to achieve so that passive methods, including early warning, evacuation, reservoir drawdown, or provision of adequate freeboards of dam reservoirs, are applied. A prediction of wave features is essential for the success of these methods, which must be conducted frequently, particularly in lakes [8,31] and during the planning and operational phases of reservoirs [6,31,32].

A common method for the assessment of this hazard is to conduct a physical model study in the laboratory environment. Generic physical model studies [20–24,26–29] systematically vary parameters (slide properties, hill slope angle, water depth) and express the unknown wave parameters (amplitude, height, period) as a function of these "known" parameters. The most relevant parameters for subaerial landslide tsunamis (impulse waves) are shown in Figure 1. The developed empirical equations allow for first estimates of future events [8,31,33], and their application is often the most straightforward method if time is limited. The results also determine whether a more accurate prototype specific numerical [7,11,19] or physical [6,32] model study is required; these latter methods are costly and require considerable more time and resources.

Herein, the significance of the impulse product parameter P is reviewed, which is believed to be the most universal parameter for generic landslide tsunami (impulse wave) predictions. Useful analytical derivations based on P are deduced and it is illustrated how P greatly simplifies hazard assessment through real-world predictions.

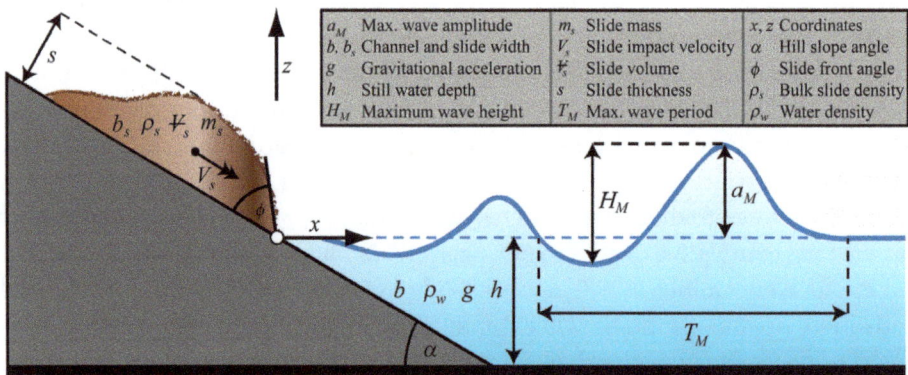

a_M	Max. wave amplitude	m_s	Slide mass	x, z	Coordinates
b, b_s	Channel and slide width	V_s	Slide impact velocity	α	Hill slope angle
g	Gravitational acceleration	Ψ_s	Slide volume	ϕ	Slide front angle
h	Still water depth	s	Slide thickness	ρ_s	Bulk slide density
H_M	Maximum wave height	T_M	Max. wave period	ρ_w	Water density

Figure 1. Definition sketch of subaerial landslide tsunami (impulse wave) generation (adapted from Heller and Hager [26], with permission from © 2010 American Society of Civil Engineers).

2

2. The Impulse Product Parameter

2.1. Derivation

The semi-empirical impulse product parameter P was developed by Heller and Hager [26] from subaerial landslide generated impulse wave model tests in a laboratory wave channel (2D) (see description of method in Appendix A). It is defined as

$$P = FS^{1/2}M^{1/4}\{\cos[(6/7)\alpha]\}^{1/2} \tag{1}$$

Equation (1) includes the slide Froude number $F = V_s/(gh)^{1/2}$, the relative slide thickness $S = s/h$, the relative slide mass $M = m_s/(\rho_w b_s h^2)$ and the hill slope angle α. Figure 1 shows all parameters required for these dimensionless numbers, namely the slide impact velocity V_s, gravitational acceleration g, still water depth h, slide thickness s, slide mass m_s, water density ρ_w and slide width b_s. The parameter P is based on the square root of the streamwise slide momentum flux component, involving the bulk slide density ρ_s and slide discharge Q_s, as [24]

$$(\rho_s Q_s V_s \cos\alpha)^{1/2} \approx (\rho_s s b_s V_s^2 \cos\alpha)^{1/2} = \rho_s^{1/2}s^{1/2}b_s^{1/2}V_s\cos^{1/2}\alpha \tag{2}$$

The expression on the right hand side in Equation (2) was further refined with a multiple regression data analysis based on hundreds of 2D experiments, resulting in the establishment of P [26]. This refinement was conducted to minimize the data scatter in the prediction of the wave parameters. The analysis revealed that the relative effects of V_s and s are correctly retained in P, as predicted in Equation (2). Since a vertically impacting slide with $\cos(90°) = 0$ would result in no impulse wave action, the term $\cos^{1/2}\alpha$ in Equation (2) was replaced by the empirical parameter $\{\cos[(6/7)\alpha]\}^{1/2}$ resulting in the smallest data scatter. The relative effect of $\rho_s^{1/2}$ in the slide mass $m_s = \rho_s \Psi_s$, with bulk slide volume Ψ_s, was with $\rho_s^{1/4}$ found to be less pronounced in the data analysis than predicted in Equation (2).

Figure 2 shows P *versus* the relative maximum wave height H_M/h resulting from the multiple regression data analysis. The average data scatter is ±30%, and the maximum scatter in the order of ±40% is considerable smaller than in previous studies [21,23], involving +100/−50% maximum scatter relative to the relative wave amplitude a/h or height H/h. In addition, the parameter P considers wider parameter ranges thereby applying to landslides and ice masses, to hill slope angles $30° \leqslant \alpha \leqslant 90°$, and to further conditions (Table A1). Heller and Spinneken [29] (see description of method in Appendix A) demonstrated that P, originally developed for granular slides, applies also to waves generated by block slides. Further, the study of Fuchs et al. [34] used P to describe underwater landslide characteristics. These diverse conditions under which P applies may establish P as the most universal parameter for landslide tsunami (impulse wave) hazard assessment.

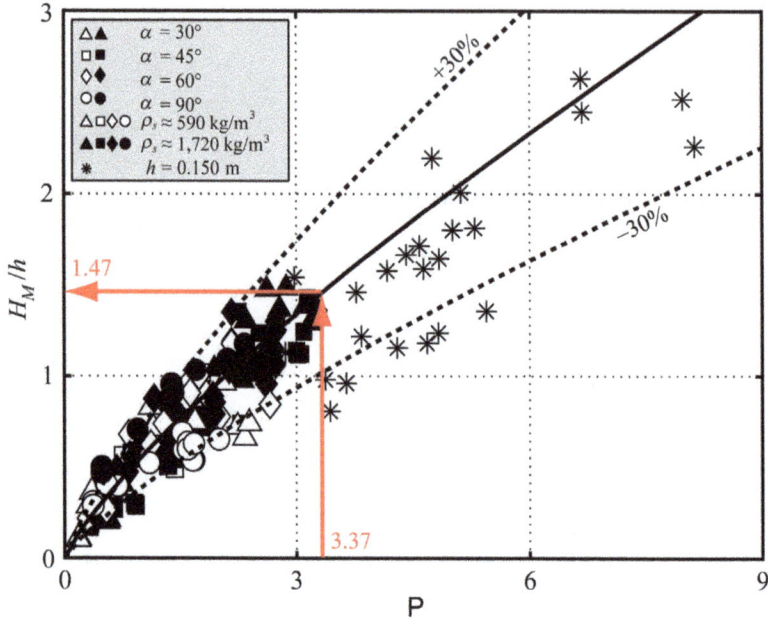

Figure 2. Example of prediction diagram based on P: Relative maximum wave height H_M/h *versus* P for granular slides and best fit $H_M/h = (5/9)P^{4/5}$ ($R^2 = 0.82$, Table 1) with $\pm 30\%$ lines including most data points; the red arrows show the prediction for the 1958 Lituya Bay case resulting in $H_M = 1.47h = 179$ m; data indicated with * are too low due to non-negligible scale effects (adapted from Heller and Hager [26], with permission from © 2010 American Society of Civil Engineers).

Table 1. Empirical equations based on P derived in 2D physical model studies by Heller and Hager [26] and Heller and Spinneken [29]; maximum wave height H_M and period T_M correspond to the identical wave and location x_M where the maximum wave amplitude a_M was measured; for block model slides the blockage ratio $B = b_s/b$ (0.88 − 0.98), the expression $\phi = \sin^{1/2}\varphi$ (0.71 − 1.00) considering the slide front angle φ and the expression $T_s = t_s/\{[h + V_s/(sb_s)]/V_s\}$ (0.34 − 1.00) considering the transition type are relevant, with $t_s =$ characteristic time of submerged landslide motion.

Wave parameter	Heller and Hager [26]		Heller and Spinneken [29]	
Slide type	granular		block	
Maximum amplitude	$a_M = (4/9)P^{4/5}h$	($R^2 = 0.88$)	$a_M = (3/4)[PB\phi T_s^{1/2}]^{9/10}h$	($R^2 = 0.88$)
Streamwise distance at a_M	$x_M = (11/2)P^{1/2}h$	($R^2 = 0.23$)	—	
Maximum height (Figure 2)	$H_M = (5/9)P^{4/5}h$	($R^2 = 0.82$)	$H_M = [PB\phi T_s^{1/4}]^{9/10}h$	($R^2 = 0.93$)
Maximum period	$T_M = 9P^{1/2}(h/g)^{1/2}$	($R^2 = 0.33$)	$T_M = (19/2)[PT_s^{1/2}]^{1/4}(h/g)^{1/2}$	($R^2 = 0.24$)
Amplitude evolution	$a(x) = (3/5)[P(x/h)^{-1/3}]^{4/5}h$	($R^2 = 0.81$)	$a(x) = (11/10)[P(x/h)^{-1/3}B\phi T_s^{3/4}]^{9/10}h$	($R^2 = 0.85$)
Height evolution	$H(x) = (3/4)[P(x/h)^{-1/3}]^{4/5}h$	($R^2 = 0.80$)	$H(x) = (3/2)[P(x/h)^{-1/3}B\phi T_s^{1/2}]^{9/10}h$	($R^2 = 0.89$)
Period evolution	$T(x) = 9[P(x/h)^{5/4}]^{1/4}(h/g)^{1/2}$	($R^2 = 0.66$)	$T(x) = (13/2)[P(x/h)^{5/4}T_s^{1/3}]^{1/4}(h/g)^{1/2}$	($R^2 = 0.53$)

Table 1 summarizes the most relevant empirical equations based on P, all derived from wave channel (2D) tests. This includes the maximum wave amplitude a_M (with its location x_M), height H_M and period T_M (Figure 1) and their evolutions $a(x)$, $H(x)$, $T(x)$, with distance x from the slide impact point for both granular [26] and block slides [29]. All these wave parameters simply contain P and the majority of these empirical equations result in large coefficients of determination $R^2 > 0.80$.

2.2. Analytical Aspects

The parameter P allows for the derivation of theoretical aspects which are as universally applicable as P itself. The slide centroid impact velocity V_s on a constant slope defined by α and the dynamic bed friction angle δ may be approximated with an energy balance between the slide release and impact location [30] as

$$V_s = [2g\Delta z_{sc}(1 - tan\delta cot\alpha)]^{1/2} \tag{3}$$

The parameter Δz_{sc} is the slide centroid drop height distance between slide release and the impact location. Equations (1) and (3) result with $A = (2\Delta z_{sc})^{1/2}s^{1/2}[m_s/(\rho_w b_s)]^{1/4}/h^{3/2}$ and $f(\alpha) = (1 - tan\delta cot\alpha)^{1/2}\{cos[(6/7)\alpha]\}^{1/2}$ in

$$P = A(1 - tan\delta cot\alpha)^{1/2}\{cos[(6/7)\alpha]\}^{1/2} = Af(\alpha) \tag{4}$$

Figure 3 shows $f(\alpha)$, for a typical dynamic bed friction angle range of $10° \leqslant \delta \leqslant 35°$, *versus* the hill slope angle α. The function $f(\alpha)$ is proportional to P and as such directly proportional to the maximum wave amplitude a_M, height H_M and period T_M and their evolutions $a(x)$, $H(x)$, $T(x)$ with distance x (Table 1). Figure 3 reveals that the hill slope angles resulting in maximum P are in the range $39.1° \leqslant \alpha_{max} \leqslant 65.1°$. As expected, no impulse wave is generated for $\alpha < \delta$ where the slide remains at rest. Further, α_{max} corresponding to the maximum P value increases with increasing δ. For a hill slope angle $\alpha \to 90°$ (glacier calving, rock fall), the effect of the friction angle δ is negligible so that $f(\alpha) = 0.47$, irrespective of the value of δ.

5

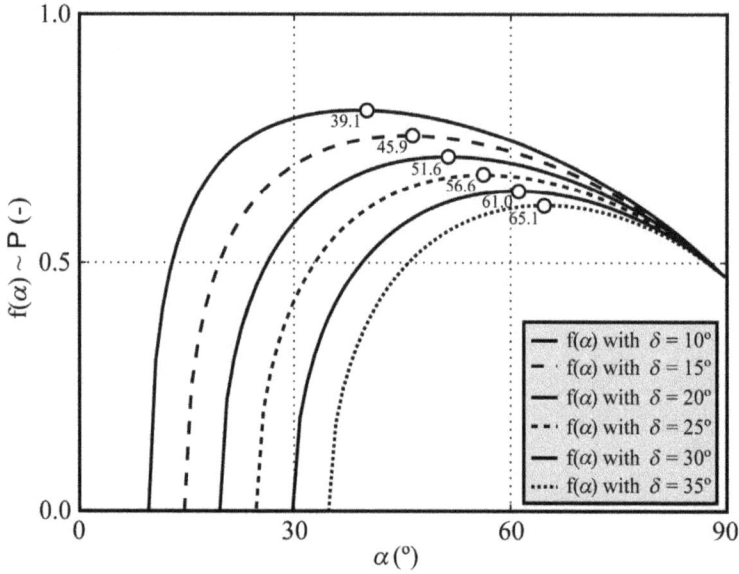

Figure 3. $f(\alpha) \sim P$ according to Equation (4) for different dynamic bed friction angles δ (°); the typical value $\delta = 20°$ results in a slope angle $\alpha_{max} = 51.6°$.

The values for α_{max} are analytically derived by differentiating Equation (4) with respect to α and set to zero, resulting in

$$(\tan\delta/\sin^2\alpha) = (1 - \tan\delta\cot\alpha)\tan[(6/7)\alpha]6/7 \qquad (5)$$

For the typical value $\delta = 20°$, Equation (5) results in $\alpha_{max} = 51.6°$ under otherwise constant conditions, which is in agreement with Figure 3. Impulse waves at $\alpha \approx 50°$ were indeed witnessed, namely in the Alps [8]. Mass movements on such steep mountain flanks, with α close to α_{max}, in combination with the confined water body geometries of lakes and reservoirs, contribute to the high relevance of landslide generated impulse waves in mountainous regions. The probability for landslides is typically highest at $\alpha = 36°$ to $39°$ [35]. Even then, only about 10% smaller waves may be expected (for $\delta = 20°$) as compared with the maximum due to the moderate change of $f(\alpha)$ with α between $35°$ and $75°$ (Figure 3).

3. Real-World Applications

The well documented 1958 Lituya Bay case [1,16,36] shown in Figure 4 is used to provide an example of a real-world prediction based on P. The T-shaped Lituya Bay is located near the St. Elias Mountains in Alaska, where the main bay is about 12 km long and 1.2 km to 3.3 km wide, except for the 300 m wide exit to the Pacific Ocean. On 9 July 1958, an 8.3 moment magnitude earthquake initiated a rock slide with a grain density of $\rho_g = 2700$ kg/m^3 sliding from a maximum altitude of 914 m above sea level on a slope of $\alpha = 40°$ (Figure 4). The parameters in P are taken from Heller and Hager [26], namely the slide impact velocity $V_s = 92$ m/s, still water depth $h = 122$ m, maximum slide thickness $s = 92$ m, mean slide width $b_s = 823$ m and slide mass $m_s = 82.62 \times 10^9$ kg. These result in a slide Froude number F = 2.66, a relative slide thickness $S = 0.75$, a relative slide mass $M = 6.74$, a hill slope angle $\alpha = 40°$, so that P = 3.37. The rock slide at Lituya Bay generated an impulse wave with a maximum run-up height of $R = 524$ m on the opposite shore at a distance $x \approx 1350$ m and a run-up angle $\beta = 45°$.

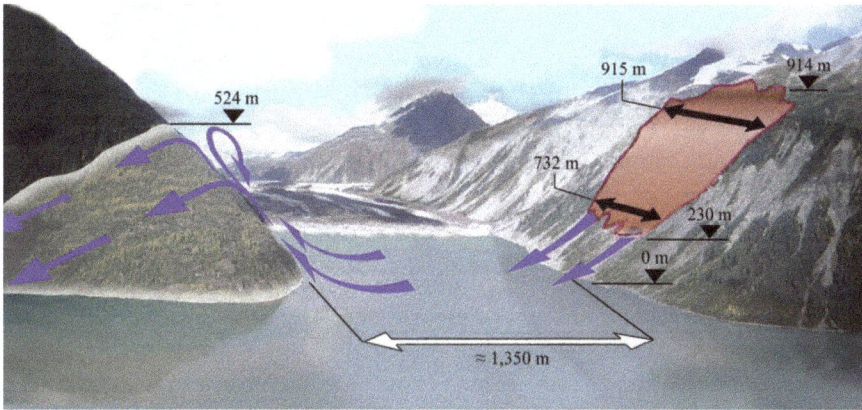

Figure 4. Artist's impression of 1958 Lituya Bay rockslide generating a tsunami of ~162 m in height destroying forest up to maximum run-up height of 524 m (adapted from Heller and Hager [26], with permission from © 2010 American Society of Civil Engineers).

Fritz *et al.* [36] investigated the Lituya Bay case in a physical model study and measured an up-scaled wave height of $H = 162$ m and a wave amplitude of $a = 152$ m close to the opposite shore at $x = 885$ m. This wave height is in excellent agreement with the inversely computed value $H = 162$ m in front of the opposite shore computed with the solitary wave run-up equation of Hall and Watts [37] and $R = 524$ m.

The prediction of H_M shown in Figure 2 results in a maximum wave height of $H_M = 179$ m just in front of the opposite shore ($x_M = 1232$ m based on empirical equation in Table 1) compared to the "observed" wave height of $H = 162$ m. The small difference is explained with the lateral spread of wave energy (Figure 4), in contrast to the laboratory tests, where any spread was prevented by the side boundaries of the 2D setup. Further, the empirical equation in Table 1 for a_M results in $a_M = 143$ m, which is in good agreement with $a = 152$ m measured at $x = 885$ m in the physical model [36].

The following examples show that empirical predictions based on P are often conclusive enough to replace expensive prototype specific physical or numerical model studies, or at least to provide well founded recommendations on whether a more expensive investigation is required. Heller *et al.* [31] provided a generic hazard assessment methodology based on P applicable in both 2D (Table 1) and 3D, including wave generation, propagation and run-up on shores or dams. Fuchs and Boes [8] calibrated and validated this method [31] with a rock fall generated impulse wave observed in 2007 at Lake Lucerne, Switzerland, and predicted then potential future waves at the same location. Preliminary estimates based on P for the planned Kühtai reservoir and hydropower dam in Austria suggested that an impulse wave due to a snow avalanche may overtop the dam so that a detailed prototype specific study was recommended. This was realized by Fuchs *et al.* [32] at scale 1:130, indicating that the wave only moistens the dam crest without overtopping. Cannata *et al.* [33] implemented the method of Heller *et al.* [31] in the open source GIS software GRASS, along with a considerable more time consuming shallow-water equation approach. A comparison of the run-up based on the two approaches for a case study at Lake Como, Italy, resulted in a general agreement of the wave height magnitude. BGC [38] predicted the wave parameters in the slide impact zone based on P for slope failure scenarios in the Mitchell Pit Lake, Canada. The wave propagation and potential dam overtopping were then numerically modelled with TELEMAC-2D with these initial predictions as input values. These examples provide evidence that the parameter P is an efficient tool for first estimates in real-world predictions and for hazard assessment in general.

4. Limitations

Whereas P was developed for subaerial landslide tsunamis (impulse waves), it does not apply to submarine landslide tsunamis [5,9,25,39–41]. Although some slide parameters are significant for both phenomena, other parameters not considered in P, including the initial slide submergence, are only relevant for underwater landslide tsunamis. Further, parameters applicable for both cases may not necessarily be relevant in the same ranges and their relative importance for submarine slides may also not be reflected by Equation (1). Finally, subaerial slides often result in considerable air entrainment, in contrast to underwater slides. The application of P to partially submerged slides is currently not investigated.

The parameter P was thus far mainly tested for wave channel (2D) rather than for wave basin (3D) geometries. Even though 2D geometries can reflect real-world cases (e.g. narrow reservoirs, lakes or fjords), the wave propagation is commonly of 3D nature. Transformation methodologies of results from 2D to 3D were presented by Huber and Hager [20], Heller *et al.* [31] and Heller *et al.* [42] in which the two latter studies already included P. Several 3D real-world predictions were based on this transformed form of P [8,32,33]. An ongoing project aims to further investigate the transformation of 2D results based on P to other idealized geometries, and future real-world applications are hoped to further confirm the applicability of P to cases with more complex water body geometries and bathymetries.

5. Conclusions

The relevance of the semi-empirical impulse product parameter P was reviewed, which is believed to be the most universal parameter to predict subaerial landslide tsunamis (impulse waves). The parameter P includes all relevant slide parameters affecting the wave generation process in wide test ranges such as densities heavier and lighter than water or slide impact angles between 30° and 90° (Table A1). The parameter P is based on the streamwise slide momentum flux component and it was refined with a multiple regression data analysis of granular slide tests conducted in a laboratory channel (2D) and further confirmed with 2D block model slide tests. Empirical equations based on P allow for a simple prediction of the maximum wave amplitude a_M, height H_M and period T_M and their evolutions $a(x)$, $H(x)$ and $T(x)$ with propagation distance (Table 1). Analytical evidence based on P revealed that the highest waves occur for a slide impact angle of $\alpha = 51.6°$. The landslide probability is highest for hill slope angles of $\alpha = 36$ to $39°$ where only about 10% smaller waves may be expected.

Despite the fact that P was derived under idealized conditions, it is considered a useful and effective parameter for estimates in real-world cases. This was demonstrated for the 1958 Lituya Bay case, where a good agreement between the "observed" and predicted wave heights resulted. Four further real-world studies conducted by other authors involving rock falls at Lake Lucerne, Switzerland; snow avalanches in the planned Kühtai reservoir in Austria; rock falls at Lake Como, Italy; and potential slope failures in Mitchell Pit Lake in Canada, provided evidence that first estimates based on P are often conclusive enough to replace expensive prototype specific physical or numerical model studies, or at least provide well founded recommendations on whether a more expensive investigation is required.

The parameter P applies to subaerial and potentially to partially submerged landslide tsunamis (impulse waves); however, it does not apply to submarine slides. A further limitation of P is its derivation for wave channel (2D) tests. Equations for waves propagating in 3D were proposed based on 2D to 3D transformation methods including P, and real-world applications showed that these equations result in realistic values. In the light of this success, the application of P to more complex water body geometries is the subject of an ongoing research effort.

Acknowledgments: The authors thank H.M. Fritz and A. Zweifel for having provided their experimental data. J. Spinneken is acknowledged for comments on an earlier version of this review. The work was supported by the Swiss National Science Foundation (Grant No. 200020-103480/1) and an Imperial College London Junior Research Fellowship.

Author Contributions: Conceived and designed the experiments: VH of experiments at Imperial College London, WHH of experiments at ETH Zurich. Performed the experiments: VH. Analyzed the data: VH of data obtained at Imperial College London, VH WHH of data obtained at ETH Zurich. Wrote the review: VH. Commented on and improved the review: WHH.

Conflicts of Interest: The authors declare no conflict of interest.

Appendix A: Experimental Methodology

The 2D experiments were conducted in two prismatic wave channels, namely the granular slide tests in an 11 m (L) × 0.500 m (W) × 1 m (H) wave channel at ETH Zurich [26] and the block model slide tests in a 24.5 m (L) × 0.600 m (W) × 1.0 m (H) wave channel at Imperial College London [29]. The channel bottoms consisted of glass sheets and the walls of glass and steel sheets (ETH Zurich, Zurich, Switzerland) or only glass sheets (Imperial College London, London, UK). The parameter ranges of both studies are reported in Table A1 and the coordinate origin $(x; z)$ is defined at the intersection of the still water surface with the hill slope ramp (Figure 1).

Table A1. Limitations of P: Parameter ranges of physical model studies of Heller and Hager [26] and Heller and Spinneken [29].

Name	Symbol	Dimension	Heller and Hager [26]	Heller and Spinneken [29]
Slide model type	—	—	granular	block
Channel width	b	(m)	0.500	0.600
Still water depth	h	(m)	0.150–0.675	0.300, 0.600
Slide thickness	s	(m)	0.050–0.249	0.120
Grain diameter	d_g	(mm)	2.0–8.0	—
Streamwise distance	x	(m)	0–8.90	0–17.7
Slide impact velocity	V_s	(m/s)	2.06–8.77	0.59–3.56
Bulk slide volume	$\mathcal{V}_s$	(m³)	0.0167–0.0668	0.0373
Bulk slide density	ρ_s	(kg/m³)	590–1,720	1,534
Slide mass	m_s	(kg)	10.09–113.30	57.23
Slide width	b_s	(m)	0.500	~0.588, ~0.578, 0.526
Slide front angle	φ	(°)	not systematic investigated	30, 45, 60, 90
Transition type	—	(−)	none	none and circular shaped
Hill slope angle	α	(°)	30–90	45
Slide Froude number	F	(−)	0.86–6.83	0.34–2.07
Relative slide thickness	S	(−)	0.09–1.64	0.20–0.40
Relative slide mass	M	(−)	0.11–10.02	0.27–1.21
Relative streamwise distance	x/h	(−)	0–59	0–40
Impulse product parameter	P	(−)	0.17–8.13	0.16–1.19
Number of tests	i	(−)	434	144

The granular slide material in the 434 tests at ETH Zurich (Zurich, Switzerland) was accelerated in a box with up to 8 bar air pressure with a pneumatic landslide generator [22]. Once the box reached its maximum velocity, its front flap opened, the slide left the box, slid down on a 3 m long hill slope ramp and generated the impulse waves. The whole pneumatic landslide generator was adjustable to various

10

still water depths h and slide impact angles α and the box height and length was also adjustable such that various bulk slide volumes V_s and slide heights s could be investigated (Table A1). The pneumatic landslide generator allowed for a systematic and independent variation of all governing parameters included in P. The granular slide materials, made of barium-sulphate (BaSO$_4$) and polypropylene (PP), consisted of four cylindrically shaped grains of diameter of d_g = 2 to 8 mm of densities heavier ($\rho_s \approx 1720$ kg/m^3) and lighter ($\rho_s \approx 590$ kg/m^3) than water. Mixtures of different grain diameters and densities were also included in the test program. However, the grain diameter and grain size distribution were found to have a negligible effect on the wave features and are not included in P. A small fraction of the tests was conducted at a small water depth h = 0.150 m where scale effects relative to the wave amplitude and height, in comparison with the remaining model tests, may be up to about 15% [43]. Figure 2 supports this statement showing that these tests resulted in a slightly smaller maximum wave height than predicted by the corresponding empirical equation in Table 1. The main measurement techniques included two Laser Distance Sensors (LDS) [44] to scan the slide profiles (estimated accuracy $\pm$ 0.5 mm) at 100 Hz, seven capacitance wave gauges to record the wave profiles ($\pm$1.5 mm) at 500 Hz at relative distances of up to x/h = 59.0 and Particle Image Velocimetry (PIV) [45] for determining the velocity vector fields in the slide impact zone. All three measurement systems were triggered simultaneously with the start bottom of the pneumatic landslide generator [26].

The 144 block model slide experiments at Imperial College London (London, UK) involved a hill slope ramp of constant front angle α = 45° (Table A1). The ramp's front surface consisted of PVC sheets and a stainless steel guide in the center matching a groove in the slide bottom to assure that the slide stayed in the channel center during impact. The four slides, one for each slide front angle φ, were also made of PVC. They were moved in the raised position with a pulley system and released with a mechanism fitted to the slide surface. The glass bottom in the immediate slide impact was protected with a 1 m long rubber sheet covered with a thin stainless steel plate. The slide was either brought to an immediate rest at the slope bottom with mastic sealant or it run out further over a circular-shaped transition made of an aluminum sheet bent to an eighth of a circle of radius 0.400 m. The test program included three specific block model parameters, namely three blockage ratios $B = b_s/b$ = 0.88, 0.96 and 0.98 (varied with PVC additions mounted at the sides of the slides), four slide front angles φ = 30, 45, 60 and 90° and two transition types, in addition to different slide parameters (Table A1). The slide impact velocity V_s was measured with a LDS [44]. A PVC strip with holes at constant intervals was bounded on the surface of each slide and this strip was scanned at 128 Hz with the LDS. The slide velocity was then calculated with the information about the spatial and temporal intervals between neighboring holes. The wave features ($\pm$1.5 mm) were

11

measured with seven resistance type wave gauges at 128 Hz at a relative distances of up to x/h = 40.0 (Table A1) [29].

References

1. Miller, D.J. *Giant Waves in Lituya Bay, Alaska*; Geological Survey Professional Paper No. 354-C; U.S. Government Printing Office: Washington, DC, USA, 1960.
2. Slingerland, R.L.; Voight, B. Occurrences, properties and predictive models of landslide-generated impulse waves. In *Rockslides and Avalanches*; Voight, B., Ed.; Elsevier: Amsterdam, The Netherlands, 1979; Volume 2, pp. 317–397.
3. Ward, S.N. Landslide tsunami. *J. Geophys. Res.* **2001**, *106*, 11201–11215. [CrossRef]
4. Synolakis, C.E.; Bardet, J.-P.; Borrero, J.C.; Davies, H.L.; Okal, E.A.; Silver, E.A.; Sweet, S.; Tappin, D.R. The slump origin of the 1998 Papua New Guinea Tsunami. *Proc. R. Soc. Lond. A* **2002**, *458*, 763–789. [CrossRef]
5. Masson, D.G.; Harbitz, C.B.; Wynn, R.B.; Pedersen, G.; Løvholt, F. Submarine landslides: Processes, triggers and hazard prediction. *Philos. Trans. R. Soc. A* **2006**, *364*, 2009–2039.
6. WCHL. *Hydraulic Model Studies—Wave Action Generated by Slides into Mica Reservoir—British Columbia*. Report; Western Canada Hydraulic Laboratories: Vancouver, Canada, 1970.
7. Løvholt, F.; Pedersen, G.; Gisler, G. Oceanic propagation of a potential tsunami from the La Palma Island. *J. Geophys. Res.* **2008**, *113*, C09026. [CrossRef]
8. Fuchs, H.; Boes, R. Berechnung felsrutschinduzierter Impulswellen im Vierwaldstättersee. *Wasser Energ. Luft* **2010**, *102*, 215–221.
9. Watt, S.F.L.; Talling, P.J.; Vardy, M.E.; Heller, V.; Hühnerbach, V.; Urlaub, M.; Sarkar, S.; Masson, D.G.; Henstock, T.J.; Minshull, T.A.; *et al.* Combinations of volcanic-flank and seafloor-sediment failure offshore Montserrat, and their implications for tsunami generation. *Earth Planet. Sci. Lett.* **2012**, *319–320*, 228–240. [CrossRef]
10. Kremer, K.; Simpson, G.; Girardclos, S. Giant Lake Geneva tsunami in AD 563. *Nat. Geosci.* **2012**, *5*, 756–757. [CrossRef]
11. Ward, S.N.; Day, S. Ritter Island Volcano—Lateral collapse and the tsunami of 1888. *Geophys. J. Int.* **2003**, *154*, 891–902. [CrossRef]
12. Monaghan, J.J.; Kos, A.; Issa, N. Fluid motion generated by impact. *J. Waterw. Port C-ASCE* **2003**, *129*, 250–259. [CrossRef]
13. Quecedo, M.; Pastor, M.; Herreros, M.I. Numerical modeling of impulse wave generated by fast landslides. *Int. J. Numer. Methods Eng.* **2004**, *59*, 1633–1656. [CrossRef]
14. Liu, P.L.-F.; Wu, T.-R.; Raichlen, F.; Synolakis, C.E.; Borrero, J.C. Runup and rundown generated by three-dimensional sliding masses. *J. Fluid Mech.* **2005**, *536*, 107–144. [CrossRef]
15. Lynett, P.; Liu, P.L.-F. A numerical study of the run-up generated by three-dimensional landslides. *J. Geophys. Res.* **2005**, *110*, C03006. [CrossRef]
16. Schwaiger, H.F.; Higman, B. Lagrangian hydrocode simulations of the 1958 Lituya Bay tsunamigenic rockslide. *Geochem. Geophy. Geosyst.* **2007**, *8*, 1–7.

17. Bascarini, C. Computational fluid dynamics modelling of landslide generated water waves. *Landslides* **2010**, *7*, 117–124. [CrossRef]
18. Abadie, S.M.; Morichon, D.; Grilli, S.; Glockner, S. Numerical simulation of waves generated by landslides using a multiple-fluid Navier-Stokes model. *Coast. Eng.* **2010**, *57*, 779–794. [CrossRef]
19. Abadie, S.M.; Harris, J.C.; Grilli, S.T.; Fabre, R. Numerical modeling of tsunami waves generated by the flank collapse of the Cumbre Viejo Volcano (La Palma, Canary Islands): Tsunami source and near field effects. *J. Geophys. Res.* **2012**, *117*, C05030. [CrossRef]
20. Huber, A.; Hager, W.H. Forecasting impulse waves in reservoirs. In Proceedings of 19th Congrès des Grands Barrages, Florence, Italy, 26–30 May 1997; pp. 993–1005.
21. Walder, J.S.; Watts, P.; Sorensen, O.E.; Janssen, K. Tsunamis generated by subaerial mass flows. *J. Geophys. Res.* **2003**, *108*. [CrossRef]
22. Fritz, H.M.; Hager, W.H.; Minor, H.-E. Near field characteristics of landslide generated impulse waves. *J. Waterw. Port C-ASCE* **2004**, *130*, 287–302. [CrossRef]
23. Panizzo, A.; De Girolamo, P.; Petaccia, A. Forecasting impulse waves generated by subaerial landslides. *J. Geophys. Res.* **2005**, *110*, C12025. [CrossRef]
24. Zweifel, A.; Hager, W.H.; Minor, H.-E. Plane impulse waves in reservoirs. *J. Waterw. Port C-ASCE* **2006**, *132*, 358–368. [CrossRef]
25. Enet, F.; Grilli, S.T. Experimental study of tsunami generation by three-dimensional rigid underwater landslides. *J. Waterw. Port C-ASCE* **2007**, *133*, 442–454. [CrossRef]
26. Heller, V.; Hager, W.H. Impulse product parameter in landslide generated impulse waves. *J. Waterw. Port C-ASCE* **2010**, *136*, 145–155. [CrossRef]
27. Heller, V.; Hager, W.H. Wave types of landslide generated impulse waves. *Ocean Eng.* **2011**, *38*, 630–640. [CrossRef]
28. Mohammed, F.; Fritz, H.M. Physical modeling of tsunamis generated by three-dimensional deformable granular landslides. *J. Geophys. Res.* **2012**, *117*, C11015. [CrossRef]
29. Heller, V.; Spinneken, J. Improved landslide-tsunami predictions: Effects of block model parameters and slide model. *J. Geophys. Res.* **2013**, *118*, 1489–1507. [CrossRef]
30. Körner, H.J. Reichweite und Geschwindigkeit von Bergstürzen und Fliessschneelawinen. *Rock Mech.* **1976**, *8*, 225–256. [CrossRef]
31. Heller, V.; Hager, W.H.; Minor, H.-E. Landslide generated impulse waves in reservoirs—Basics and computation. In *VAW-Mitteilung 211*; Boes, R., Ed.; Swiss Federal Institute of Technology (ETH) Zurich: Zurich, Switzerland, 2009.
32. Fuchs, H.; Pfister, M.; Boes, R.; Perzlmaier, S.; Reindl, R. Impulswellen infolge Lawineneinstoss in den Speicher Kühtai. *Wasserwirtschaft* **2011**, *101*, 54–60.
33. Cannata, M.; Marzocchi, R.; Molinari, M.E. Modeling of landslide-generated tsunamis with GRASS. *Trans. GIS* **2012**, *16*, 191–214.
34. Fuchs, H.; Winz, E.; Hager, W.H. Underwater landslide characteristics from 2D laboratory modeling. *J. Waterw. Port C-ASCE* **2013**, *139*, 480–488. [CrossRef]
35. Larsen, I.J.; Montgomery, D.R. Landslide erosion coupled to tectonics and river incision. *Nat. Geosci.* **2012**, *5*, 468–473. [CrossRef]

36. Fritz, H.M.; Hager, W.H.; Minor, H.-E. Lituya Bay case: Rockslide impact and wave run-up. *Sci. Tsunami Hazards* **2001**, *19*, 3–22.

37. Hall, J.V.; Watts, G.M. *Laboratory Investigation of the Vertical Rise of Solitary Wave on Impermeable Slopes*; Technical Memo Report No. 33. U.S. Army Corps of Engineers, Beach Erosion Board: Washington, DC, USA, 1953.

38. BGC. *Mitchell Pit Landslide Generated Wave Modelling. Appendix 4-E*; BGC Engineering Inc.: Vancouver, BC, Canada, 2012.

39. Watts, P. Tsunami features of solid block underwater landslide. *J. Waterw. Port C-ASCE* **2000**, *126*, 144–152.

40. Bardet, J.-P.; Synolakis, C.E.; Davies, H.L.; Imamura, F.; Okal, E.A. Landslide tsunamis: Recent findings and research directions. *Pure Appl. Geophys.* **2003**, *160*, 1793–1809. [CrossRef]

41. Najafi-Jilani, A.; Ataie-Ashtiani, B. Estimation of near-field characteristics of tsunami generation by submarine landslide. *Ocean Eng.* **2008**, *35*, 545–557.

42. Heller, V.; Moalemi, M.; Kinnear, R.D.; Adams, R.A. Geometrical effects on landslide-generated tsunamis. *J. Waterw. Port C-ASCE* **2012**, *138*, 286–298.

43. Heller, V.; Hager, W.H.; Minor, H.-E. Scale effects in subaerial landslide generated impulse waves. *Exp. Fluids* **2008**, *44*, 691–703. [CrossRef]

44. Dorsch, R.G.; Häusler, G.; Herrmann, J.M. Laser triangulation: Fundamental uncertainty in distance measurement. *Appl. Opt.* **1994**, *33*, 1306–1314. [CrossRef]

45. Raffel, M.; Willert, C.E.; Kompenhans, J. *Particle Image Velocimetry—A Practical Guide*; Springer: Berlin, Germany, 1998.

Impulse Wave Generation: Comparison of Free Granular with Mesh-Packed Slides

Frederic M. Evers and Willi H. Hager

Abstract: Slides generating impulse waves are currently generated using either block models or free granular material impacting a water body. These procedures were mainly developed to study plane impulse waves, *i.e.*, wave generation in a rectangular channel. The current VAW, ETH Zurich, research is directed to the spatial impulse wave features, *i.e.*, waves propagating in a wave basin. The two wave generation mechanisms mentioned above complicate this process for various reasons, including experimental handling, collection of slide material in the wave basin, poor representation of prototype conditions for the block model, and excessive temporal duration for free granular slides. Impulse waves originating from slides with free granular material and mesh-packed slides are compared in this paper. Detailed test series are presented, so that the resulting main wave features can be compared. The results highlight whether the simplified procedure involving mesh-packed slides really applies in future research, and specify advantages in terms of impulse wave experimentation.

Reprinted from *J. Mar. Sci. Eng.* Cite as: Evers, F.M.; Hager, W.H. Impulse Wave Generation: Comparison of Free Granular with Mesh-Packed Slides. *J. Mar. Sci. Eng.* **2015**, *3*, 100–110.

1. Introduction

Mass wasting including rockfalls, landslides, or avalanches may cause large water waves in oceans, bays, lakes, and reservoirs. As the kinetic energy is transferred from the slide mass to the water body these waves are referred to as impulse waves. They may run-up the shoreline several meters or overtop a dam, endangering thereby adjacent settlements and infrastructure. Therefore, procedures for assessing the generation and propagation of landslide-generated impulse waves form the integral part of an effective risk management strategy [1].

The generation of landslide-generated impulse waves is a complex process encompassing the interaction of the phases slide material, water, and air. To reproduce this process within a hydraulic scale model subaerial slides have so far been mainly represented by either a free granular slide or a rigid block [2]. Free granular slide material was used e.g., by Fritz [3], Heller [4], Mohammed and Fritz [5], and Viroulet *et al.* [6]; while e.g., Di Risio *et al.* [7], Heller and Spinneken [2], Kamphuis and Bowering [8], Noda [9], Panizzo *et al.* [10], Sælevik *et al.* [11], and Viroulet *et al.* [12] conducted experiments with block models. Ataie-Ashtiani and

Nik-Khah [13] and Zweifel [14] presented results comprising both approaches. Block models are unable to account for the granular slide matrix, whereas free granular slides imply a significant procedural and temporal effort for the experimental execution.

Fritz [3], Zweifel [14], and Heller [4] conducted a total of 434 experiments involving free granular material and a pneumatic landslide generator using the two-dimensional (2D) wave channel at the Laboratory of Hydraulics, Hydrology and Glaciology (VAW). The relevant wave characteristics were among others the maximum wave amplitude and wave height, as well as their decay along the propagation distance [4]. The observed wave characteristics were correlated to the measured slide parameters to establish general design information. These empirical equations for significant wave parameters are used as a reference for the experiments presented hereinafter, since they were established on a sound data basis.

As described by Slingerland and Voight [15], Davidson and Whalin [16] also applied bags containing loose iron and lead elements for generating impulse waves. Yet their results were not compared to free granular slides. As a prerequisite to investigate the generation and propagation of spatial impulse waves in a wave basin, experiments with mesh-packed granular material in a 2D wave channel were conducted. The focus of these tests comprised only selected wave features as the wave amplitudes and heights, their decay, and the wave crest celerity. However, no comprehensive test program involving further test parameters was attempted, because this will be the purpose of future research. The measured wave characteristics are compared with the empirical equations derived from the 2D data resulting from the free granular material to assess whether mesh-packed slides provide a sufficient reproducibility. The following therefore involves a comparison between free and mesh-packed slides to investigate whether the simplifications offered by the latter approach are justified by the experimental data. If mesh-packed slides would be able to adequately reproduce the former approach, substantial simplifications in terms of experimental effort, test duration including material package and collecting from the test facility, among others, would result.

2. Previous Research and Experimental Setup

2.1. Impulse Product Parameter

2D experiments involving free granular material were conducted by systematically varying the slide parameters to study 2D impulse waves. Heller and Hager [17] identified the slide (subscript s) impact velocity V_s, the slide thickness s, the slide mass m_s, the still water depth h, and the slide impact angle α as the relevant parameters for impulse wave generation (Figure 1). A set of dimensionless quantities, namely the slide Froude number F, the relative slide thickness S, and the relative slide mass M, entirely define the physics of impulse waves, namely

$$F = V_s/(gh)^{1/2} \tag{1}$$

$$S = s/h \tag{2}$$

$$M = m_s/(\rho_w bh^2) \tag{3}$$

Based on these dimensionless quantities, Heller and Hager [17] developed the so-called impulse product parameter P for describing the 2D characteristics of landslide generated impulse waves as

$$P = FS^{1/2}M^{1/4}\{\cos[(6/7)\alpha]\}^{1/2} \tag{4}$$

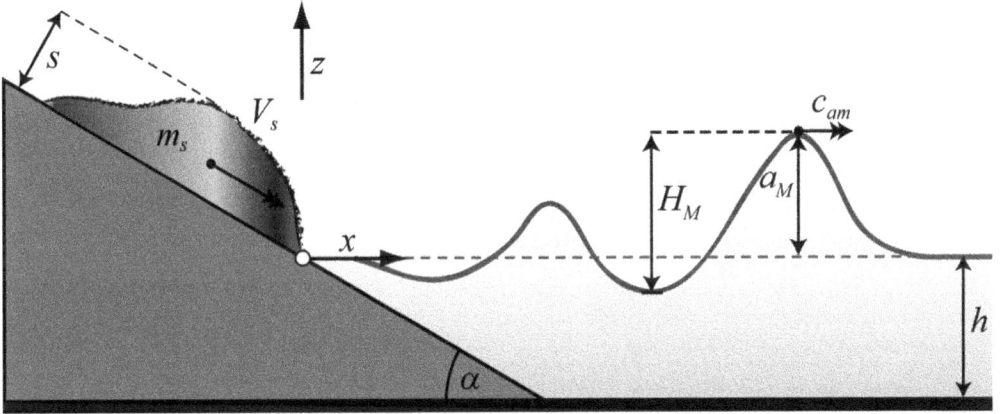

Figure 1. Relevant slide parameters and wave characteristics (adapted from Heller [4], with permission from VAW).

2.2. Wave Characteristics

To study the reproducibility of waves generated by mesh-packed slides compared with free granular slides, the maximum (subscript M) wave amplitude a_M and the maximum wave height H_M as shown in Figure 1, as well as their decay along the propagation distance x were analyzed. The governing 2D wave characteristics include the relative maximum wave amplitude $A_M = a_M/h$ and height $Y_M = H_M/h$ plus their decay $A(X)$ and $Y(X)$ along the relative distance $X = x/h$ measured from the

17

location of the free water surface at the slide plane (Figure 1). Heller and Hager [17] empirically derived these wave characteristics as

$$A_M = (4/9)P^{4/5} \tag{5}$$

$$Y_M = (5/9)P^{4/5} \tag{6}$$

$$A(X) = (3/5)\left(PX^{-1/3}\right)^{4/5} \tag{7}$$

$$Y(X) = (3/4)\left(PX^{-1/3}\right)^{4/5} \tag{8}$$

Further, the mean (subscript m) wave crest celerity c_{am} of the mean wave amplitude a_m is [4]

$$c_{am}/(gh)^{1/2} = \left(1 + 2(a_m/h)^2\right)^{1/2} \tag{9}$$

2.3. Experimental Setup and Procedure

For the present experiments with mesh-packed slides the identical wave channel was used as for these involving free granular material. The instrumentation consisted of six Capacitance Wave Gauges (CWG) (Figure 2). The slide impact velocity V_s and the slide thickness s were measured by laser distance sensors mounted perpendicularly to the slide plane. In contrast to the previous experiments, V_s of the mesh-packed slides was not determined as a slide centroid velocity, but as the velocity of the slide front under dry conditions, given that the loose mesh bag does not allow for a correct capturing of the slide profile.

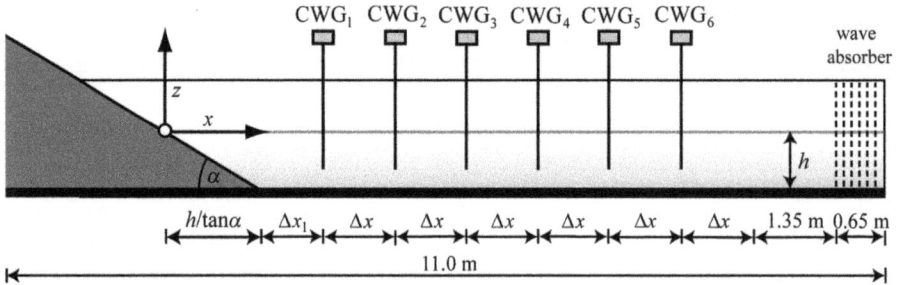

Figure 2. Capacitance Wave Gauge (CWG)$_{1-6}$ positions with $\Delta x_1 = 0.71$ m and $\Delta x = 1.00$ m for $\alpha = 30°$; $\Delta x_1 = 1.13$ m and $\Delta x = 1.00$ m for $\alpha = 45°$; $\Delta x_1 = 1.27$ m and $\Delta x = 1.06$ m for $\alpha = 60°$ (adapted from Heller [4], with permission from VAW).

Table 1. Overview of experimental parameters and dimensionless quantities.

Parameter	Free Granular Slides [4]	Mesh-Packed Slides
h [m]	0.15, 0.20, 0.30, 0.45, 0.60, 0.675	0.20, 0.30, 0.40
s [m]	0.05–0.249	0.062–0.145
V_s [m/s]	2.06–8.77	1.2–9.2
m_s [kg]	10.09–113.30	19.5–20.1
α [°]	30, 45, 60	30, 45, 60
F [-]	0.86–6.83	0.70–5.36
S [-]	0.09–1.64	0.16–0.65
M [-]	0.11–10.02	0.24–1.01
P [-]	0.17–8.13	0.26–2.78

The granular slide material used for the mesh-packed experiments corresponded to that used by Heller [4]. It has a grain (subscript g) diameter of d_g = 8 mm and a grain density of ρ_g = 2,429 kg/m^3. The granular material is loosely packed into mesh bags made of sifting media (SEFAR NYTAL® PA-38GG-500, Sefar AG, Heiden, Switzerland) with a mesh opening of 500 μm and a porosity of 47%. The bags were accelerated with the pneumatic landslide generator [18]. The main parameters and dimensionless quantities of the mesh-packed slides considered herein are compared with these of the free granular slides in Table 1. The total number of mesh-packed experiments evaluated is 42.

3. Results and Discussion

3.1. Slide Impact and Wave Generation Process

In Figure 3 the slide impact and wave generation processes are documented by means of high-speed photography at various times for slide impact angles of α = 30° and α = 60°. Free granular slides are affected by compaction and strong deformation processes during the impact onto the still water and the underwater movement to the channel bottom [3]. The mesh-packed slides are both bended and lifted upwards after impacting the water body, depending on the slide impact angle α. For α = 30° these effects are larger than for α = 60°. These effects significantly increase the slide thickness s resembling the mechanisms of compaction and deformation of free granular slides. Also the process of flow separation and the formation of an impact crater along with air entrainment are observed in both cases.

Figure 3. Photographs at various stages of wave generation process with mesh-packed slides. Left column: $\alpha = 30°$ with F = 2.27, S = 0.23, M = 0.45, and P = 0.84 at t = 0.00, 0.14, 0.34, and 0.90 s. Right column: $\alpha = 60°$ with F = 3.03, S = 0.21, M = 0.45, and P = 0.90, at t = 0.00, 0.11, 0.31, and 0.76 s.

3.2. Maximum Wave Amplitude and Height

The relative maximum wave amplitude A_M and wave height Y_M of waves generated with mesh-packed slides versus the impulse product parameter P including Equations (5) and (6) are shown in Figure 4. The maxima are recorded independently from their position within the wave train and along their propagation distance. In most wave trains the wave maxima were observed at the first wave crest and were fully developed at CWG_1. The data of both wave maxima predominantly scatter within $\pm 30\%$ of Equations (5) and (6), as do these for free granular slides. A concentrated undercut of the -30% curve for $P \leqslant 1$ is also detected in the corresponding plots of Heller [4]. The coefficients of determination are $R^2 = 0.82$ and $R^2 = 0.85$ for maximum wave amplitude and height, respectively, compared to $R^2 = 0.89$ and $R^2 = 0.85$ for the 434 free granular slide experiments [4].

Note the effect of the slide impact angle α on the relative maximum wave amplitude A_M and height Y_M. For $\alpha = 60°$ the maxima are predominantly scattered within the area between the curves of Equations (5) and (6) and their corresponding -30% curves; at $\alpha = 45°$ the wave maxima are narrowly scattered along Equations (5) and (6), while the maxima of $\alpha = 30°$ are located in the upper half above the equations up to the $+30\%$ curves. In summary, a good overall reproducibility of maximum impulse wave amplitudes and heights results by using mesh-packed slides.

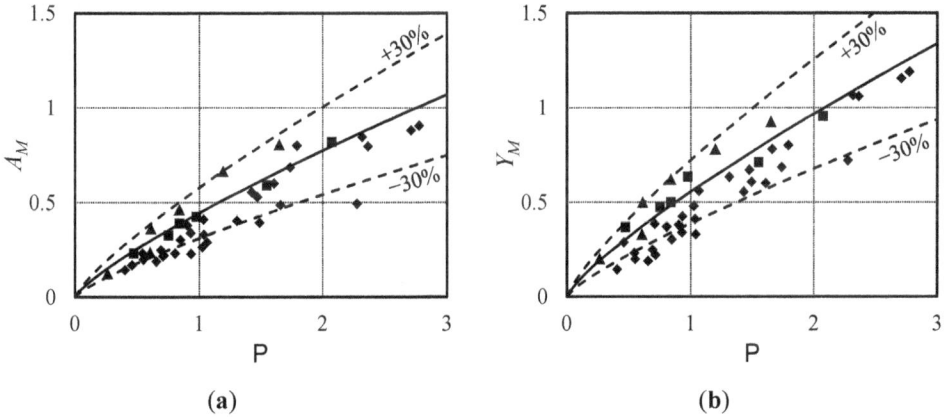

(a)

(b)

Figure 4. Waves generated by mesh-packed slides (a) relative maximum wave amplitude $A_M(P)$ with (–) Equation (5); (b) wave height $Y_M(P)$, with (–) Equation (6) for (▲) $\alpha = 30°$, (■) $\alpha = 45°$, (◆) $\alpha = 60$.

3.3. Wave Amplitude and Height Decay

To evaluate the wave amplitude and height decays, only the first wave crest was considered. This applies also to wave trains, where these maxima occur at the second wave crest. The relative wave amplitude $A(X)$ and height $Y(X)$ decays generated with mesh-packed slides versus $PX^{-1/3}$ as well as the Equations (7) and (8) are shown in Figure 5, from where good overall agreement results. In contrast to the maxima of wave amplitude and height, no immediate effects of the slide impact angle α

result. For wave trains with $PX^{-1/3} \leqslant 0.75$ at X_{CWG1} an increase in wave amplitude and height is observed. These wave trains developed their maxima at CWG_2. An increased undercut of the -30% curve of Equation (7) of $A(X)$ for $PX^{-1/3} \leqslant 1$ applies also to the data of Heller [4]. This statement is valid for the data exceeding the $+30\%$ curves of $A(X)$ as well as $Y(X)$ for $0.5 \leqslant PX^{-1/3} \leqslant 1$. Despite the values of wave amplitude and height decay that undercut and exceed the $\pm30\%$ curves of Equations (7) and (8) in certain ranges of $PX^{-1/3}$, they reproduce the results of free granular slides well. The coefficients of determination are $R^2 = 0.71$ and $R^2 = 0.78$ for wave amplitude and height decay, respectively, compared to $R^2 = 0.83$ and $R^2 = 0.84$ for the 434 free granular slide experiments [4].

3.4. Wave Crest Celerity

The wave crest celerity was evaluated by averaging the wave amplitudes of the first crests of CWG_1 to CWG_6 to a mean wave amplitude a_m. By accounting for the runtime of the first wave crest between CWG_1 to CWG_6 the mean wave crest celerity c_{am} was determined in analogy to [4]. The relative celerity $c_{am}/(gh)^{1/2}$ of waves generated with mesh-packed slides shown in Figure 6 lies within the experimental scatter of free granular material and reproduces Equation (9) well. The coefficient of determination is $R^2 = 0.95$, compared to $R^2 = 0.91$ for the 434 free granular slide experiments [4].

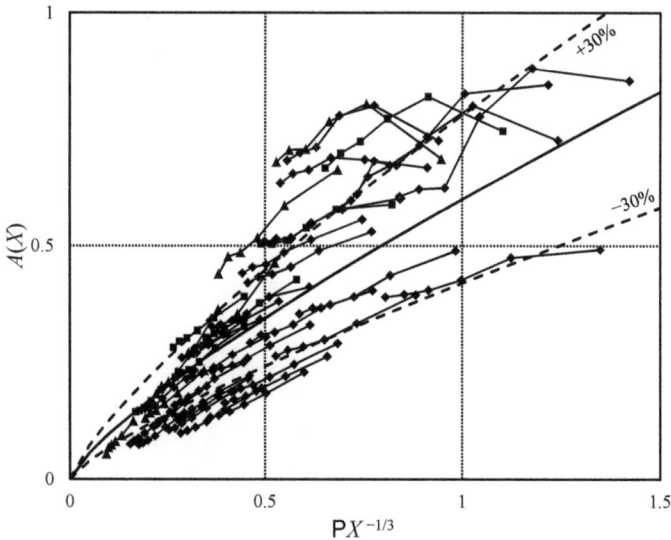

(a)

Figure 5. *Cont.*

22

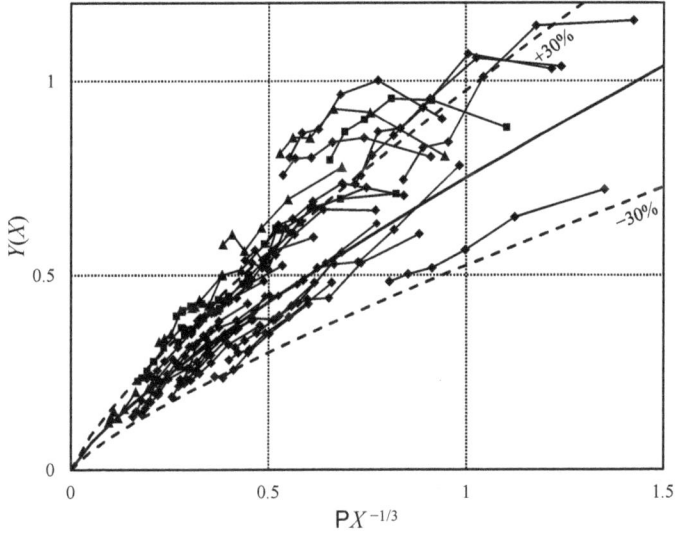

(b)

Figure 5. Mesh-packed slides (**a**) relative wave amplitude decay $A(X)$ with (–) Equation (7); (**b**) relative wave height decay $Y(X)$ with (–) Equation (8) versus $PX^{-1/3}$ for $\alpha = 30°$ (▲), $\alpha = 45°$ (■), $\alpha = 60°$ (♦).

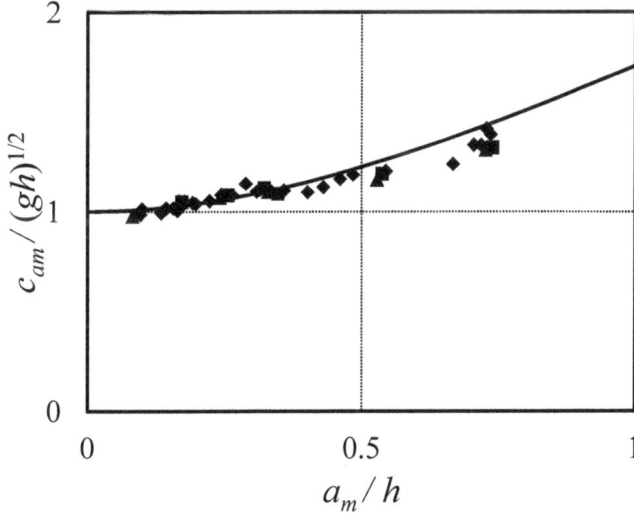

Figure 6. Waves generated by mesh-packed slides, mean relative wave crest celerity $c_{am}/(gh)^{1/2}$ versus mean relative wave amplitude a_m/h for (▲) $\alpha = 30°$, (■) $\alpha = 45°$, (♦) $\alpha = 60°$, (–) Equation (9).

23

4. Conclusions

This research explores the question whether impulse waves can be generated by mesh-packed slides as an alternative to free granular slides mainly in view of experimental effort and reduction of test preparation. Therefore, the prominent wave features induced by mesh-packed slides were investigated in a 2D wave channel. The impulse waves generated were analyzed regarding their main characteristics, including the maximum wave amplitude and height, the distance related decay, and the wave crest celerity. These features were compared with the previously established empirical equations derived from experiments involving free granular material to assess their reproducibility. The main findings are:

- The impulse product parameter P describes adequately both waves generated by mesh-packed and free granular slides;
- Waves generated by mesh-packed slides follow a $\pm 30\%$ scatter around the equations derived from experiments with free granular material. This scatter applies equally to free granular slides;
- For small values of P, the present data may undercut the -30% range. This behavior applies also for the corresponding ranges of free granular slides; and
- For values of $PX^{-1/3}$ ranging between 0.5 and 1, the present data may exceed the $+30\%$ range, similar as for free granular slides.

The present experiments evidence that mesh-packed slides suit for model experimentation of landslide-generated impulse waves which are physically similar regarding selected wave parameters to these generated with free granular material. This implies a substantial improvement of efficiency for the standard test procedure and is especially beneficial for future experiments in a wave basin.

Acknowledgments: This work was supported by the Swiss National Science Foundation (Project No 200021-143657/1).

Author Contributions: The experiments were designed by WHH and FME. The experiments and the data analysis were conducted by FME. The article was written by FME and improved by WHH.

Conflicts of Interest: The authors declare no conflict of interest.

References

1. Heller, V.; Hager, W.H.; Minor, H.-E. Landslide generated impulse waves in reservoirs: Basics and computation. In *VAW-Mitteilung 211*; Minor, H.-E., Ed.; ETH Zurich: Zürich, Switzerland, 2009.
2. Heller, V.; Spinneken, J. Improved landslide-tsunami prediction: Effects of block model parameters and slide model. *J. Geophys. Res.-Oceans* **2013**, *118*, 1489–1507.
3. Fritz, H.M. Initial phase of landslide generated impulse waves. In *VAW-Mitteilung 178*; Minor, H.-E., Ed.; ETH Zurich: Zürich, Switzerland, 2002.

4. Heller, V. Landslide generated impulse waves: Prediction of near field characteristics. In *VAW-Mitteilung 204*; Minor, H.-E., Ed.; ETH Zurich: Zürich, Switzerland, 2008.

5. Mohammed, F.; Fritz, H.M. Physical modeling of tsunamis generated by three-dimensional deformable granular landslides. *J. Geophys. Res.* **2012**, *117*, C11015.

6. Viroulet, S.; Sauret, A.; Kimmoun, O. Tsunami generated by a granular collapse down a rough inclined plane. *Europhys. Lett.* **2014**, *105*, 34004.

7. Di Risio, M.; De Girolamo, P.; Bellotti, G.; Panizzo, A.; Aristodemo, F.; Molfetta, M.G.; Petrillo, A.F. Landslide-generated tsunamis runup at the coast of a conical island: New physical model experiments. *J. Geophys. Res.* **2009**, *114*, C01009.

8. Kamphuis, J.W.; Bowering, R.J. Impulse waves generated by landslides. In Proceedings of the 12th Coastal Engineering Conference, Washington, DC, USA; 1970; pp. 575–588.

9. Noda, E. Water waves generated by landslides. *J. Waterway Div.-ASCE* **1970**, *96*, 835–855.

10. Panizzo, A.; de Girolamo, P.; Petaccia, A. Forecasting impulse waves generated by subaerial landslides. *J. Geophys. Res.* **2005**, *110*, C12025.

11. Sælevik, G.; Jensen, A.; Pedersen, G. Experimental investigation of impact generated tsunami; related to a potential rock slide, Western Norway. *Coast. Eng.* **2009**, *56*, 897–906.

12. Viroulet, S.; Cébron, D.; Kimmoun, O.; Kharif, C. Shallow water waves generated by subaerial solid landslides. *Geophys. J. Int.* **2013**, *193*, 747–762.

13. Ataie-Ashtiani, B.; Nik-Khah, A. Impulsive waves caused by subaerial landslides. *Environ. Fluid Mech.* **2008**, *8*, 263–280.

14. Zweifel, A. Impulswellen: Effekte der Rutschdichte und der Wassertiefe (Impulse waves: Effects of slide density and water depth). In *VAW-Mitteilung 186*; Minor, H.-E., Ed.; ETH Zurich: Zürich, Switzerland, 2004.

15. Slingerland, R.L.; Voight, B. Evaluating hazard of landslide-induced water waves. *J. Waterw. Port C. Div.* **1982**, *108*, 504–512.

16. Davidson, D.D.; Whalin, R.W. *Potential landslide-generated water waves, Libby Dam and Lake Koocanusa, Montana*; Technical Report H-74–15; Waterway Experiment Station, U.S. Army Corps of Engineers: Vicksburg, MO, USA, 1974.

17. Heller, V.; Hager, W.H. Impulse product parameter in landslide generated impulse waves. *J. Waterw. Port C-ASCE* **2010**, *136*, 145–155.

18. Fritz, H.M.; Moser, P. Pneumatic landslide generator. *Int. J. Fluid Power* **2003**, *4*, 49–57.

Tsunamigenic Earthquakes at Along-dip Double Segmentation and Along-strike Single Segmentation near Japan

Junji Koyama, Motohiro Tsuzuki and Kiyoshi Yomogida

Abstract: A distinct difference of the earthquake activity in megathrust subduction zones is pointed out, concerning seismic segmentations in the vicinity of Japan—that is, the apparent distribution of earthquake hypocenters characterized by Along-dip Double Segmentation (ADDS) and Along-strike Single Segmentation (ASSS). ADDS is double aligned seismic-segmentation of trench-ward seismic segments along the Japan Trench and island-ward seismic segments along the Pacific coast of the Japan Islands. The 2011 Tohoku-oki megathrust earthquake of Mw9.0 occurred in ADDS. In the meantime, the subduction zone along the Nankai Trough, the western part of Japan, is the source region of a multiple rupture of seismic segments by the 1707 Houei earthquake, the greatest earthquake in the history of Japan. This subduction zone is narrow under the Japan Islands, which is composed of single aligned seismic-segmentation side by side along the Nankai Trough, which is typical of ASSS. Looking at the world seismicity, the 1960 and 2010 Chile megathrusts, for example, occurred in ASSS, whereas the 1952 Kamchatka and the 1964 Alaska megathrusts occurred in ADDS. These megathrusts in ADDS result from the rupture of strong asperity in the trench-ward seismic segments. Since the asperity of earthquakes in ASSS is concentrated in the shallow part of subduction zones and the asperity of frequent earthquakes in ADDS is in deeper parts of the island-ward seismic segments than those of ASSS, there must be a difference in tsunami excitations due to earthquakes in ADDS and ASSS. An analysis was made in detail of tsunami and seismic excitations of earthquakes in the vicinity of Japan. Tsunami heights of ASSS earthquakes are about two times larger than those of ADDS earthquakes with the same value of seismic moment. The reason for this different tsunami excitation is also considered in relation to the seismic segmentations of ADDS and ASSS.

Reprinted from *J. Mar. Sci. Eng.* Cite as: Koyama, J.; Tsuzuki, M.; Yomogida, K. Tsunamigenic Earthquakes at Along-dip Double Segmentation and Along-strike Single Segmentation near Japan *J. Mar. Sci. Eng.* **2015**, *3*, 1178–1193.

1. Introduction

Devastating tsunamis in the last two decades required a new paradigm of the earthquake occurrences and tsunami generations on a geologically extended time-span. The extraordinary tsunami in the Indian Ocean of 2004 was generated by the earthquake of more than 1000 km in fault length, which includes historical fault ruptures of smaller scales along the Sumatra-Andaman subduction zone [1]. The 2011 Tohoku tsunami in Japan resulted from a fault rupture developed from the Japan Trench to the coastline of the Tohoku prefecture of Japan along the dip-direction of the Pacific plate as well as along the strike-direction of the Japan Trench. This wide-spread source region of the 2011 megathrust earthquake of Mw9.0 includes trench-ward seismic segments, and the segments have been believed to be an interface of aseismic segments [2,3] without any stress accumulation.

These tsunamis were due to megathrust earthquakes much larger than anticipated sizes in their respective seismological histories, although some geological evidence demonstrated past earthquakes and tsunamis larger than those in historical documents and instrumentally observed records. It has been pointed out recently that observations of their seismic activity in recent years revealed that the megathrust earthquakes occurred in seismic segments of either Along-strike Single Segmentation (ASSS) or Along-dip Double Segmentation (ADDS) [4–6]. The different seismic segmentations—ASSS and ADDS—have been identified, referring to their regional seismic-activity, focal mechanisms, rupture patterns, geometry of subduction zones, types of overriding plates and back-arc activity [5].

Previous studies discuss the seismic segmentation along the subduction zones as a more complex structure, being composed of aseismic-zone, unstable zone, conditionally stable zone, and stable sliding zone, e.g. [2]. The present ADDS/ASSS hypothesis would be one where the segmentation relates directly to observable seismic activity in the source regions of great (Mw~8) and megathrust earthquakes (Mw~9). The main purpose of this study is to discuss the tsunami excitation in relation to these segmentations so as to make a comprehensive understanding of source regions of megathrust earthquakes.

Previous studies showed that tsunami excitation in the Japan Sea is larger than that in the Pacific Sea side of Japan [7–9]. These studies attributed the difference to the dip angle of earthquake faults or to the rigidity of earthquake source-regions. In this study, a further discussion is made about the tsunami excitation referring to the recent understanding of earthquake sources in different segmentations of ADDS and ASSS in the vicinity of Japan.

2. Diversity of Megathrust Earthquakes in the World

Seismic activity off the Pacific coast of the Tohoku district, Japan has been extensively investigated from historical and instrumental records. It is characterized into regional seismic segments; there is a double aligned seismic-segmentation along the dip-direction of the subduction zone. The segments along the island-arc side of Japan frequently generated earthquakes as large as Mw8, but the segments along the Japan Trench had been considered to be aseismic. This double aligned seismic-segmentation is called Along-dip Double Segmentation (ADDS). The 2011 megathrust earthquake of Mw9.0 ruptured many segments along not only strike- but also dip-directions of the Japan Trench, covering an area of about 200×500 km^2, as shown in Figure 1.

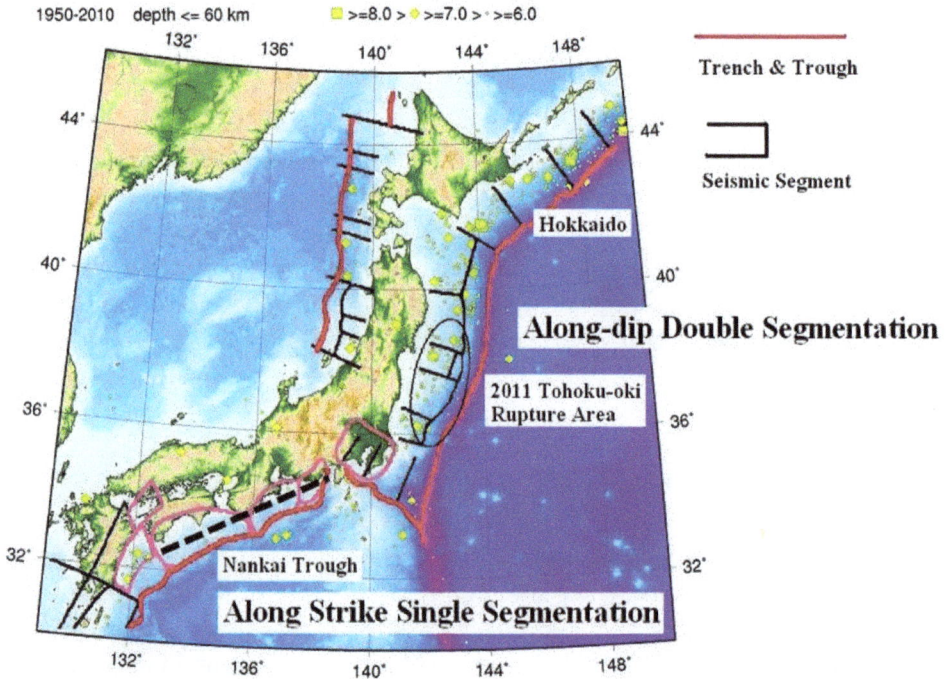

Figure 1. Seismic activity in Japan and in its vicinity. Seismic segments have been used in the official earthquake forecasting [10] of Evaluation of Major Subduction Zone Earthquakes by the Headquarters for Earthquake Research Promotion. Epicenters of earthquakes are plotted by yellow symbols from 1950 to 2010 with magnitudes larger than 5.9 and their focal depths shallower than 61 km determined by the Japan Meteorological Agency [11]. Trench and trough near the Japan Islands are illustrated by red curves. The 2011 Tohoku-oki megathrust earthquake ruptured the area circled by a solid ellipse, where is Along-dip Double Segmentation (ADDS). Along-strike Single Segmentation (ASSS) can be found in the Nankai Trough, where little recent seismic activity has been observed. Such regions are often called seismic gaps [12]. The source extent of the 1707 Houei great earthquake in this segmentation is added by a broken line along the Nankai Trough to the original figure on [13].

The best-known megathrust earthquake in Japan is the 1707 Houei great earthquake along the Nankai Trough with three major segments in Figure 1. This historical event exhibits an interaction in the trench-axis direction among adjacent segments. The 1707 Houei great earthquake took place at a very different site from that of the 2011 megathrust event, where little seismic activity is observed in the single aligned seismic-segmentation along the axis of the Nankai Trough (Figure 1). The inactive seismicity in this region not only applies to the period analyzed in Figure 1 but also to the whole period from 1924 to the present, according to the Japan Meteorological Agency [11], except for the enhanced aftershock activity following the 1944 Tonankai earthquake Mw8.2 and the 1946 Nankaido earthquake Mw8.2, which successively occurred along the Nankai Trough. The Houei earthquake of multi-segment rupture is referred to an earthquake in Along-strike Single Segmentation (ASSS), which contrasts to the 2011 Tohoku-oki megathrust earthquake in ADDS.

The reason for a gigantic megathrust earthquake in ADDS to grow up to such the scale is due to the rupture of strongly-coupled asperity in the trench-ward segment with a longer recurrence time (e.g., millennial) after large earthquakes in the land-ward segments repeated with a shorter recurrence time (e.g., centennial). The activity in ASSS is characterized by almost 100% coupled areas of shallow subduction zones, which finally gives rise to a great earthquake. In other words, the difference between these two types of segmentations appears in a seismic gap [12] along the subduction zone in ASSS and in a doughnut pattern [14] of seismic activity in the subduction zone prior to a gigantic megathrust earthquake in ADDS.

These two types of segmentations (*i.e.*, ASSS and ADDS) can be found not only in the vicinity of Japan but also elsewhere in the world. The 1952 Kamchatka earthquake of Mw 9.0 and the 1964 Alaska earthquake of Mw9.2 are pointed out to be of ADDS type, while the 1960 Chile earthquake of Mw9.5 and the 2010 Maule earthquake of Mw8.8 are of ASSS [4,5]. Speaking about the 2004 Sumatra-Andaman earthquake of Mw9.3, the faulting process is peculiar. Judging from the detailed analysis on the seismicity around the source region of the 2004 earthquake [1], the 2004 event started in the ADDS region and extended into the Andaman-Nicobar ASSS region [13]. The latter region is a typical oblique subduction zone, and is quite similar to the source areas of great earthquakes of the 1957 Andreanof Mw8.6 and the 1965 Rat Island Mw8.7 in the Aleutian arc.

General descriptions of proposed ADDS and ASSS are summarized in Table 1, indicating the distinction from the previous seismic segmentations e.g., [2,3]. Seismic segments in ASSS are usually characterized by a narrow subduction zone from the oceanic trench to the island arc. The evidence has been presented, showing the asperity of the 2010 Maule earthquake in accordance with the strongly-coupled plate interface identified by GPS observation beforehand of the 2010 event [15]. Seismic segments in ADDS are, on the other hand, rather wide. The 2011 Tohoku-oki megathrust earthquake revealed that the strongly-coupled segments exist in the trench-ward of ADDS, where the millennial seismic-asperity of the event was observed [16,17]. This is also true in the 1964 Alaska earthquake [18], showing a very strong asperity in the Alaska Bay close to the Aleutian Trench. Large (not as large as gigantic) earthquakes occur repeatedly in the landward segments of ADDS with their foci in the deep part of the subduction zone. As a result, the asperity of repeated large earthquakes in ASSS is distributed in a shallow part of the subduction zone, while that of the repeated large earthquakes in ADDS stay in a rather deeper part. This would result in a different tsunami excitation between ADDS and ASSS earthquakes.

Figure 2 illustrates an image of earthquake cycles in ADDS and ASSS. An earthquake cycle in ASSS is commonly restricted in the shallow part of the subduction zone, where the contact of the plate interface is strong, forming the seismic gap all over the particular subduction zone. Successive rupture of multiple seismic segments along the trench direction results in a megathrust earthquake (metaphoric centennial event) in ASSS. On the other hand, an earthquake cycle in ADDS is characterized by repeated earthquakes as large as Mw8 and many smaller earthquakes are commonly found in the island-ward segments of the subduction zone, though the asperity in the trench-ward stays still, which eventually induces a megathrust earthquake (metaphoric millennial event) in ADDS. At the time of an ADDS megathrust event, the rupture extends in the dip-direction as well as in the strike-direction of the subduction zone.

Table 1. Characteristics of the seismic activity in Along-strike Single Segmentation (ASSS) and Along-dip Double Segmentation (ADDS).

	ASSS	ADDS
Alignment	Single Aligned	Double Aligned
Seismic Zone	Narrow	Wide
Width/Length	1:4	1:2
Interface Contact	Single Whole Contact	Island-ward/Trench-ward
Seismic Activity	Quiet Everywhere Seismic Gap	Active/Quiet Doughnut Pattern
Recurrence Time	A Few Hundred	A Few Hundred/A Thousand
Interface Contact	Whole Zone	Contact/Strong Contact
Previous Events	1960 Chile, 1707 Houei	1964 Alaska, 2011 Tohoku
Possible Region	Cascade, Canada	Hokkaido, Japan [21]

Although the segmentation of ADDS/ASSS is our proposed understanding on the seismic source region of megathrust/great earthquakes, we need to deepen our understanding of these segmentations in order to distinguish megathrust earthquakes in the future and also to reduce the disasters due to such different types of earthquakes and tsunamis. This study is an attempt to enlarge our knowledge on the tsunami excitations in relation to the different seismic segmentations.

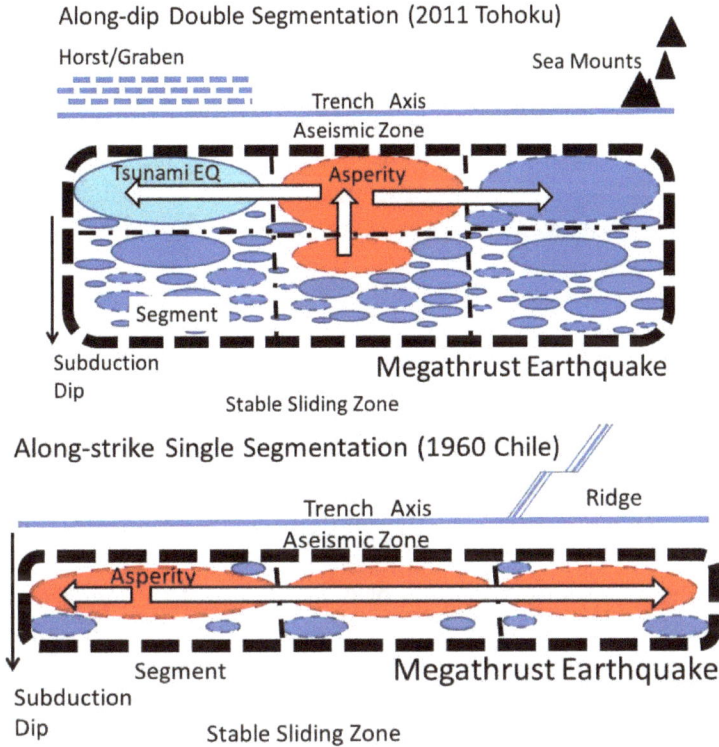

Figure 2. Schematic illustration of Along-dip Double Segmentation (ADDS) and Along- strike Single Segmentation (ASSS). ADDS is characterized by a double aligned seismic-segments along dip direction of subduction, one of which is a strongly-coupled asperity in the trench-ward segments, which eventually ruptures induced by an asperity in the island-ward segment(s) where repeated large earthquakes as large as Mw8 and many smaller-sized earthquakes (solid ellipses) occur. Rupturing all these segments results in a millennial megathrust earthquake (square by thick broken line) like the 2011 Tohoku event. ASSS shows a seismic gap of subduction interface along the trench axis in single aligned seismic-segments side by side. The asperity stays in a narrow part of the subduction interface in ASSS. Horst and Graben and seamounts are considered to be responsible for generating tsunami earthquakes [19–21]. As well as Horst and Graben, surface roughness such as seamounts and oceanic ridges plays an important role in blocking seismic segments (dotted broken lines) along the subduction zone [2].

3. Tsunami Magnitudes for Tsunamigenic Earthquakes near Japan

Tsunamis generated in the vicinity of Japan have been quantified from -1 to 4 using tsunami heights and disaster distribution near respective source regions [22,23]. This is called the Imamura-Iida scale of tsunamis, and it has been determined from historical documents and recorded evidence in Japan since 176 A.D. This is available at present, as listed in major tsunami catalogs, such as Historical Tsunami Data Base for the World Ocean [24]. This scale for tsunamis is similar to earthquake intensity, such as the Mercalli intensity scale. Local tsunami magnitude m is numerically defined [25,26], extending the Imamura-Iida scale. The definition of m is, using tsunami heights H (peak to trough in meter) recorded on tide-gauges and correction for tsunami travel distances $\triangle$ (km) as

$$m = 2.7 \log H + 2.7 \log \triangle - 4.3 \qquad (1)$$

Since this is the tsunami magnitude scale investigated well in Japan, at first we analyzed this local tsunami magnitude m comparing with seismic moment Mo of each earthquake that occurred near Japan.

Another magnitude scale, Mt is the far-field tsunami magnitude [27], of which definition is

$$Mt = \log H + 9.1 + \triangle C \qquad (2)$$

where H is the maximum amplitude of tsunami heights in meter and $\triangle C$ is an empirical station correction or regionally-averaged correction. The correction term and the constant 9.1 are introduced so that Mt agrees with moment magnitude Mw [28] of each earthquake. Mt scale is extended to estimate tsunami magnitude Mt for local earthquakes near Japan [7] as

$$Mt = \log H + \log \triangle + 5.80 \qquad (3)$$

The same analysis is made to study the relation between Mt and Mo of tsunamigenic earthquakes near Japan, taking into account the seismic segmentations.

Historical tsunami heights until recent years were measured in tide gauge stations at best located in developed harbors, most of which face in the opposite direction to their respective open seas. Others are eyewitness records and traces of inundations and up-streams. All these data and their sources are listed on the worldwide tsunami catalogs [23,24]. Quantification of tsunami excitation and scaling relations of tsunamis to earthquake source parameters have been made based on those observations. After the recent installation of ocean-bottom pressure gauges, the observation makes it possible to study the tsunami generation free from the contaminations of detail oceanic bathymetry and coastal topography. Unfortunately or fortunately, only few records are available now e.g., [17]. Therefore, it is necessary to apply historical tsunami data to describe the general scaling relation of tsunami excitation to earthquakes and to study the generation mechanism of tsunamis.

In the above equations of (1) and (2), the dependence of tsunami magnitudes on tsunami heights are different. This is because local tsunami magnitude m is based on the energy of tsunami waves and far-field tsunami magnitude Mt on the amplitude of tsunami waves. It is true that these tsunami magnitudes are empirical parameters and needed to be rigorously quantified; however, the present study is interested only in their regional bias due to regional seismic segmentation. The analysis would be granted, since the discussion here is made in a relative manner based on the tsunami magnitudes, which would be free from the absolute uncertainty of tsunami magnitude determination.

4. Tsunami Magnitude and Seismic Moment of Tsunamigenic Earthquakes near Japan

Local tsunami magnitude m and seismic moment Mo of 59 tsunamigenic earthquakes in the vicinity of Japan since 1923 have been determined. Table 2 summarizes those data with their references. There is a variety of earthquakes occurring near Japan, which cannot be classified into ADDS nor ASSS, although tsunamis were excited and recorded. Some of them were intra-plate and outer-rise earthquakes of normal-fault type and the others are earthquakes near the shore of the Japan Islands of the strike-slip type [13]. Therefore, we classify them into ADDS, ASSS and NFSS (Normal Fault-Strike Slip). Figure 3 shows the locations of all the earthquakes in Table 2.

Figure 4 shows the least-squares regressions between m and Mo for ADDS (m_D; 22 events in Table 2) and ASSS (m_S; 10 events in Table 2).

$$m_D = 1.26 \log Mo - 24.4 \ (\pm 0.17)$$
$$m_S = 1.19 \log Mo - 22.3 \ (\pm 0.39)$$

(4)

The data of which m is smaller than 0 in Table 2 is not included in the regression analyses, because the tsunami height is smaller than 15 cm at the distance of 100 km for $m = -0.5$ and smaller than -0.5, which we considered large uncertainty in observations. The regression relation for the NFSS category is not calculated, because it is irrelevant in this study. However, it should be noted that the relationship looks similar to that for ADDS in general and deviates from ASSS.

The purpose of above analyses is not to introduce other empirical relationships but to quantify the difference in tsunami excitation due to the seismic segmentation. We find that m of ASSS is about 0.7 larger than that of ADDS for the earthquakes with Mo of 10^{20} Nm and 0.6 for 10^{21} Nm, which are within the 95% confidence interval. Considering the scatter of the data in Figure 4 and the basic data of tsunami heights to determine m, the difference of m would be about 1.0 at most.

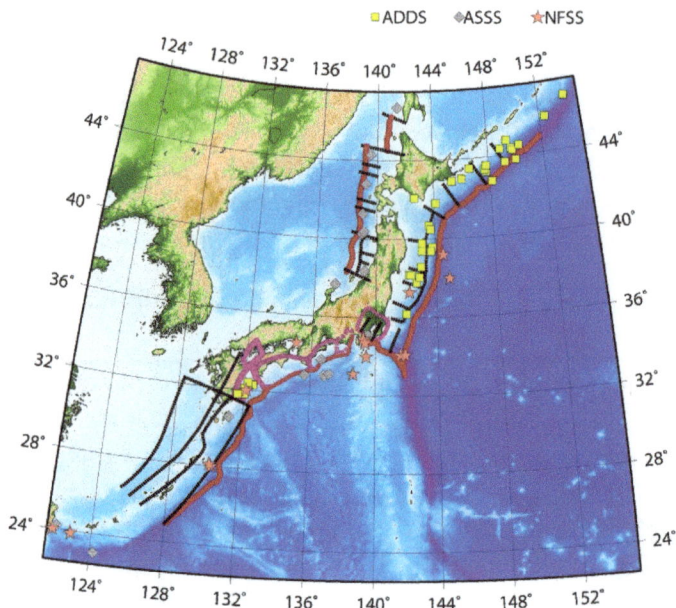

Figure 3. Epicenter locations of tsunamigenic earthquakes near Japan from 1923. Earth-quakes are categorized into ADDS, ASSS and Normal Fault/Strike Slip (NFSS), which are plotted by different symbols. Seismic segments in Figure 1 are also drawn.

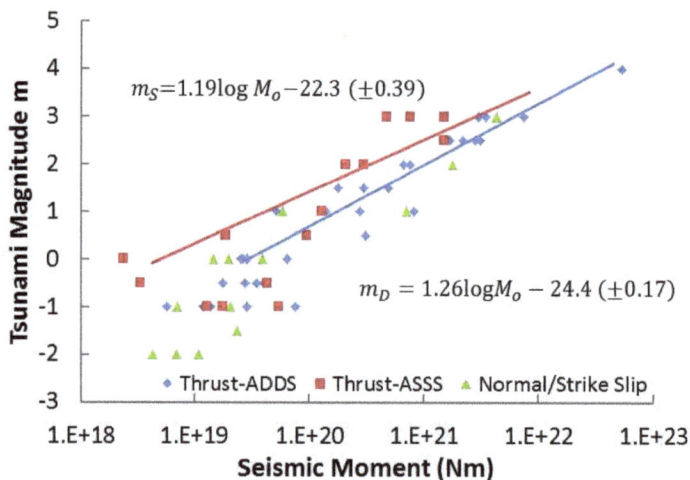

Figure 4. Tsunami magnitude m in relation to seismic moment Mo of corresponding earthquakes near Japan. Different symbols represent different category of earthquakes. Least squares regressions between m (m_D and m_S) and Mo for ADDS and ASSS earthquakes, respectively are derived.

Table 2. Tsunami magnitude and seismic moment of earthquakes near Japan since 1923.

Year	Month	Day	Location	$Mw^{\#}$	$Mt^{\$}$	$m^{\%}$	Mo (N·m)	Dip^{+}	DipSlip*	Segment	Ref
1923	9	1	Kanto	7.9	8.0	2.0	7.60×10^{20}	0.56	0.17	ADDS	[29]
1933	3	3	Sanriku	8.4	8.3	3.0	4.30×10^{21}			NF	[30]
1938	11	5	Fukushima	7.8	7.6	1.0	7.00×10^{20}			NF	[31]
1940	8	2	W.Hokkaido	7.5	7.7	2.0	2.10×10^{20}	0.72	0.72	ASSS	[32]
1944	12	7	Tonankai	8.1	8.1	2.5	1.50×10^{21}	0.17	0.17	ASSS	[33]
1946	12	21	Nankaido	8.1	8.1	3.0	1.50×10^{21}	0.17	0.17	ASSS	[33]
1952	3	4	Tokachi-oki	8.1	8.2	2.5	1.70×10^{21}	0.34	0.33	ADDS	[34]
1963	10	13	Kurile	8.5	8.4	3.0	7.50×10^{21}	0.37	0.37	ADDS	[34]
1964	5	7	Oga-oki	7.0	7.1	−0.5	4.30×10^{19}			ASSS	[32]
1964	6	16	Niigata	7.6	7.9	2.0	3.00×10^{20}	0.94	0.94	ASSS	[35]
1968	4	1	Hyuganada	7.4	7.7	1.5	1.80×10^{20}	0.29	0.29	ADDS	[36]
1968	5	16	Tokachi-oki	8.2	8.2	2.5	2.80×10^{21}	0.34	0.21	ADDS	[37]
1968	6	12	Iwate-oki	7.1	7.4	1.0	5.10×10^{19}	0.50	0.24	ADDS	[38]
1969	8	12	Kurile	8.2	8.2	2.5	2.20×10^{21}	0.27	0.27	ADDS	[39]
1970	7	26	Hyuganada	7.0	7.1	−0.5	4.10×10^{19}			ADDS	[36]
1971	9	6	Sakhalin	7.3	7.5	0.5	9.50×10^{19}	0.63	0.62	ASSS	[32]
1973	6	17	Nemuro-oki	7.8	8.1	2.0	6.70×10^{20}	0.45	0.42	ADDS	[40]
1975	6	10	Kurile	7.0	7.9	1.5	3.00×10^{20}	0.22	0.22	ADDS	[41]
1978	1	14	Oshima	6.6	6.7	−2.0	1.10×10^{19}			SS	[42]
1978	6	12	Miyagi-oki	7.6	7.4	0.5	3.10×10^{20}	0.24	0.21	ADDS	[43,44]
1980	6	29	E. Izu	6.4	6.3	−2.0	7.00×10^{18}			SS	[43,44]
1982	3	21	Urakawa	6.9	7.1	0.0	2.60×10^{19}	0.47	0.47	ADDS	[43,44]
1982	7	23	Ibaraki-oki	7.0	7.0	−0.5	2.80×10^{19}			ADDS	[43,44]
1983	5	26	C.Nihonkai	7.9	8.1	3.0	7.60×10^{20}	0.45	0.45	ASSS	[43,44]
1983	6	21	W. Aomori	7.0	7.3	0.5	1.90×10^{19}	0.68	0.68	ASSS	[43,44]
1984	3	24	Etorof-oki	7.1	7.1	0.0	6.40×10^{19}	0.29	0.28	ADDS	[43,44]
1984	8	7	Hyuganada	6.9	6.9	−1.0	2.90×10^{19}			ADDS	[43,44]
1984	9	19	Boso-oki	6.8	7.3	0.0	2.00×10^{19}			NF	[43,44]
1986	11	15	Taiwan-oki	7.3	7.6	1.0	1.30×10^{20}	0.54	0.54	ASSS	[43,44]
1989	10	29	Sanriku-oki	6.9	6.8	−1.0	5.80×10^{18}			ADDS	[43,44]
1989	11	2	Sanriku-oki	7.2	7.5	1.0	1.40×10^{20}	0.23	0.23	ADDS	[43,44]
1990	2	20	N. Oshima	6.2	6.5	−2.0	4.30×10^{18}			SS	[43,44]
1990	9	24	Tokai-oki	6.5	6.8	−1.0	7.10×10^{18}			SS	[43,44]
1991	12	22	Uruppu Isl	7.5	7.5	1.0	2.80×10^{20}	0.28	0.27	ADDS	[43,44]
1992	7	18	Sanriku-oki	6.8	7.2	0.0	2.70×10^{19}	0.19	0.18	ADDS	[43,44]
1993	2	7	Noto Pen.	6.6	6.7	−0.5	3.40×10^{19}			ASSS	[43,44]
1993	7	12	SW Hokkaido	7.7	8.1	3.0	4.70×10^{20}	0.57	0.57	ASSS	[43,44]
1994	10	4	E Hokkaido	8.1	8.2	3.0	3.00×10^{21}	0.66	0.27	ADDS	[43,44]
1994	12	28	Sanriku-oki	7.7	7.7	1.5	4.90×10^{20}	0.21	0.19	ADDS	[43,44]
1995	1	17	S Hyougo	6.8	6.4	−1.5	2.40×10^{19}			SS	[43,44]
1995	10	18	Kikaijima	6.9	7.6	1.0	5.90×10^{19}			NF	[43,44]
1995	10	19	Kikaijima	6.7	7.3	0.0	1.50×10^{19}			NF	[43,44]
1995	12	4	Etorof-oki	7.6	7.6	1.0	8.20×10^{20}	0.21	0.21	ADDS	[43,44]
1996	10	19	Hyuganada	6.6	6.9	−1.0	1.40×10^{19}			ADDS	[43,44]
1996	12	3	Hyuganada	6.7	6.7	−1.0	1.20×10^{19}			ADDS	[43,44]
2001	12	18	Yonagunijima	6.8	6.8	−1.0	2.10×10^{19}			NF	[43,44]
2002	3	31	E Taiwan	7.1	7.2	−1.0	5.40×10^{19}			ASSS	[43,44]
2003	9	26	Tokachi-oki	8.1	8.1	2.5	3.10×10^{21}	0.19	0.14	ADDS	[43,44]
2003	10	31	Fukushima-oki	6.8	7.0	−0.5	3.50×10^{19}			ADDS	[43,44]
2005	8	16	Miyagi-oki	7.1	7.0	−1.0	7.60×10^{19}			ADDS	[43,44]
2005	11	15	Sanriku-oki	6.9	7.3	0.0	3.90×10^{19}			NF	[43,44]
2006	11	15	Kurile	7.9	8.2	3.0	3.50×10^{21}	0.26	0.26	ADDS	[43,44]
2007	1	13	Kurile	8.2		2.0	1.80×10^{21}			NF	[43,44]
2007	3	25	Noto Pen.	6.9		−1.0	1.30×10^{19}			ASSS	[43,44]
2007	8	2	Sakhalin	6.2		0.0	2.40×10^{18}	0.67	0.67	ASSS	[43,44]
2008	7	19	Fukushima-oki	6.9		0.0	2.90×10^{19}	0.28	0.28	ADDS	[43,44]
2008	9	11	Tokachi-oki	7.1		−0.5	1.80×10^{19}			ADDS	[43,44]
2009	10	30	Amamioshima	6.8		−1.0	1.80×10^{19}			ASSS	[43,44]
2011	3	11	Tohoku-oki	9.0	9.1	4.0	5.31×10^{22}	0.17	0.17	ADDS	[43,44]

$Mw^{\#}$; Moment magnitude [45], $Mt^{\$}$; Tsunami magnitude [46], $m^{\%}$; Tsunami magnitude [8,25,26]; $^{+}$Effect of fault dip; $\sin \delta$, *Effect of fault dip and slip; $\sin\delta\sin\lambda$.

Since the increase of one unit in m indicates the increase by the factor of 2.24 in tsunami heights [25,26], the above difference of 1.0 means the tsunami heights of ASSS earthquakes are about two times larger than those of ADDS earthquakes with the same seismic moment.

Figure 5 shows the relations between Mt and Mo of earthquakes in ADDS (Mt_D) and ASSS (Mt_S):

$$Mt_D = 1/1.5 \log Mo - 6.04 \pm 0.03$$
$$Mt_S = 1/1.5 \log Mo - 5.86 \pm 0.04$$

(5)

where the coefficient of $1/1.5$ of the relations is from the definition of Mw [28] with the assumption of $Mt = Mw$;

$$Mw = 1/1.5 \log Mo - 6.06 \qquad (6)$$

Although there seems to be a little difference in plots between ADDS and ASSS in Figure 5, the difference is within the estimation uncertainty. It is about 0.2 in Mt from Equation (5), which would suggest that tsunami heights by ASSS earthquakes are larger by a factor of about 1.8 than those by ADDS earthquakes with the same value of seismic moment. All the discussions here suggest that the tsunami excitations by ASSS earthquakes are larger, by more or less about two times, than those by ADDS earthquakes near Japan.

5. Discussion

The difference in tsunami excitation is obtained by earthquakes in ADDS and ASSS near Japan, as shown in Figures 4 and 5. The difference is by a factor of about two. Similar emphasis has been made for tsunamigenic earthquakes in the Japan Sea through the analyses including those of tsunamis with smaller tsunami magnitude [7–9]. Generally speaking, the larger dip angles and dip components of dislocations on the faults of earthquakes in the Japan Sea give rise to the larger tsunami excitation. This is the major reason for the difference stated in the previous studies, in addition to the effect of a soft ocean-bottom structure. However, the previous emphasis should not be restricted to tsunamis in the Japan Sea, since ASSS earthquakes outside of the Japan Sea show larger tsunami excitation similar to those in the Japan Sea (Table 3). Therefore, the difference may not be attributed to the local ocean-bottom structures nor to faulting parameters but to the global characteristics of seismic segmentation or to the geological activity which makes such the surface structures.

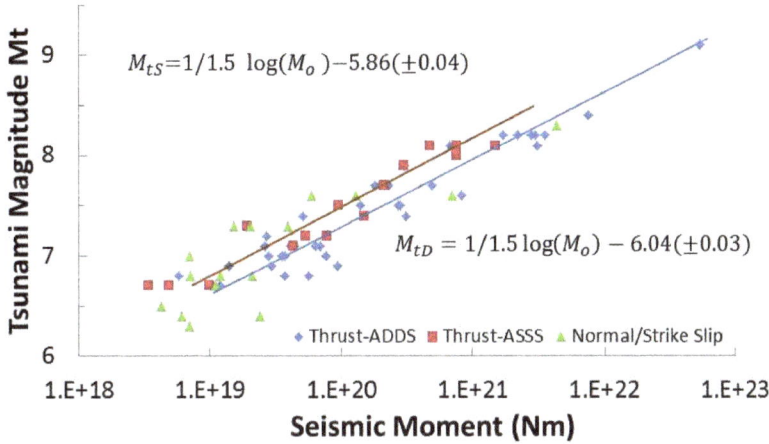

Figure 5. Tsunami magnitude Mt [27] in relation to seismic moment Mo of each earth-quake near Japan. Others are the same as those in Figure 4.

Table 3. Effect of dip angles and slip angles of dislocations on faults resposible for ocean-bottom vertical deformation. Average values are calculated from Table 2.

	Dip Angle $(\sin\delta)$	Sigma $^{\%}$	Dip and Slip Angle $(\sin\delta\sin\lambda)$	Sigma $^{\%}$
ADDS	0.32	0.13	0.26	0.08
ASSS	0.55	0.24	0.55	0.24
ASSSexJapan *	0.44	0.25	0.44	0.25

$^{\%}$ Standard deviation, * Earthquakes classified ASSS but not in the Japan Sea.

Tsunami height is basically responsible for the ocean-bottom vertical deformation due to earthquake faults [47], so that the difference obtained indicates larger ocean-bottom vertical deformation for ASSS earthquakes than that for ADDS earthquakes of the same seismic moment. There would be many reasons to lead to such the difference as follows;

(A) Since the focal depth of ASSS earthquakes is in general shallower than those of ADDS earthquakes, the dislocation by ASSS earthquakes on particular underground faults produces larger effect on ocean-bottom deformation.

(B) Since the dip angle of faults by ASSS earthquakes is larger than those by ADDS earthquakes, the focal mechanism affects larger ocean-bottom deformation.

(C) Slip angle on the faults specifies the partition of along dip and along strike components of the dislocation on a fault. Focal mechanism of ASSS earthquakes provides more along dip components than that of ADDS earthquakes.

(D) Due to three dimensional structures of ocean-bottom topography, such as trench and continental shelf, tsunami excitations become large in ASSS.

37

(E)   Focusing effect of tsunamis due to ocean-bottom topography near ASSS earthquake sources results in larger tsunamis.

(F)   Soft skin layer in the ocean bottom enhances the vertical deformation, therefore, the soft sediment in the Japan Sea side would enhance the tsunami excitation there.

The reason (A) is the result of the character of ASSS asperities and/or coupled zones existing in the shallower depth of plate interfaces than that of ADDS asperities, as described in Section 2 and Figure 2. The reason (B) is related to (F), that is, dip angles of earthquake faults in the Japan Sea are generally larger than those of the subduction zone earthquakes in the Pacific Ocean side (Table 3). However, ASSS earthquakes in this study include earthquakes in the Nankai Trough and the Okinawa Trough along the Philippine Sea Plate subduction as well as earthquakes in the west coast of Sakhalin. It is also apparent from the result in Table 3 that the effect of dip and slip angles of faults classified into ASSS earthquakes excluding the events in the Japan Sea is consistent with that of ASSS earthquakes. Smaller tsunamis ($m \leqslant -0.5$) which were not included in the present analysis do show the similar tendency of the larger tsunami excitations in ASSS compared to those in ADDS (Figure 4). Furthermore, a theoretical study on the deformation due to earthquake faults gives that the difference in the maximum deformation on the surface between dip angles with 30 and 60 degrees is within about 30% [48–50]. That is too small to comprehend the observed difference in Equation (4) between the Japan Sea and the Pacific Sea side of Japan. For (B) and (C), dip angles and slip angles of faults in ADDS and ASSS are evaluated and listed in Table 2. It is found that there are larger values for the earthquakes in the Japan Sea side than those in the Pacific Ocean side, but these earthquakes are not restricted in the Japan Sea as is mentioned in (B).

Regarding reason (D), ASSS earthquakes include those occurring in the Japan Sea side, in the back-arc basin without apparent trenches, as well as along the trench with developed continental shelves in the Nankai Trough. They generate larger tsunamis than those by ADDS earthquakes elsewhere. This eliminates the possibility of (D) for the different tsunami excitation. Speaking about (E), the results from both the local and far-field tsunami magnitude give consistent results in Equations (4) and (5). Therefore, the effect of local ocean-bottom structures alone is not the reason. Considering the reason (F) in the above, we should remind readers that tsunami earthquakes, earthquakes with slow-slip characteristics along the Sanriku Coast of the Tohoku, have occurred, where the source region of the 2011 megathrust earthquake located. The subduction zone shows a developed Horst and Graben structure [19–21], which is formed by sedimentation and related to the generation of tsunami earthquakes. The source region is not similar to the sedimentary layer in the Japan Sea.

6. Conclusions

The relationship has been derived between moment magnitude Mw of earthquake sources and local tsunami magnitude m [25,26] for the tsunamigenic earthquakes near Japan, and it turns out to be similar to the relation between Mt and Mw [27]. These two tsunami magnitude scales suggest that the tsunami excitations in ASSS are about two times larger than those of ADDS near Japan. Observations of tsunamis are of special importance in measuring the strength of historical large earthquakes before the dawn of instrumental seismometry in the 19th century. Previous studies pointed out that there are larger tsunami excitations in the Japan Sea compared to those in the Pacific Ocean side of Japan; however, it was found in this study that the larger excitation of tsunamis is not only restricted for earthquakes in the Japan Sea but also for those in the extended seismic region of Along-strike Single Segmentation (ASSS). The reason inducing this difference has been discussed in detail and the most probable reason is suggested that the asperity of earthquakes in ASSS spreads over the whole shallow area of the subduction interface. The rupture of the shallow asperity causes large ocean-bottom deformation. On the other hand, an asperity in the deeper seismic segments of the subduction interface of ADDS exists as well as an everlasting asperity in the shallower seismic segments, which form the double aligned seismic segments along the dip direction of the subduction. In the deeper part, repeated large earthquakes occur. A rupture of such asperity induces a smaller amount of the ocean-bottom deformation compared to that by ASSS events. The asperity in the shallower part of ADDS ruptures only in the case of a megathrust event as a millennial event. This rupture process has been observed in the 2011 Tohoku-oki megathrust and the 1964 great Alaska earthquake, and the other candidate for this type of a gigantic tsunami is the 17th century tsunami in Hokkaido, Japan [21]. The global importance of different tsunami excitation due to the seismic segmentation of ADDS or ASSS is stressed and is now open to future studies.

It is necessary to take into account the different excitation of tsunamis due to the seismic segmentation pointed out in this study and to consider the effect of dip and slip angles to retrieve the information on the source size of historical earthquakes. This is also important in estimating the strength of tsunamis for future earthquakes judging from so-called seismic gaps and/or doughnut patterns as well as from geodetic deformations by GPS observations.

Acknowledgments: A.R. Gusman kindly discussed with us on the tsunami excitation, showing many of his synthetic calculations to us. We thank K.T. Ginboi for his thorough review of the galley proof of the manuscript.

Author Contributions: All co-authors contributed to the present study. J. Koyama performed the data analysis and prepared the manuscript. M. Tsuzuki performed the data collection and analysis. K. Yomogida took part in discussions regarding the analyses, reviewed and edited the manuscript.

Conflicts of Interest: The authors declare no conflict of interest.

References

1. Lay, T.; Kanamori, H.; Ammon, C.J.; Nettles, M.; Ward, S.N.; Aster, R.; Beck, S.L.; Bilek, S.L.; Brudzinski, M.R.; Butler, R.; *et al.* The great Sumatra-Andaman earthquake of 26 December, 2004. *Science* **2005**, *308*, 1127–1138. PubMed]
2. Bilek, S.L. Invited review paper: Seismicity along the South American subduction zone: Review of large earthquakes, tsunamis, and subduction zone complexity. *Tectonophysics* **2010**, *495*, 2–14.
3. Peng, Z.; Gomber, J. An integrated perspective of the continuum between earthquakes and slow-slip phenomena. *Nature Geosci.* **2010**, *3*, 599–607.
4. Yomogida, K.; Yoshizawa, K.; Koyama, J.; Tsuzuki, M. Along-dip segmentation of the 2011 off the Pacific coast of Tohoku earthquake and comparison with other megathrust earthquakes. *Earth Planets Space* **2011**, *63*, 697–701.
5. Koyama, J.; Yoshizawa, K.; Yomogida, K.; Tsuzuki, M. Variability of megathrust earthquakes in the world revealed by the 2011 Tohoku-oki earthquake. *Earth Planets Space* **2012**, *64*, 1189–1198.
6. Kopp, H. The control of subduction zone structural complexity and geometry on margin segmentation and seismicity. *Tectonophysics* **2013**, *589*, 1–16.
7. Abe, K. Quantification of major earthquake tsunamis of the Japan Sea. *Phys. Earth Planet. Inter.* **1985**, *38*, 214–223.
8. Hatori, T. Characteristics of tsunami magnitude near Japan. *Rep. Tsunami Eng.* **1996**, *13*, 17–26.
9. Watanabe, H. Regional variation of the formulae for the determination of tsunami magnitude in and around Japan. *J. Seism. Soc. Japan* **1995**, *48*, 271–280.
10. Available online: http://www.j-shis.bosai.go.jp/map/?lang=en (accessed on 21 September 2015).
11. Available online: http://www.jma.go.jp/jma/en/Activities/earthquake.html (accessed on 21 September 2015).
12. Kelleher, J.; Savino, J. Distribution of seismicity before large strike slip and thrust-type earthquakes. *J. Geophys. Res.* **1975**, *80*, 260–271.
13. Koyama, J.; Tsuzuki, M. Activity of significant earthquakes before and after megathrust earthquakes in the world. *J. Seism. Soc. Jpn.* **2014**, *66*, 83–95.
14. Mogi, K. Seismicity in western Japan and long-term earthquake forecasting. In *Earthquake Prediction*, Maurice Ewing Series; 4th ed.; Simpson, D.W., Richards, P.G., Eds.; AGU: Washington, DC, USA, 1981; pp. 43–51.
15. Moreno, M.; Rosenau, M.; Oncken, O. Maule earthquake slip correlates with pre-seismic locking of Andean subduction zone. *Nature* **2010**, *467*, 198–202. PubMed]
16. Kohketsu, K.; Yokota, Y.; Nishimura, N.; Yagi, Y.; Miyazaki, S.I.; Satake, K.; Fujii, Y.; Miyake, H.; Sakai, S.; Yamanakae, Y.; *et al.* A unified source model for the 2011 Tohoku earthquake. *Earth Planet. Sci. Let.* **2011**, *310*, 480–487.

17. Maeda, T.; Furumura, T.; Sakai, S.; Shinohara, M. Significant tsunami observed at ocean-bottom pressure gauges during the 2011 off the Pacific coast of Tohoku earthquake. *Earth Planet Space* **2011**, *63*, 803–808.

18. Ruff, L.; Kanamori, H. The rupture process and asperity distribution of three great earthquakes from long-period diffracted P-waves. *Phys. Earth Planet. Inter.* **1983**, *31*, 202–230.

19. Tanioka, Y.; Ruff, L.; Satake, K. What controls the lateral variation of large earthquake occurrence along the Japan trench. *The Island Arc* **1998**, *6*, 261–266.

20. Bell, R.; Holden, C.; Power, W.; Wang, X.; Downes, G. Hikurangi margin tsunami earthquake generated by slow seismic rupture over a subducted seamount. *Earth Planet. Sci. Let.* **2014**, *397*, 1–9.

21. Ioki, K.; Tanioka, Y. Re-evaluated fault model of the 17th century great earthquake off Hokkaido using tsunami deposit data. *Earth Planet. Sci. Lett.* **2015**. submitted.

22. Imamura, M. Chronological table of tsunamis in Japan. *J. Seism. Soc. Japan* **1949**, *2*, 23–28.

23. Iida, K. Magnitude and energy of earthquakes associated by tsunami, and tsunami energy. *J. Earth Sci. Nagoya Univ.* **1958**, *6*, 101–112.

24. Available online: http://www.ngdc.noaa.gov/nndc/struts/form?t=101650&s=70&d=7 (accessed on 21 September 2015).

25. Hatori, T. Relation between tsunami magnitude and wave energy. *Bull. Earthq. Res. Inst. Tokyo Univ.* **1979**, *54*, 531–541.

26. Hatori, T. Classification of tsunami magnitude scale. *Bull. Earthq. Res. Inst. Tokyo Univ.* **1986**, *61*, 473–515.

27. Abe, K. Size of great earthquakes of 1837–1974 inferred from tsunami data. *J. Geophys. Res.* **1979**, *84*, 1561–1568.

28. Hanks, T.; Kanamori, H. A moment magnitude scale. *J. Geophys. Res.* **1979**, *84*, 2348–2350.

29. Kanamori, H. Faulting of the great Kanto earthquake of 1923 as revealed by seismological data. *Bull. Earthq. Res. Inst. Tokyo Univ.* **1971**, *49*, 13–18.

30. Kanamori, H. Seismological evidence for a lithospheric normal faulting—The Sanriku earthquake of 1933. *Phys. Earth Planet. Int.* **1971**, *4*, 289–300.

31. Abe, K. Tectonic implications of the large Shioya-oki earthquakes of 1938. *Tectonophysics* **1977**, *41*, 269–289.

32. Fukao, Y.; Furumoto, M. Mechanism of large earthquakes along the eastern margin of the Japan Sea. *Tectonophysics* **1975**, *25*, 247–266.

33. Kanamori, H. Tectonic implications of the 1944 Tonankai and the 1946 Nankaido earthquakes. *Phys. Earth Planet. Int.* **1972**, *5*, 129–139.

34. Kanamori, H. Synthesis of long-period surface waves and its application to earthquake source studies—Kurile Islands earthquake of October 13, 1963. *J. Geophys. Res.* **1970**, *75*, 5011–5027.

35. Aki, K. Generation and propagation of G waves from the Niigata earthquake of June 16, 1964, Part 1. A statistical analysis. *Bull. Earthq. Res. Inst. Tokyo Univ.* **1966**, *44*, 33–72.

36. Shiono, K.; Mikumo, T.; Ishikawa, Y. Tectonics of the Kyushu-Ryukyu arc as evidenced from seismicity and focal mechanism of shallow to intermediate-depth earthquakes. *J. Phys. Earth* **1980**, *28*, 17–43.

37. Kanamori, H. Focal mechanism of the Tokachi-Oki earthquake of May 16, 1968: Contortion of the lithosphere at a junction of two trenches. *Tectonophysics* **1971**, *12*, 1–13.

38. Yoshioka, N.; Abe, K. Focal mechanism of the Iwate-oki earthquake of June 12, 1968. *J. Phys. Earth* **1976**, *24*, 251–262.

39. Abe, K. Tsunami and mechanism of great earthquakes. *Phys. Earth Planet. Int.* **1973**, *7*, 143–153.

40. Shimazaki, K. Nemuro-oki earthquake of June 17, 1973: A lithospheric rebound at the upper half of the interface. *Phys. Earth Planet. Int.* **1974**, *9*, 315–327.

41. Takemura, M.; Koyama, J.; Suzuki, Z. Source process of the 1974 and 1975 earthquakes in Kurile Islands in special relation to the difference in excitation of tsunami. *Tohoku Geophys. J.* **1977**, *24*, 113–132.

42. Shimazaki, K.; Sommerville, P. Static and dynamic parameters of the Izu-Oshima, Japan, earthquake of June 14, 1978. *Bull. Seism. Soc. Am.* **1979**, *69*, 1343–1378.

43. Available online: http://earthquake.usgs.gov/earthquakes/eqarchives/sopar/ (accessed on 21 September 2015).

44. Available online: http://www.globalcmt.org/CMTsearch.html (accessed on 21 September 2015).

45. Available online: http://earthquake.usgs.gov/earthquakes/world/historical_country. php (accessed on 21 September 2015).

46. Available online: http://wwweic.eri.u-tokyo.ac.jp/tsunamiMt.html (accessed on 21 September 2015).

47. Gusman, A.R.; Tanioka, Y. W phase inversion and tsunami inundation modeling for tsunami early warning: Case study for the 2011 Tohoku event. *Pure Appl. Geophys.* **2013**, *170*.

48. Matsu'ura, M.; Sato, R. Displacement fields due to the faults. *J. Seism. Soc. Jpn.* **1975**, *28*, 429–434.

49. Himematsu, Y. Hokkaido Univ. prepared numerical calculations on the displacements fields based on Okada, Y. *Bull. Seism. Soc. Am.* **1992**, *82*, 1018–1040.

50. Available online: http://www.bosai.go.jp/study/application/dc3d/DC3Dhtml_J.html (accessed on 21 September 2015).

Time Evolution of Man-Made Harbor Modifications in San Diego: Effects on Tsunamis

Aggeliki Barberopoulou, Mark R. Legg and Edison Gica

Abstract: San Diego, one of the largest ports on the U.S. West Coast and home to the largest U.S. Navy base, is exposed to various local and distant tsunami sources. During the first half of the twentieth century, extensive modifications to the port included but were not limited to dredging, expansion of land near the airport and previous tidal flats, as well as creation of jetties. Using historical nautical charts and available Digital Elevation Models, this study gives an overview of changes to San Diego harbor in the last 150+ years due to human intervention and examines the effects of these changes on tsunamis. Two distant and two local scenarios were selected to demonstrate the impact of modified nearshore topography and bathymetry to incoming tsunamis. Inundation pattern, flow depths, and flooded localities vary greatly from year to year in the four scenarios. Specifically, flooded areas shift from the inner harbor to outer locations. Currents induced by the distant tsunamis intensify with modifications and shift from locations primarily outside the harbor to locations inside. A new characteristic in tsunami dynamics associated with port modifications is the introduction of high current spots. Numerical results also show that the introduction of high currents could threaten navigation, vessels, and facilities at narrow openings and also along the harbor "throat"—therefore, at an increased number of locations. Modifications in the port show that changes could have a negative but also a positive impact through constraint of flooding outside of the harbor and shifting of high currents to locations of minimal impact. The results of this study may be used as a first step toward future harbor design plans to reduce tsunami damages.

Reprinted from *J. Mar. Sci. Eng.* Cite as: Barberopoulou, A.; Legg, M.R.; Gica, E. Time Evolution of Man-Made Harbor Modifications in San Diego: Effects on Tsunamis. *J. Mar. Sci. Eng.* **2015**, 3, 1382–1403.

1. Introduction

Harbor modifications and redevelopment plans (see current plans for the Port of San Diego in [1]) are quite common to enhance or increase space in ports and, thus, to improve operations. In fact, harbor modifications at San Diego Bay decreased the water area by filling shoals around the main channels to add more land, including the area where Lindbergh Field—now known as San Diego Airport—was built. Space

43

was developed for mooring boats, including private yachts and Navy vessels, but the reduction of the water surface area increased tsunami amplitudes and currents by constricting channels between Bay margins. Deepening of channels by dredging to improve navigability also enables greater penetration of tsunami energy into the harbor. Similar problems occurred along major rivers, e.g., the Mississippi River, when construction of levees restricted the channel width, negating the purpose of natural flood plains to distribute and absorb the excess water volume during major floods (e.g., severe flooding in the Midwest and South in 1973 [2]).

Ports are designed to be protected against vertical water fluctuations, but the subject of tsunami currents—a newly recognized problem—is not typically considered in the design of harbor modifications. Tsunami currents form a common threat to ports, as seen in previous tsunamis (22 May 1960, Chile; 27 March 1964, Alaska; 27 February 2010, Chile; 11 March 2011, Japan; e.g., [3,4]); usually, they are not accompanied by significant (>1 m) tsunami amplitudes. Tsunami amplitudes are related to the tsunamigenic potential of a source, but tsunami duration and tsunami currents—even if they relate to relatively small amplitude waves—are also important. The long duration of major trans-Pacific tsunamis (from M_w 8+ subduction events) creates a persistent hazard that may produce the greatest damage many hours after the initial tsunami arrival.

In order to understand how anthropogenic changes to ports affect tsunamis, we studied the time evolution of man-made modifications in San Diego Bay since the late nineteenth century and its influence on tsunami amplitudes and currents. San Diego Bay is one of the largest harbors on the U.S. West Coast, and it has been significantly transformed in the twentieth century.

Most of the changes in San Diego Bay have included dredging, enlargement of the North Island/Coronado land area, and creation of new marinas by enhancing already existing sand bars and shoals with fill and creating breakwaters. As done along major rivers, dredging increases the channel depth, but addition of levees, breakwaters, and other shore protection structures combined with narrowing of the channel results in the increase of flow depths and current speeds during inundation (flooding or surge). Tsunami currents can cause damage to structures (e.g., by erosion or scouring; [5]).

We examined the effects of harbor modifications on tsunami hazard by simulating major tsunami events from both local and distant sources. The changes primarily occurred during the first half of the twentieth century. Four specific years were selected as representative of major changes in San Diego Bay: 1892, 1938, 1945 and present day. Post-1945, the bay continued to undergo changes but maintained a similar appearance and, therefore, the bathymetry/topography in and around the bay is represented by a Digital Elevation Model (DEM) showing the harbor shape as we know it today (dated 2014). To reflect these changes, historical nautical charts were used to show the state of the bay during that period.

2. San Diego Bay

San Diego Harbor is located on the southwestern corner of the United States. It has a long crescent-like shape of approximately 24 km in length and 1–3 km width. The harbor is the fourth largest in California and is home to a U.S. Navy base offering shelter to several aircraft carriers and other naval vessels[1]. Naval Air Station North Island (NASNI; [6]) is located at the north end of the Coronado peninsula in San Diego Bay. It is the home port of several aircraft carriers of the U.S. Navy, and is part of the largest aerospace-industrial complex in the U.S. Navy, Naval Base Coronado, California ([6]). A cruise ship terminal is located near downtown San Diego, a regular ferry service runs from Coronado to San Diego, and several marinas, for both private and commercial boats, exist throughout the harbor. Although it is naturally well protected from large wave amplitudes, the same does not apply for tsunami currents.

3. Time Evolution of San Diego Bay

San Diego Bay is one of the best natural harbors in the world, protected by the Point Loma Peninsula at the western harbor entrance, and by North Island, Coronado, and the Silver Strand enclosing the inner harbor. Coronado is an affluent resort town located in San Diego County, California, just over 8 km (5.2 mi) SSW of downtown San Diego, with a population of 24,697 ([7]). Coronado Island lies on a peninsula connected to the mainland by a 16-km (10 mi) isthmus called the Silver Strand (locally, The Strand; [8]). North Island was an uninhabited sand flat in the nineteenth century (Figure 1A). It was also referred to as North Coronado Island, because it was separated from South Coronado (now the city of Coronado) by a shallow bay, known as the Spanish Bight, which was later filled in during World War II (1945; see Figures 1 and 2). South Coronado, which is not an island but the terminus of a peninsula known as the Silver Strand, became the town of Coronado.

We present nautical charts that portray the changes that have occurred in San Diego Bay since the nineteenth century. Because modifications to the bay happened quite frequently, especially during the first half of the twentieth century, we summarized major changes over selected years. A small set of original nautical charts are presented next. The major changes primarily included modifications to Coronado Island (also referred to as The Island, The North Island, and Coronado,

---

[1] There are also some submarines, an explosives loading area, and mooring facilities and docks for many other naval vessels throughout the bay ([6]).

or North Beach Island and Coronado separated by Spanish Bight[2] in older charts) and the area around the Dutch Flat (see Figure 1). We use Coronado Island to refer to the entire island of Coronado and North Island after the Spanish Bight was filled in at the post-1945 configuration, and North Island to the pre-1945 configuration to distinguish the northern part of Coronado island separated by Spanish Bight in older charts (see Figure 1).

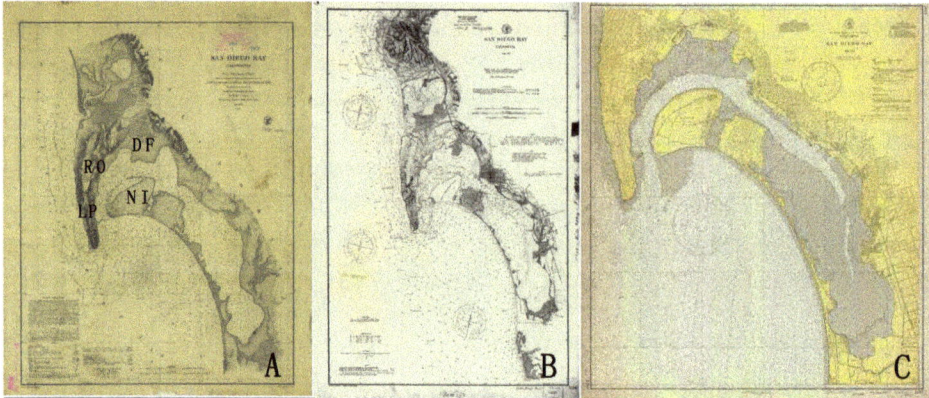

Figure 1. Maps of San Diego Bay. Depths are in feet. (**A**) Nautical chart of 1857. Approximate locations of Dutch Flat (DF), North Island (NI), Roseville (RO), and La Playa (LP) are shown in bold capitals; (**B**) Nautical chart of 1892. The major difference in this chart compared to (A) is the increase of depth at the entrance of the bay, east and northeast of Point Loma (between Roseville and LaPlaya) and between Ballast Point and North Island as a result of dredging. The change appears at least five-fold when compared to the earlier map of 1857; (**C**) Nautical chart of 1938. The major modifications include filling west and north of the North Island (compared to maps A and B) to unify with existing sand dunes and sand spits into North Island and Coronado, with a more regular shape and increased land space. The maps were downloaded from the historical map and chart collection of the Office of Coast Survey, NOAA [9].

3.1. Pre-Nineteenth Century to 1895

During this period, as shown in Figure 1B, the bay remained largely unmodified, with some dredging performed in the late nineteenth century. The depths inside the

---

2   Spanish Bight is a graben associated with the Rose Canyon fault zone, e.g., [10]. San Diego Bay lies within the releasing stepover of the Rose Canyon fault zone, which created the bay and uplift of Point Loma. The Silver Strand is a narrow sand spit that connects North Island/Coronado to the mainland at Imperial Beach to the south. San Diego Bay was a natural lagoon formed behind this barrier beach ([11]).

bay were fairly shallow at the time (characteristic depths 15–54 ft or 4.5–16.5 m near the entrance; Figure 1; [9]). The major characteristics of the bay were the irregularly shaped North Island[3] and the width of the bay around the Dutch Flat (North East of the Bay entrance; see Figure 1) to the northeast. The first tide gauge for San Diego was installed in 1854 at La Playa (near Roseville). It was removed in 1872 and in 1906, it was relocated to the south side of Navy Pier in downtown San Diego. The Dutch Flat area was part of the fan delta complex (depositional area) of the San Diego River, which was diverted to the west through Mission Bay (originally called False Bay) via levee and jetty construction after major floods in the mid-nineteenth century (*ca.* 1862; [12,13]).

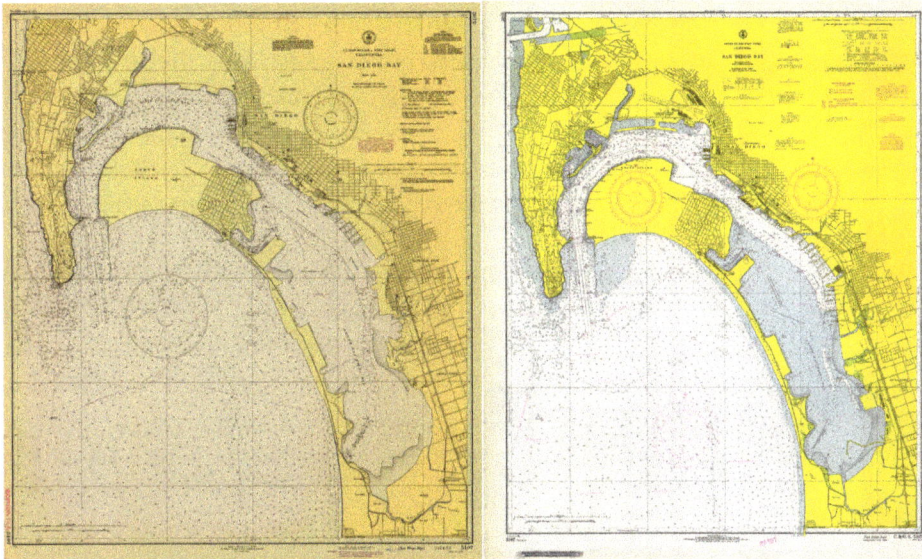

Figure 2. (**Left**) Nautical chart of San Diego Bay of 1945; (**Right**) Nautical chart of San Diego Bay of 1966. The maps were downloaded from the historical map and chart collection of the Office of Coast Survey, National Oceanic and Atmospheric Administration (NOAA; [9]).

3.2. Post-1895 and Early Twentieth Century

Most of the modifications for the enhancement of space in the bay happened during this period and prior to the 1950s. The modifications included dredging at and around the entrance to the bay and expansion around the Dutch Flat, probably

---

[3]   North Island was the source of fresh water for the lighthouse keeper on Point Loma during the nineteenth century.

to enable better use of the Naval Training Station and the U.S. Marine Corps base near the last location[4]. In addition to the previously noted modifications, sand spits and sand dunes close to North Island were connected through filling, for the creation of the more regularly shaped North Island[5] in the 1930s (Figure 1C).

3.3. Circa 1945 (1945 to 1960s)

By 1945, San Diego Bay had changed considerably, looking very close to how we know it today. The major changes to the harbor configuration resulted in narrow channels between islands and mainland shorelines, specifically from the harbor entrance to New San Diego (downtown San Diego). The area between San Diego and Coronado has always been narrow and the bathymetry fairly shallow, although it has increased over the years from 2.5 m (8 ft) in 1857 to about 13 m (40 ft) today; North Island and Coronado Island may also be uplifts flanking the Spanish Bight and Coronado fault segments of the Rose Canyon fault zone [10]. In fact, uplift exists along the west side of the north-trending Coronado Fault, which passes from the east side of Coronado Island to the mainland shore along the Pacific Highway (note how straight this section of coast was in the nineteenth century maps; Figure 1; west of New San Diego).

The Spanish Bight was a narrow channel that separated Coronado from North Island up to the early 1940s (Figure 1). The development of North Island by the U.S. Navy prior to and during World War II led to the filling of the Bight in 1943, combining the land areas into a single body. The Navy still operates NASNI on Coronado. On the southern side of the town is Naval Amphibious Base Coronado, a training center for Navy SEALs. Both facilities are part of the larger Naval Base Coronado complex [14].

---

4   Originally, this area was the fan delta for the San Diego River, which was flooded in the 1861/62 "Noachian Deluge"—a severe flood event. Levees (restraining wall on 1892, no 5106 chart) were built to redirect the river flow to False Bay (Mission Bay), which finally succeeded after a couple more floods broke through to San Diego Bay. Dutch Flat was expanded, and adjoining basins were subsequently filled as Lindbergh Field (the municipal airport) grew and adjacent land was developed. Finally, Shelter Island and Harbor Island were constructed to provide additional protection for marinas and land for hotels, restaurants, *etc.* (e.g., Figure 3).

5   Spanish Bight was mostly filled to connect North Island to Coronado, and additional dredging created local bays, such as Glorietta Bay, for marinas and naval vessels. Piers were constructed at San Diego for commercial boats and naval shipyards to the south at National City where major shipyards were built. The southern end of the bay had salt evaporation ponds, although a marina was also built for the City of Chula Vista. The Silver Strand was enhanced to reduce flooding during storms, and marinas were constructed in some locations (other locations were restricted to military operations including an amphibious base and radio station).

Sand bars (shoals) to the north of the bay were further enhanced to create new marina locations (between Quarantine Station, now known as La Playa[6], and Roseville) in North San Diego Bay (Figures 1 and 2). By 1966, Shelter Island and Harbor Island were created by dredging and filling the area east of Roseville and Dutch Flat, providing more space for marinas (Figure 3). These islands are actually connected to the mainland by causeways on landfill. The depths in the bay are more uniform and larger than before. The area around Dutch Flat, Fisherman's Point and the municipal airport were filled, with the exception of a channel east of Roseville (Figure 2A). This channel provided boat access to the Marine Corps base. Small, elongated boat basins now exist between Shelter Island, Harbor Island and the mainland.

Figure 3. View of Shelter Island in 1937, the 1950s and the 1990s, left to right. Photos courtesy of Unified Port of San Diego [1]. An example of extreme transformation showing how the sand spit, or "mud island" as it was referred to, now offers home to numerous vessels. View is southwest to northeast. The entrance to Shelter Island has also been subject to high currents in past tsunamis (e.g., 22 May 1960, Chile and 11 March 2011, Japan; [4]).

4. Numerical Grids of San Diego

Four different DEMs for San Diego were used to depict the evolution of the port due to man-made modifications. Three were digitized from historical charts of the historical map and chart collection of the Office of Coast Survey, NOAA [9] for the purpose of our numerical simulations. The present day (2014) DEM originated from

---

6   La Playa is the location of the original tide gauge for San Diego. It was first installed in 1854. A strong local earthquake, 27 May 1862 M~6, created a wave in the bay that was observed by the tide gauge engineer (Andrew Cassidy) who was repairing the pier for the tide gauge at the time of the earthquake. He observed a runup of about 1–1.2 m (3–4 ft) on the beach at La Playa but the water returned to its normal level. We do not know if this runup was a vertical or horizontal distance [12,13]. The Rose Canyon fault was likely the source of the earthquake but data are inadequate to determine the epicenter with certainty.

the National Geophysical Data Center (now National Centers for Environmental Information, or NCEI) [15]. The quality of the paper chart (*i.e.*, detail) and the years representing the major changes in the port were the primary factors that affected our choice of charts. The years representing the major changes to the port are: 1892, 1938, 1945, and present day for the current harbor configuration (Figure 4).

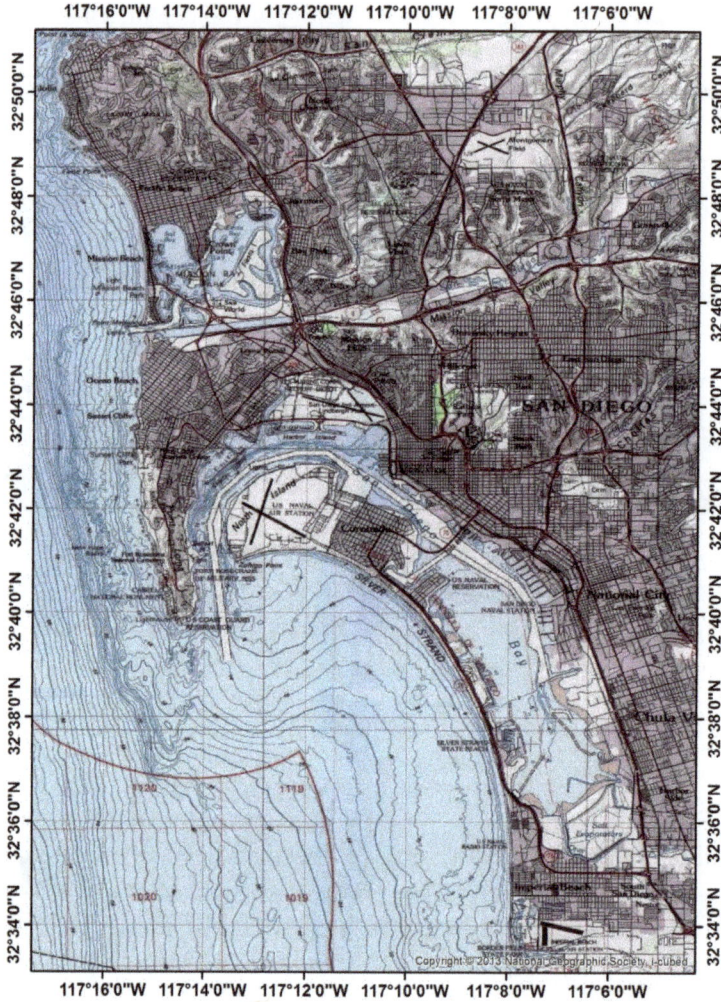

Figure 4. Map of San Diego Bay, showing the state of the harbor today with locations that are referenced frequently throughout the text.

All numerical grids (1892, 1938, 1945, and present day) share two outer grids (usually referred to as A and B grids) that form a nested system. The A grid, or outermost grid, was created from interpolated ETOPO1 data to 30 arc-sec (Figure 5;

~ 900 m [16]). The B grid was also created from interpolated ETOPO1 data to 12 arc-sec [16], while the C grid uses a 3 arc-sec resolution, subsampled from a 1/3 arc-sec resolution dataset (~ 90 m; Figure 5; [17]).

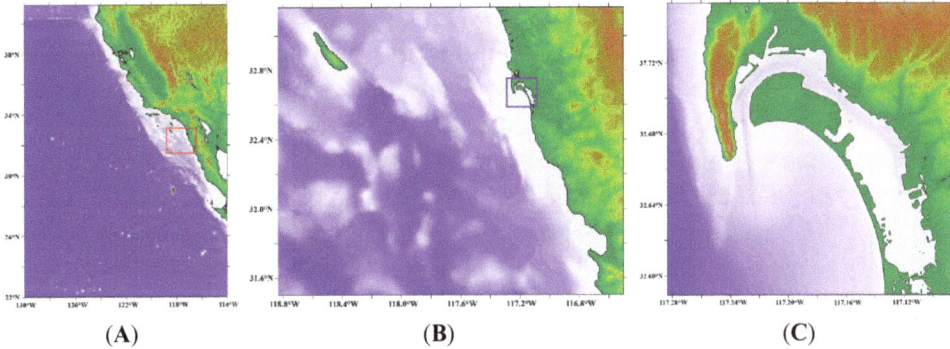

Figure 5. Extent of numerical grids **A (left)**, **B (middle**; shown as red rectangle within the A grid) and **C (right**; shown in blue colour within the B grid). C grid here only shows the current configuration but the extent of all four C grids (1892, 1938, 1945 and 2014) is the same.

Digitization of old nautical charts is challenging, especially in cases where not only the coastline and bathymetry around a port have changed, but also the topography has changed due to urbanization. Nautical charts are meant for navigation, and therefore, elevation information provided in them is limited. Elevation data for past years are also generally difficult to find so a number of assumptions were made for the creation of the DEMs. Elevation data of San Diego for the high elevation areas to the north of San Diego toward Mission Bay were assumed to have changed little in the last 100 years. This may be a reasonable assumption since construction of the restraining wall (levee) to redirect the San Diego River helped to maintain topography in the low area between Mission Bay and San Diego Bay. This allowed us to use topography from the United States Geological Survey (USGS) with datum NAD83[7] (North American Vertical Datum 1988, or NAVD88; [18]). Flat areas close to shores usually have large uncertainties and were manually digitized to match available contour data (Figure 6). The coastline was digitized from the charts in addition to bathymetric points (Figure 6). All these data were combined to create the final DEMs at 90 m resolution and mean high water (MHW) depth.

---

[7] Although contours were matching fairly well with old charts an additional small shift ensured a better fit to the contours in the nautical charts and the coastline.

Topographic maps (e.g., from USGS) generally have elevations referenced to orthometric datums, either the NAVD 88 or the older National Geodetic Vertical Datum 1929 (NGVD 29). All GPS positioning data are referenced to one of many 3-D/ellipsoid datums. NOAA's nautical charts have depths referenced to mean lower low water (MLLW), and bridge clearances are referenced to MHW. The legal shoreline in the U.S., which is the shoreline represented on NOAA's nautical charts, is the MHW shoreline; that is, the land-water interface when the water level is at an elevation equal to the MHW datum. For the construction of DEMs a tidal datum—representing a reference plane—is necessary. A tidal datum refers to an average height of the water level at different phases of the tidal cycle ([19]). The MLLW line is also depicted on NOAA's charts[8] [19]. Mean Sea Level (MSL) is usually the average between MLLW and MHHW[9] as provided on the charts. MHW is more challenging, but we assumed the value of 1.52 m (4.98 ft) above MLLW as estimated by NOAA for San Diego for the years between 1993 and 2003 [20].

We also assumed that the highest elevations on Coronado Island have not changed considerably over time (a 6 m (20 ft) contour is the highest elevation contour that appears in early charts; Figure 1, [9]). The plane of reference used in the charts is the average of MLLW (Figures 1 and 2).

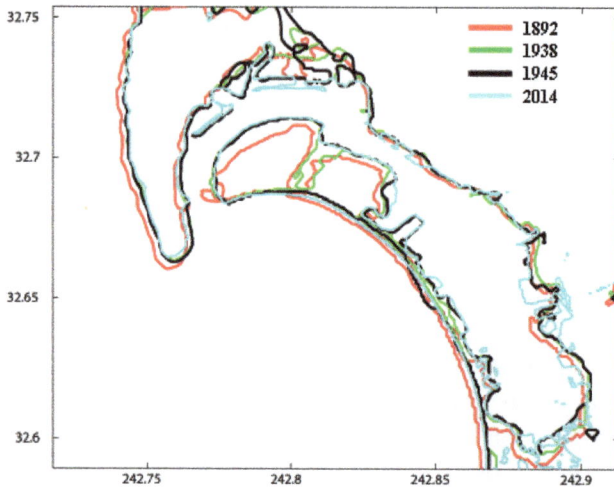

Figure 6. Coastlines of the four digital elevation models (DEMs) used in the numerical simulations.

---

[8] MLLW is the lower low water height average of each tidal day calculated over a tidal datum epoch (19 year period adopted by the National Ocean Service).
[9] MHHW is the higher high water height average of each tidal day calculated over a tidal datum epoch (19 year period adopted by the National Ocean Service).

5. Numerical Modeling

For the numerical modeling of tsunamis, we use the well-known and tested MOST (Method of Splitting Tsunami) code [21], which solves the depth-averaged, nonlinear shallow-water wave equations that allow wave evolution over variable bathymetry and topography [21,22]. This is a suite of numerical codes that model all phases of the tsunami simulation (generation, propagation, and runup). The initial displacement on the surface of the water is obtained directly from the terminal coseismic deformation of the ocean floor using Okada's formulation for the surface expression of shear and tensile faults in a homogeneous, elastic half space [23]. The equations of motion are then solved numerically to propagate the water surface disturbance across the computational domain [21,22,24]. A 0.0009 Manning's roughness coefficient was used in all simulations.

MOST has been tested extensively, complies with the standards and procedures outlined in [25], and has shown reliability in forecasting inundation. The Tohoku 2011 source used in this article was also selected due to a really good fit to DART® buoy data from the Pacific Ocean [26]. More recently it was also shown to compare fairly well against tsunami currents measured by recently installed acoustic Doppler current profilers (ADCP; [27]).

A set of four scenarios were considered on all numerical grids. A common A and B grid was used for all four DEMs; those are the outermost and intermediate grids used in the simulations for computational efficiency, respectively. We applied two distant scenarios and two local/regional scenarios to investigate the role of modified bathymetry and nearshore topography on our results. Distant scenarios were run for approximately 24 h while local scenarios for 12 h each.

6. Tsunami Sources

San Diego is exposed to tsunamis from both local and distant sources. A small but balanced set of sources was selected (two distant, two local) to allow for the arrival of waves from different directions in order to investigate, among other factors, the contribution of source location to tsunami impact in San Diego. Specifically, two distant tsunamis generated in the south and in the northwest Pacific were considered. With respect to near-field sources, one local tectonic source and one landslide source were included for locations to the southwest, where the waves' approach is directly into the bay.

The distant sources are major historical subduction zone earthquakes (in opposite hemispheres): the 22 May 1960 Chile earthquake (M_w 9.5) and the 11 March 2011 Tohoku, Japan (M_w 9.0) earthquake. The local sources are theoretical and include a large transpressional earthquake (M_w 7.4) on the San Clemente fault (SCF), located about 70 km southwest of San Diego Bay, and a large submarine landslide located in Coronado Canyon about 25 km southwest of San Diego. For a comprehensive tsunami hazard assessment for San Diego harbor, considering all types of local, regional, and distant sources capable of generating tsunamis that may be damaging, see [28]. A more detailed discussion of the sources with impact to San Diego, as well as the most significant tsunami records in San Diego, can also be found in the same article.

6.1. Chile 1960

A simple rectangular source (800 km × 200 km) was used to represent the fault surface that produced the 22 May 1960 M_w 9.5 earthquake and transpacific tsunami [29]. The collision zone of Nazca and South American plates (NazSA[10]) is an important source of tsunamis and hazard studies for the U.S. West Coast, and California in particular. The 22 May 1960 Chile earthquake is known as the largest recorded earthquake and was preceded and followed by large earthquakes within a few minutes (~M_w 8.0 or greater; see [30,31]). Rupture of the NazSA zone is likely more complex and possibly longer, but far-field studies are not affected greatly by such variation. We consider a Chile 1960 type of source to represent a source similar to the 22 May 1960 event, represented by a single fault plane [29] with a slip of 24 m. More details are provided in Table 1.

Table 1. Table of sources.

Source Name (Far-field)	Fault Length (km)	Fault Width (km)	Slip (m)	Dip (degrees)	Rake (degrees)	Strike (degrees)	Depth[11] (km)
Chile 1960	800	200	24	10	90	10	87.73
Chile 1960[12]	800	200	24	10	90	10	87.73
Tohoku 2011[13]							
Segment 1	100	50	4.66	19	90	185	5
Segment 2	100	50	12.23	19	90	185	5
Segment 3	100	50	26.31	21	90	188	21.28
Segment 4	100	50	21.27	19	90	188	5
Segment 5	100	50	22.75	21	90	198	21.28
Segment 6	100	50	4.98	19	90	198	5
Source Name (Near-field)	Fault Length (km)	Fault Width (km)	Slip (m)	Dip (degrees)	Rake (degrees)	Strike (degrees)	Depth (km)
San Clemente Fault Bend Region							
Segment 1	18.52	10	1.6	89	−161.6	330.4	0.5
Segment 2	8.86	10	2.7	89	158.2	310.2	0.5
Segment 3	6.96	10	7.0	80	135.0	307.8	0.5
Segment 4	23.65	10	4.2	89	166.0	304.9	0.5
Segment 5	12.92	10	5.8	85	149.0	314.5	0.5
Segment 6	6.9	10	3.2	89	161.6	312.4	0.5
Segment 7	19.5	10	1.6	70	−161.6	311.8	0.5
Landslide Case	**Lat**	**Lon**	**Depth (m)**	**Uplift (m)**	**Subsidence (m)**	**Direction (deg)**	
Coronado Canyon	32.519	−117.338	370	5	5	260	

---

[10] We abbreviated the source simply for easy reference in this article.
[11] Depth is measured from ground surface to top of the fault.
[12] [29].
[13] Source from NOAA Center for Tsunami Research.

6.2. Tohoku 2011

The 11 March 2011 M_w 9.0 Tohoku Japan tsunami caused wide devastation on the Japanese coast, damaged facilities and vessels on the U.S. West Coast [4] and was recorded by multiple sea level stations across the Pacific. As a result of the tsunami, at least 20,000 perished [24] and the nearshore topography was completely redefined [32,33]. The tsunami source used in this study is based on real-time forecasting done by NOAA Center for Tsunami Research. Tsunami wave data from two DART® stations near Japan (21418 and 21401; [34]) were used by an inversion algorithm [26] that selected six unit sources from a pre-computed propagation database and its corresponding coefficients [35]. Parameters of each unit source and their corresponding coefficients are listed in Table 1.

6.3. San Clemente Fault (SCF)—Bend Region

We modeled a major left bend in the SCF offshore northern Baja California, Mexico (Table 1). This is a zone of oblique shortening, where the fault impedes the general northwest movement of the Pacific plate. Seafloor uplift resulting from this restraining bend covers a region of about 875 km² and represents as much as 720 m of tectonic uplift [36–38]. Recent submersible investigations of the SCF revealed vertical seafloor scarps, representative of seafloor uplift during the most recent large earthquakes on the SCF [39]. The largest scarps are 1–3 m high and would be associated with earthquakes of magnitude 6.5 or greater. Three moderate earthquakes (M6) along the SCF have been recorded by modern seismographs in the past 83 years.

6.4. Coronado Canyon Landslide

Sediment and sedimentary rock slope failures may occur along the Coronado Escarpment within Coronado and La Jolla submarine canyons. A large failure in Coronado Canyon was modeled to represent a submarine landslide close to San Diego Bay. Our landslide case was represented as a dipole (uplift pattern; see Table 1; [40,41]).

7. Results

7.1. Chile 1960

The inundation extent and corresponding inundation pattern varies greatly from year to year in this scenario for San Diego (Figure 7). Inundation for the present-day (2014) configuration of San Diego is primarily constrained to the south. For earlier years we see large inundation on the inner side of the harbor and NNE in the area of what used to be large tidal lands. Large inundation is evident for the 1945 DEM, especially north of the San Diego region and close to Lindbergh Field (San Diego airport), but that was reversed with the expansion of the Dutch Flat. The early bay, mostly unmodified by human activity, suffered relatively minor wave amplitudes beyond the harbor entrance. This is probably a result of shallow water at the entrance, which reduces the surge into the harbor. Inundation was increased locally at narrow embayments north and west of San Diego where dredging created deeper channels for boats, and fill areas further restricted the channel width. In contrast, the low-lying Silver Strand was overtopped in earlier configurations, but widening and increased elevation reduced the area of potential inundation in more recent years.

Generally, the pattern of wave amplitude distribution away from (or outside) the harbor has remained the same over the years but wave amplitudes inside the harbor show a gradual increase as a result of modifications. Currents also show a gradual increase as a function of further modifications for this event. A high current curve following the shape of the harbor is evident for all DEMs in this South American scenario. Additional spots of high current activity also appear in later years.

7.2. 11 March 2011 Japan Tsunami

Similar to the previous distant (Chile 1960) scenario, smaller wave amplitudes appear in older configurations. More specifically, at the earliest year (1892), numerical modeling shows energy spots constrained outside the harbor along the coastline and at the entrance of the harbor. Further development in the bay introduced larger waves inside the harbor around La Playa (Figure 8, 1938). In later years, the creation of the jetty near La Playa changed this pattern and redistributed waves away from the original coastline and along the jetty. Currents also show an increase, but the overall trend is not so clear. Generally, in earlier years, currents were more widely distributed compared to more recent years, while spots of high currents were introduced. The high current spots within the harbor correspond to a new characteristic introduced in the second half of the twentieth century. This is common in both distant scenarios.

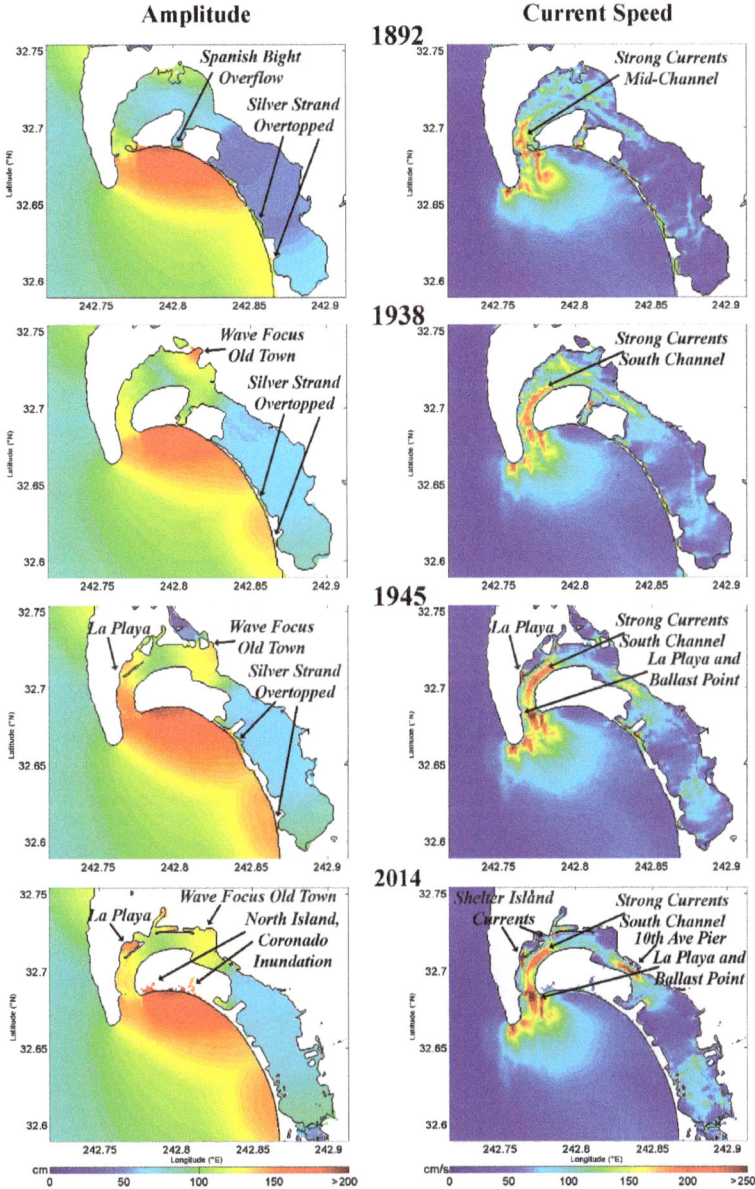

Figure 7. Maximum wave amplitude and maximum current distribution in San Diego harbor for four different configurations (1892, 1938, 1945, and 2014). A Chile 1960 type of scenario was used to investigate changes in wave dynamics as a result of anthropogenic changes. Years indicate time tagging of large changes in the harbor. **Left column** shows wave heights in cm; **right column** tsunami currents in cm/s.

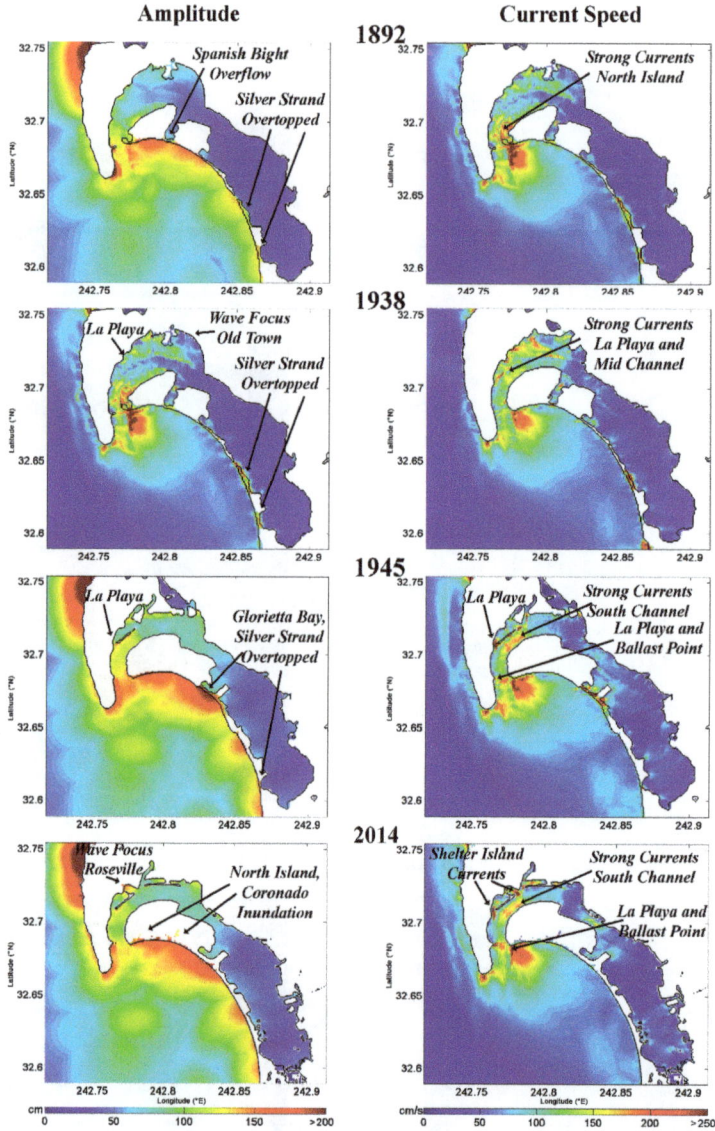

Figure 8. Maximum wave amplitude and maximum current distribution in San Diego harbor for four different configurations (1892, 1938, 1945, and 2014). A tsunami scenario resembling the 11 March 2011 Tohoku tsunami was used to investigate changes in wave dynamics as a result of anthropogenic changes. Years indicate time tagging of large changes in the harbor. **Left column** shows wave heights in cm; **right column** tsunami currents in cm/s.

The tsunami current distribution from both distant scenarios is worth comparing. In the case of Chile 1960, high currents appear to be more widely distributed along the harbor, while for the Japan tsunami, high currents appear more concentrated near the entrance. This would likely be attributed to the location of the tsunami source and directivity of waves. The maximum amplitudes occur on the south side of Coronado/North Island for the Chile event, showing the effect of the southern source (directivity). In contrast, high amplitudes occur west of Point Loma for the Tohoku event, which arrives from the north and west. Multiple wave arrivals due to refraction along Pacific Ocean waveguides, e.g., Emperor–Hawaiian Seamounts produce strong energy many hours after the first wave arrivals ([42]). Later wave arrivals may involve local resonant effects, so large amplitudes and strong currents were observed during both events. The long period character of these distant subduction megathrust events is very important for emergency planning. Major surges arriving several hours after the first waves caused serious and unexpected damage in 2011 (the surges also persisted for several hours in 1960).

We notice high currents in all plots, but it is clear that the creation of narrow islands and boat basins changes the distribution pattern in the harbor. Although there is an increasing trend with the modifications, enlargement of land space and dredging appear also to have contributed to reduction of currents in some cases while introducing high currents elsewhere (see 1892 and 1938 results; Figures 5 and 6). Currents are high in narrow openings as one might expect, and more such spots of high currents appear in later years as San Diego land expansion led to a narrower opening in the port. The development of a narrow opening combined with channel deepening to the harbor appears also to be responsible for the intensification of the currents that follow a curved wider line along the harbor in 1938 as opposed to 1892. This implies increased danger to vessels not only at the entrances of narrow openings, but also along the "throat[14]" of the San Diego harbor. Strong currents were indeed observed in Harbor Island and Shelter Island during the Tohoku tsunami and in other harbors in California where narrow channels exist (e.g., Catalina harbor, King Harbor in Redondo beach *etc.*).

---

[14] Numerous piers and anchorages exist in the area between San Diego and Coronado. Strong currents in these areas were damaging to Navy ships and other boats, while the ferry service was disrupted during the 1960 event. Docks were damaged and some boats dragged from their moorings [3].

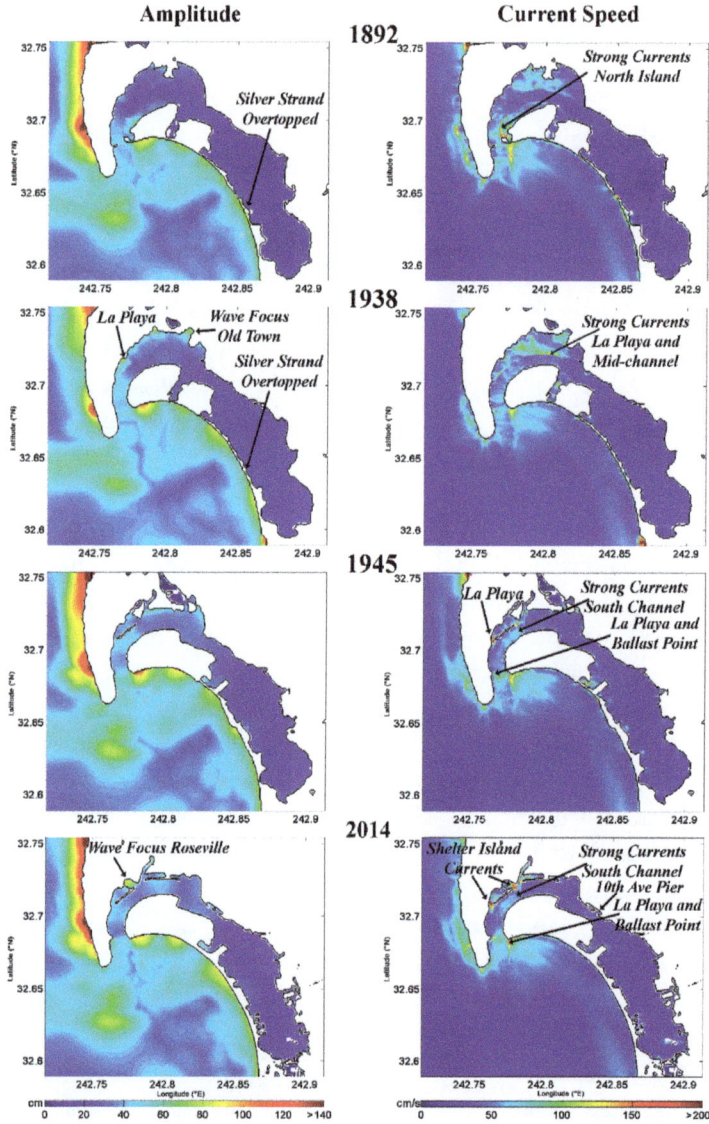

Figure 9. Maximum wave amplitude and maximum current distribution in San Diego harbor for four different configurations (1892, 1938, 1945, and 2014). A San Clemente fault rupture type of tsunami scenario was used to investigate changes in wave dynamics as a result of anthropogenic changes. Years indicate time tagging of large changes in the harbor. **Left column** shows wave heights in cm; **right column** tsunami currents in cm/s.

60

Coronado Canyon - Landslide

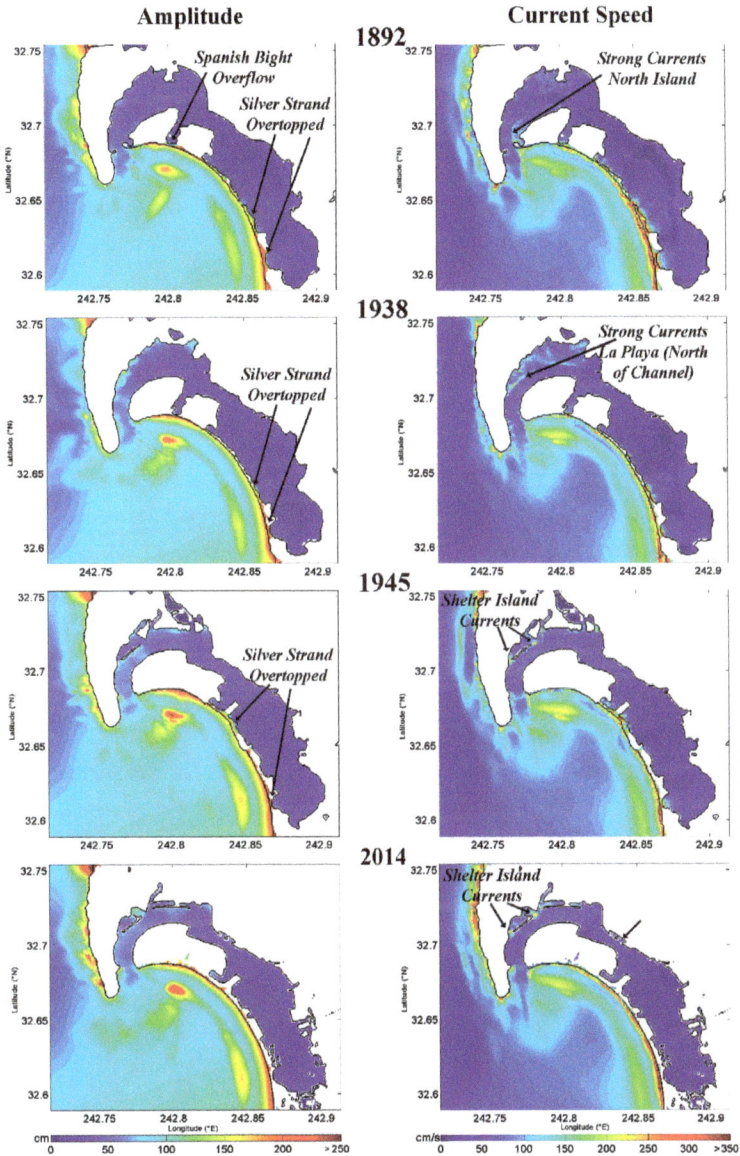

Figure 10. Maximum wave amplitude and maximum current distribution in San Diego harbor for four different configurations (1892, 1938, 1945, and 2014). A hypothetical tsunami scenario generated by a landslide on Coronado Canyon was used to investigate changes in wave dynamics as a result of anthropogenic changes. Years indicate time tagging of large changes in the harbor. **Left column** shows wave heights in cm; **right column** tsunami currents in cm/s.

61

7.3. San Clemente Fault—Bend Region

Wave amplitudes for this scenario appear relatively low (predominantly 1–2 m). The pattern does not change with modifications. Tsunami currents also appear low for all years, but isolated spots of high energy are introduced after the extended modifications at the entrance of boat basins (e.g., La Playa; see Figure 9). The small changes for this event likely are attributed to waves of shorter wavelength.

7.4. Coronado Canyon Landslide

Wave amplitude pattern does not appear vastly changed during the years for this scenario. Currents also generally maintain their distribution pattern while some spots of high activity get introduced. High currents appear constrained mainly along Coronado Island and Silver Strand outside the harbor (Figure 10). Specifically, we notice similar patterns to the other scenarios with increased currents where channels narrow. Most importantly, for these scenarios, amplitudes are similar to distant events or greater (2 m or higher) due to proximity of the source and the landslide mechanism (larger displacement). The Coronado Canyon landslide source has a nearly direct path into San Diego Bay coming from the south.

8. Discussion/Conclusions

San Diego harbor has undergone extensive modifications and development. Four scenarios were run on four different DEMs to investigate the effects of the major harbor changes to tsunamis. The set of scenarios was selected to ensure a balanced set of scenarios that included both distant and local sources with varying geographical locations. The numerical code MOST was used to model all phases of the tsunami (generation to inundation).

Results show varying inundation patterns for the scenarios with some common characteristics for the distant events when compared to the local events. Inundation within the bay generally was reduced and constrained to locations outside the bay mainly due to expansion of land (channel narrowing) and further filling to elevate tidal flats. Introduction of jetties also caused redistribution of wave heights.

With respect to tsunami currents, it appears that spots of high energy increased in later years. This appears to be associated with the narrowing and deepening of the channel combined with the construction of jetties. In some occasions, changes had a positive effect (reduction of currents), but these positive effects appear to have been counteracted by creation of new high energy spots, due to the creation of narrow entrances to boat basins, in order to accommodate more vessels.

Altering the location of entrances to the boat basins or changing the shape of the boat basins might lead to alternative designs that attenuate currents. While the harbor modifications improve navigation, anchorage and mooring areas, the narrowed channels, especially where elongate islands were created near the harbor entrance, create intense currents during long period wave activity from major distant tsunamis. These currents have created havoc, damaging boats and docking facilities during large events in 1960 and 1988 (Chile), 1964 (Alaska), and 2011 (Japan) [4]. The short period character and shorter duration of tsunamis from local sources reduce these effects. Some narrow low-lying areas along the Silver Strand may be susceptible to overtopping and inundation during large amplitude events, depending also on tidal conditions during the tsunami arrival. In addition, strong currents along beaches may cause severe erosion and scour, and pose serious threat to swimmers and small boats.

The effects of harbor modifications on tsunami impact is a subject that merits further investigation. However, the work presented here leads to the following more general conclusions:

- Enhancement of port space increases exposure to assets within the harbor compared to previous years as more vessels are now accommodated within the port facilities.

- Numerical results also show that the introduction of high currents could threaten navigation, vessels and facilities at narrow openings and also along the harbor "throat".

- Harbor modifications can have both positive and negative effects on tsunami impact. For example, currents can be constrained to areas of minimal impact (e.g., outside of the harbor), thereby exposing less assets to dangerous wave action.

- The long duration of distant subduction megathrust tsunamis (Mw >8 source magnitude) spans several tidal cycles so that serious currents occur for several hours, or even days, after the first wave arrivals.

Acknowledgments: We would like to thank Finn Scheele for help with some of the maps in this manuscript. We would like to thank Cynthia Holder, Deputy District clerk/records manager of the Unified Port of San Diego for giving us access to the historical photographs of San Diego Bay. This publication is partially funded by the Joint Institute for the Study of Atmosphere and Ocean (JISAO) at the University of Washington under NOAA Cooperative Agreement No. NA17RJ1232, Contribution No. 2445. This is Contribution No. 4364 from the NOAA/Pacific Marine Environmental Laboratory.

Author Contributions: A.B. wrote the majority of the paper, digitized the nautical charts for the simulations and made about half of the figures appearing in the manuscript. M.R.L. had the original idea for this paper, provided information on the local sources and wrote sections of the paper relative to the sources and San Diego Bay history. E.G. ran the simulations, helped with the decision-making of the distant sources and contributed to the editing of this paper.

Conflicts of Interest: The authors declare no conflict of interest.

References

1. Unified Port of San Diego. Available online: http://www.portofsandiego.org (accessed on 7 July 2015).
2. Kemp, K. The Mississippi Levee System and the Old River Control Structure. Available online: http://www.tulane.edu/~{}bfleury/envirobio/enviroweb/FloodControl.htm (accessed on 7 July 2015).
3. Lander, J.F.; Lockridge, P.A. *United States Tsunami 1890–1988*; National Geophysical Data Center, National Environment Satellite, Data: Boulder, CO, USA, 1989.
4. Wilson, R.I.; Admire, A.R.; Borrero, J.C.; Dengler, L.A.; Legg, M.R.; Lynett, P.; McCrink, T.P.; Miller, K.M.; Ritchie, A.; Sterling, K.; *et al.* Observations and impacts from the 2010 Chilean and 2011 Japanese tsunamis in California (USA). *Pure Appl. Geophys.* **2012**, *170*, 22.
5. Shuto, N. Damages to coastal structures by tsunami-induced currents in the past. *J. Disaster Res.* **2009**, *4*, 462–468.
6. Military.com. Naval Air Station North Island. Available online: http://www.military.com/base-guide/naval-air-station-north-island (accessed on 13 July 2015).
7. 2010 Census—Census.gov, 2010. Available online: http://www.census.gov/2010census/ (accessed on 14 September 2015).
8. California Department of Parks and Recreation. About the Beach. Available online: http://www.parks.ca.gov/?page_id=984 (accessed on 13 July 2015).
9. Office of Coast Survey, NOAA. Historical Charts. Available online: http://historicalcharts.noaa.gov/ (accessed on 8 July 2015).
10. Fischer, P.J.; Mills, G.I. *The Offshore Newport-Inglewood-Rose Canyon Fault Zone, California: Structure, Segmentation and Tectonics*; Environmental Perils of the San Diego Region: San Diego, CA, USA, 1991; pp. 17–36.
11. Harper & Row. *Shepard, Submarine Geology*, 3rd ed.; Harper & Row: New York, NY, USA, 1973.
12. Gohres, S.D.H.C.-D. Tidal Marigrams. Available online: https://www.sandiegohistory.org/journal/64october/marigrams.htm#Gohres (accessed on 10 July 2015).
13. Gohres, H. Tidal Marigrams. *J. San Diego Hist.* **1964**, *10*, 4.
14. Military.com. CNIC I Naval Base Coronado Complex. Available online: http://www.cnic.navy.mil/regions/cnrsw/installations/navbase_coronado.html (accessed on 13 July 2015).
15. NOAA, National Centers for Environmental Information. NOAA Tsunami Inundation Digital Elevation Models (DEMs). Available online: http://www.ngdc.noaa.gov/mgg/inundation/ (accessed on 8 July 2015).
16. NOAA, National Centers for Environmental Information. ETOPO1 Global Relief Model. Available online: https://www.ngdc.noaa.gov/mgg/global/global.html (accessed on 8 July 2015).

17. Carignan, K.S.; Taylor, L.A.; Eakins, B.W.; Friday, D.Z.; Grothe, P.R.; Love, M. *Digital Elevation Models of San Diego, California: Procedures, Data Sources and Analysis*; National Geophysical Data Center, National Oceanic Atmospheric Administration: Silver Spring, MD, USA, 2012.

18. National Geodetic Survey, NOAA. Vertical Datums. Available online: http://www.ngs.noaa.gov/datums/vertical/ (accessed on 8 July 2015).

19. NOAA, Center for Operational Oceanographic Products and Services. *Tidal Datums and Their Applications*; US Department of Commerce: Silver Spring, MD, USA, 2000.

20. NOAA, Tides and Currents-Superseded Bench Mark Sheet for 9410170, San Diego CA. Available online: http://beta.tidesandcurrents.noaa.gov/benchmarks.html?id=9410170&type=superseded (accessed on 10 July 2015).

21. Titov, V.V.; Synolakis, C.E. Numerical Modeling of tidal waver runup. *J. Waterw. Port Coast. Ocean Eng.* **1998**, *124*, 15.

22. Titov, V.V.; Gonzalez, F.I. *Implementation and Testing of the Method of Splitting Tsunami (MOST) Model*; NOAA Technical Memorandum ERL PMEL-112; NOAA: Seattle, WA, USA, 1997.

23. Okada, Y. Surface deformation due to shear and tensile faults in a half-space. *Bull. Seismol. Soc. Am.* **1985**, *75*, 20.

24. International Tsunami Information Center (ITIC). m.yamamoto@unesco.org (accessed on 9 October 2012).

25. Synolakis, C.E.; Bernard, E.N.; Titov, V.V.; Kanoglu, U.; Gonzalez, F.I. Validation and Verification of Tsunami Numerical Models. *Pure Appl. Geophys.* **2008**, *165*, 38.

26. Percival, D.B.; Denbo, D.W.; Eble, M.C.; Gica, E.; Mofjeld, H.O.; Spillane, M.C.; Tang, L.; Titov, V.V. Extraction of tsunami source coefficients via inversion of DART® buoy data. *J. Int. Soc. Prev. Mitig. Nat. Hazards* **2011**, *58*, 567–590.

27. Admire, A.R.; Dengler, L.A.; Crawford, G.B.; Uslu, B.U.; Borrero, J.C.; Greer, S.D.; Wilson, R.I. Observed and modeled currents from the Tohoku-oki Japan and other recent tsunamis in Northern California. *Pure Appl. Geophys.* **2014**, *171*, 18.

28. Barberopoulou, A.; Legg, M.R.; Uslu, B.; Synolakis, C.E. Reassessing the tsunami risk in major ports and harbors of California I: San Diego. *Nat. Hazards* **2011**, *58*, 18.

29. Kanamori, H.; Cipar, J.J. Focal Process of the great Chilean earthquake 22 May 1960. *Phys. Earth Planet. Inter.* **1974**, *9*, 9.

30. Kanamori, H. The energy release in great earthquakes. *J. Geophys. Res.* **1977**, *82*, 7.

31. Barrientos, S.E.; Ward, S.N. The 1960 Chile earthquake: Inversion for slip distribution from surface deformation. *Geophys. J. Int.* **1990**, *103*, 9.

32. Tanaka, H.; Tinh, N.X.; Umeda, M.; Hirao, R.; Pradjoko, E.; Mano, A.; Udo, K. Coastal and estuarine morphology changes induced by the 2011 Great East Japan Earthquake Tsunami. *Coast. Eng. J.* **2012**, *54*, 25.

33. Taylor, A. Japan Earthquake, 2 Years Later: Before and After. Available online: http://www.theatlantic.com/photo/2013/03/japan-earthquake-2-years-later-before-and-after/100469/#img18 (accessed on 13 July 2015).

34. NOAA, National Data Buoy Center. Available online: http://www.ndbc.noaa.gov/dart. shtml (accessed on 10 July 2015).

35. Gica, E.; Spillane, M.C.; Titov, V.V.; Chamberlin, C.D.; Newman, J.C. *Development of the forecast propagation database for NOAA's Short-Term Inundation Forecast for Tsunamis (SIFT)*; NOAA Technical Memorandum, OAR PMEL-139; NOAA: Seattle, WA, USA, 2008.

36. Legg, M.R. Geologic Structure and Tectonics of the Inner Continental Borderland Offshore Northern Baja California, Mexico, Santa Barbara, CA. Ph.D. Dissertation, University of California, Santa Barbara, CA, USA, 1985.

37. Legg, M.R.; Borrero, J.C. Tsunami potential of major restraining bends along submarine strike-slip faults. In Proceedings of the International Tsunami Symposium 2001, NOAA/PMEL, Seattle, WA, USA, 7–10 August 2001.

38. Legg, M.R.; Goldfinger, C.; Kamerling, M.J.; Chaytor, J.D.; Einstein, D.E. Morphology, structure and evolution of California Continental Borderland restraining bends. *Tecton. Strike-Slip Restraining Releas. Bends Cont. Oceanic Settings* **2007**, *290*, 143–168.

39. Goldfinger, C.; Legg, M.R.; Torres, M.E. *New Mapping and Submersible Observations of Recent Activity on the San Clemente Fault*; EOS, Transactions of the American Geophysical Union: Washington, DC, USA, 2000.

40. Synolakis, C.E.; Borrero, J.C.; Eisner, R. Developing inundation maps for Southern California. In Proceedings of the 2002 Coastal Disasters Conference, San Diego, CA, USA, 24–27 February 2002.

41. Synolakis, C.E.; Bardet, J.P.; Borrero, J.; Davies, H.; Okal, E.; Silver, E.; Sweet, J.; Tappin, D. Slump origin of the 1998 Papua New Guinea tsunami. *Proc. R. Soc. Lond. A* **2002**, *458*, 763–789.

42. Barberopoulou, A.; Legg, M.R.; Gica, E. *Multiple Wave Arrivals Contribute to Damage and Tsunami Duration on the US West Coast*; Springer: Berlin, Germany; Heidelberg, Germany, 2014.

An Experimental and Numerical Study of Long Wave Run-Up on a Plane Beach

Ulrike Drähne, Nils Goseberg, Stefan Vater, Nicole Beisiegel and Jörn Behrens

Abstract: This research is to facilitate the current understanding of long wave dynamics at coasts and during on-land propagation; experimental and numerical approaches are compared against existing analytical expressions for the long wave run-up. Leading depression sinusoidal waves are chosen to model these dynamics. The experimental study was conducted using a new pump-driven wave generator and the numerical experiments were carried out with a one-dimensional discontinuous Galerkin non-linear shallow water model. The numerical model is able to accurately reproduce the run-up elevation and velocities predicted by the theoretical expressions. Depending on the surf similarity of the generated waves and due to imperfections of the experimental wave generation, riding waves are observed in the experimental results. These artifacts can also be confirmed in the numerical study when the data from the physical experiments is assimilated. Qualitatively, scale effects associated with the experimental setting are discussed. Finally, shoreline velocities, run-up and run-down are determined and shown to largely agree with analytical predictions.

Reprinted from *J. Mar. Sci. Eng.* Cite as: Drähne, U.; Goseberg, N.; Vater, S.; Beisiegel, N.; Behrens, J. An Experimental and Numerical Study of Long Wave Run-Up on a Plane Beach. *J. Mar. Sci. Eng.* **2016**, *4*, 1.

1. Introduction

On 26 December 2004 at 01:01:09 UTC a submarine earthquake of magnitude $M = 9.0$ hit the west coast of Northern Sumatra, Indonesia [1]. The main shock had its epicenter at 3.09° N and 94.26° E, and the fault line extended for 1200 km to 1300 km from Indonesia northward to the Andaman Islands. The generated tsunami waves caused disastrous destruction around the Indian Ocean, with approx. 220,000 casualties, and severe material losses [2]. Run-up heights around the Indian Ocean reached maximal values of about 30 m [3], where the maximum run-up of waves is commonly defined as the shoreline elevation at maximum inundation above mean sea level (the subsequent or preceding retreat of water is called run-down).

In the light of this and other disastrous flooding events, major efforts have been established to improve the understanding of generation, propagation and run-up of what can be approximated as shallow water waves [4]. These efforts can be roughly classified into three methodological categories: (a) experimental [5–7]; (b) numerical/computational [8–12]; and (c) theoretical [13,14].

In physical modeling the ability to generate a wave of a certain type determines quality and reliability of the experimental data. Even in the age of super computers, experiments are still valuable in research, as the data they produce does not result from any model simplifications and is vital for validation and calibration of numerical models. For over 40 years, the solitary wave paradigm has been assumed to yield a good model to study tsunami waves (see, e.g., [4,15–17]). However, the findings in [18] demonstrated the shortcomings of this model to represent tsunami waves since temporal and spatial scales of solitary waves are significantly shorter than those of the prototype. Instead, the current state-of-the-art model (see [14]) is the N-wave in its general form, e.g., elevation and depression are of different size.

Analogous to the improvement of the knowledge about wave types, the techniques to generate waves have evolved over the past decades. Very early experiments were carried out in 1844 by Scott Russell who used a sinking box to generate a solitary wave (see an investigation in [19]). In the 1980s, Synolakis conducted experiments in a rectangular wave flume using a piston type wave generator [20,21]. Using this technique, a vertical rectangular wall (a piston) is hydraulically driven forward and backward to transfer momentum into the water column which imitates the depth-averaged particle velocity of a passing wave and finally to generate the wave. In this way one or more solitary waves are generated. By experimental means, and by employing linear and non-linear shallow water wave theory, Synolakis [21] was able to predict the run-up of non-breaking solitary waves. A similar comparison between analytical, numerical and experimental data was conducted by Titov and Synolakis [22] to show that the non-linear shallow water wave equations successfully reproduce experimental results or geophysical tsunamis with complicated small scale bathymetric features.

Another hydraulic approach is the wave generation with a vertically moveable bottom as used by Hammack [23]. In that way it was intended to model a submarine earthquake. According to [24] this technique is not suitable for modeling waves in the vicinity of a coast since a distinction between a generation section and a downstream section has to be made. A third approach to transfer momentum into a water body is to release an amount of water from above the water surface [25]. Although a wave is generated, this dam break like mechanism has the disadvantage that significant turbulence is induced into the water. Furthermore, it is difficult to control the wave characteristics such as amplitude and period [24]. Recently, a new technique to generate long waves has been developed at the Franzius-Institute for Hydraulic, Estuarine and Coastal Engineering in Hannover, Germany. Utilizing a pump-driven wave generator, precise control over the wave characteristics (wave length, amplitude and shape) can be maintained to high accuracy [26]. In particular, different kinds of waves can be generated including single cycle sinusoidal waves, solitary waves and N-waves.

Flood research and forecasting are mainly carried out with the help of computer models that employ robust and accurate numerical techniques to solve equations suitable to describe geophysical fluid flow. Various state-of-the-art discretization techniques can be and are used for this purpose, such as finite difference, finite element and finite volume models (see, e.g., [8,9,12,22,27]). Among these models, discontinuous Galerkin models (as described in, e.g., [28,29]) have recently become popular, because they combine numerical conservation properties with geometric flexibility, high-order accuracy and robustness on structured and unstructured grids. Furthermore, the communication between elements is local making them especially suitable for parallel and high performance computing (see Kelly and Giraldo [30] for a study with a 3D model). Therefore, numerical modeling became a most valuable tool in tsunami science; particularly powerful once employed in combination with analytical and experimental methods.

While a number of analytical solutions are available [31], the present study focuses on the important theoretical results in [14], who applied the methodology originally developed by Synolakis [21] for different wave shapes. Model calibration and validation calls for further realistic experimental data, in order to link numerical modeling with realistic fore- and hindcasting results.

In order to increase confidence in our physical and numerical models, and to test the applicability of the theoretical derivations, the authors investigate the agreement of experimental and numerical modeling results. The present study therefore provides measurements of shoreline motion using an innovative pump-driven wave generator (see [24]), and may serve as a novel benchmark for leading depression sinusoidal waves. The produced data set is used to validate a numerical discontinuous Galerkin non-linear shallow water model concerning shoreline dynamics. With both, experiment and simulation, the authors reproduce the theoretical results for long periodic waves that were presented in [14].

Our main research questions ask, whether run-up, calculated with the shallow water model, is capable of representing shoreline motion adequately in terms of theoretical understanding according to [14,21], and in terms of physical experiments as well as numerical methods, for a single cycle sinusoidal wave as a very basic representation of a tsunami. Once convinced that the numerical method adequately reproduces the theoretical expression, the authors investigate deviations of the experimental results from theory by running numerical simulations with perfect and imperfect initial conditions (as taken from "imperfect" experiments), assuming that impurities in the initial wave setup lead to contaminated run-up results. The overall aim is to showcase how comparative, intermethodological work contributes to the understanding of shoreline motion of long waves.

After summarizing the theoretical results of Madsen and Schäffer [14] at the beginning of Section 2, the experimental and numerical setup of this study is introduced and the novel design of the wave flume is described. Furthermore, the authors provide background of the one-dimensional discontinuous Galerkin non-linear shallow water model. Once convinced that the experimental data are useful and sufficiently accurate (Section 3.1), the authors perform numerical simulations with analytically prescribed wave shapes to validate the numerical model in Section 3.2 in terms of the theoretical expressions. The experimental data is then compared with analytical as well as numerical results in Section 3.3. For further validation and in order to assess useful information on wave impact, maximum as well as minimum run-up and shoreline velocities are addressed in Section 3.4. Finally, an evaluation of all results, conclusions and an outlook is given in Section 4.

2. Methodology

The current study is based on the theoretical findings of Madsen and Schäffer [14], who derived explicit formulae for long wave run-up on a plane beach generated by waves of different shapes. The goal of this study was to reproduce these functional relationships for sinusoidal waves (a) experimentally using a wave flume facility at the Leibniz Universität Hannover; and (b) by numerical simulations with a one-dimensional shallow water model. After summarizing the results of Madsen and Schäffer [14] for periodic sinusoidal waves, the experimental setup and the numerical model used for this study are introduced in this section. At the end the boundary conditions (BC) are detailed, which the authors used to generate the waves in the experimental and the numerical model.

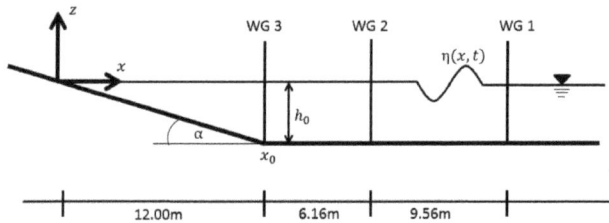

Figure 1. Schematic cross section of the considered setup: A constant depth region is attached to a linearly sloping beach with angle $\gamma = \tan \alpha$, where γ is the beach slope. Also indicated are the wave gauges (WG) from the experimental setup.

2.1. Theoretical Background

The explicit formulae for the run-up/run-down of incoming waves with fixed shape from Madsen and Schäffer [14] are given for a one-dimensional bathymetry which consists of a constant depth offshore region of depth h_0 attached to a linearly sloping beach with γ being the constant beach slope (*cf.* Figure 1). In the offshore region the waves are assumed to be solutions of the linearized shallow water

70

equations while they obey the non-linear shallow water equations on the sloping beach. Effects of wave breaking and bottom dissipation are neglected in this theory.

Analytical solutions for the non-linear shallow water equation on a plane beach are already derived by Carrier and Greenspan [13] by using a so-called hodograph transformation in which a new set of independent variables (ρ, λ) and a velocity potential $\psi(\rho, \lambda)$ are introduced. This leads to a linear differential equation in ψ, for which one can derive exact solutions. In [14] the independent variables are chosen in such a way, that ρ becomes 0 at the shoreline and λ is a modulated time. By letting $\rho \to 0$, the expressions for surface elevation η and velocity v in terms of ρ, λ and ψ lead to explicit formulae for the evolution of run-up elevation R and the associated run-up velocity V. Furthermore, the theory of Carrier and Greenspan [13] yields a theoretical breaking criterion, in which case the time derivative of V and the space derivative of η go to infinity. This corresponds to a discontinuity in these profiles.

Using some further approximations, Madsen and Schäffer [14] finally arrive at the run-up expression for periodic sinusoidal waves. Let the coordinate system have its origin at the still water shoreline with the x-axis being positive in the offshore direction (see Figure 1). The z-axis points upwards. The time series of the incoming wave at the beach toe $x_0 = h_0/\gamma$ is prescribed by

$$\eta_i(x_0, t) = A_0 \cos(\Omega [t - t_1]) \tag{1}$$

where η_i is its surface elevation, Ω is the angular frequency, A_0 is the offshore wave amplitude, and t_1 is aphase shift. The run-up velocity and elevation are then

$$\tilde{V}(\lambda) = \frac{\Omega R_0}{\gamma} \sin(\theta + \pi/4) \quad \text{and} \quad \tilde{R}(\lambda) = R_0 \cos(\theta + \pi/4) - \frac{\tilde{V}(\lambda)^2}{2g} \tag{2}$$

with

$$R_0 = 2A_0 \sqrt{\pi \Omega t_0} \quad \text{and} \quad \theta = \Omega(\lambda - t_1 - 2t_0) \tag{3}$$

and g being the gravitational constant. Further, $t_0 = x_0/\sqrt{gh_0}$, and the time is parameterized through λ, i.e.,

$$t(\lambda) = \lambda + \frac{\tilde{V}\lambda^2}{2g} \tag{4}$$

This describes the temporal variation of $R(t) = \tilde{R}(\lambda(t))$ and $V(t) = \tilde{V}(\lambda(t))$. One can also derive the maximum run-up, which occurs for $\theta = -\pi/4$. At this time $R \to R_0$ and $V \to 0$. Similar considerations can be made for the run-down and the extreme values of the velocity.

The maximum values of run-up/run-down elevation and velocity and the theoretical breaking criterion for periodic sinusoidal waves are conveniently given by introducing the non-linearity of a wave $\epsilon = A_0/h_0$ and the surf similarity parameter. The latter was originally introduced by Battjes [32] and is defined by $\xi = \gamma/\sqrt{H/L_\infty}$, where $H = 2A_0$ is the incident wave height and $L_\infty = gT^2/(2\pi)$ the deep water wavelength of small amplitude sinusoidal waves with period $T = 2\pi/\Omega$. Thus,

$$\xi = \sqrt{\pi}\left(\frac{A_0}{h}\right)^{-1/2}\left(\frac{\Omega^2 h}{g\gamma^2}\right)^{-1/2} \tag{5}$$

relates the beach slope to the wave steepness. With these definitions one obtains for a given non-linearity the extreme values

$$\frac{R_{\mathrm{up/down}}(\xi;\epsilon)}{A_0} = \pm\frac{2\pi^{3/4}}{\epsilon^{1/4}\xi^{1/2}} \quad\text{and}\quad \frac{V_{\mathrm{up/down}}(\xi;\epsilon)}{\sqrt{gA_0}} = \pm\frac{2\pi^{5/4}}{\epsilon^{1/4}\xi^{3/2}} \tag{6}$$

only depending on the surf similarity parameter. The theoretical breaking criterion is met for

$$\frac{R_{\mathrm{up/down}}^{\mathrm{limit}}(\xi)}{A_0} = \pm\frac{1}{\pi}\xi^2 \quad\text{or}\quad \frac{V_{\mathrm{up/down}}^{\mathrm{limit}}(\xi)}{\sqrt{gA_0}} = \pm\frac{1}{\sqrt{\pi}}\xi \tag{7}$$

In this study it will be shown how these results were reproduced in the laboratory experiment, $i.e.$, in a wave flume. Furthermore, they will be used later on to validate the numerical model regarding its treatment of inundation.

2.2. Physical Model

The dynamics of long sinusoidal waves approaching the sloping beach and their subsequent interaction during the run-up and run-down process was studied experimentally in a wave flume at the University of Hannover. The closed-circuit wave flume that was used to generate the sinusoidal waves including the experimental setup is already described in [24,26,33–35]. In summary, the wave generation relies on electronically controlled high-capacity pipe pumps which allow for the acceleration and deceleration of a water volume. A control loop feedback system allows the generation of arbitrary wave shapes such as sinusoidal, solitary or N-waves over a large range of wave periods and lengths.

A major advantage of this wave generation method with active control loop is that wave lengths much longer than the available propagation distance of the wave flume can be generated. This feature is accomplished by intrinsic treatment of the seaward propagating re-reflections. These re-reflections would normally limit the effective wave length to be generated to one wave flume length or less. Through inverse pump response, it is possible to compensate for the re-reflected wave components and in principle, a "clean" wave generation is provided over the entire duration of the target surface elevation time series, similar to the active wave absorption technique used in laboratory wind wave generation.

However, a disadvantage of the wave generation is the development of spurious high frequency ripples (or "riding waves") overlaid with the long wave. These are caused by the active control loop overshooting set reference values (or sometimes called *target* values) during the generation process, and emanate from excess discharge into the wave flume at short times. In the sequence, this unintentional generation of shorter waves alongside of the long waves gave rise to additional effects occurring where the run-up and run-down took place. Most prominent, these riding waves arrived somewhat delayed with respect to the long wave, broke on the shallow beach and interfered with the targeted long wave run-up process. As will be shown later, some of the presented results are attributed to this fact; a wave generation improvement useful in future studies was yet recently reported by Goseberg *et al.* [36] or Bremm *et al.* [37] to circumvent such behavior.

Figure 2 shows a sketch of the facility used for the experiments. Pump station (a), propagation section (b), reservoir section (c), sloping beach (d), and the water storage basin (e) are depicted, respectively. Walls and horizontal bottom sections are made of plasterworks and floating screed and the width of the flume is 1.0 m. A 1 in 40 sloping beach (*i.e.*, beach angle of $\alpha = 1.43°$) made of aluminum boards with small surface roughness was used to model the run-up. The effective length of the constant depth propagation section from pump station to the beach toe was 19.92 m. The undisturbed offshore water depth for this study was set to $h_0 = 0.3$ m.

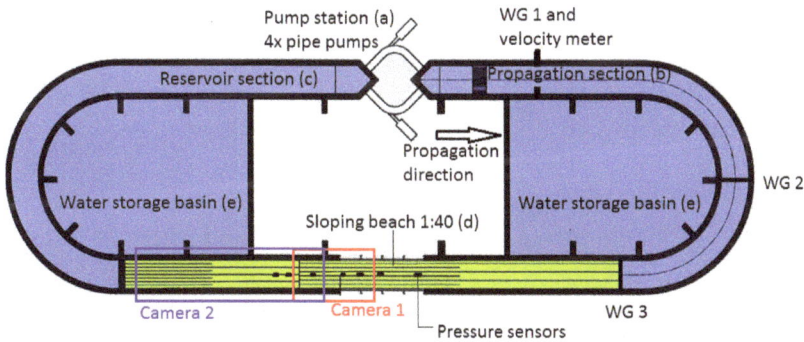

Figure 2. Schematic drawing of the wave flume and its components with position indication of instrumentation. Adapted from [38], with permission from © 2012 World Scientific Publishing Company Incorporated.

To control the generation of the waves, a pressure sensor was installed at the water inlet of the pump station. In the deep water region offshore of the beach, wave gauges (Delft Hydraulics) and velocity meters (Delft Hydraulics) completed the instrumentation during the experiments as shown in Figure 2. The surface piercing wave gauges comprised two parallel electrodes. The immersion depth (surface elevation) was determined by measuring the electric resistance between both electrodes. The measurement accuracy was $\pm 0.5\%$. Wave gauges (WG) were located close to the pump (WG1, at $x = 27.72$ m), halfway of the flume bend (WG2, at $x = 18.16$ m) and at the beach toe (WG3, at $x = 12.0$ m). All instruments were carefully tested and calibrated prior to the experiments, and were subsequently set to zero before single experimental runs.

In addition, two high definition cameras (Basler, Pilot pi-1900-32-gc) captured the wave interaction process with a sampling frequency of 32 Hz. Image processing techniques in the form of color space conversion, lens distortion correction, image rectification and projection shore-parallel, and image stitching were used in post processing on the two sets of images. The resulting processed images from the two cameras (whose field of view is indicated in Figure 2) covered a length of 9.80 m with an original overlap of 0.5 m. The time span of image recording was adapted based on the wave period of the experimental run.

Stages of the image processing process are depicted in Figure 3 which includes scene snapshots of each of the cameras and a final result after image processing routines. A manual processing was used as the amount of experimental runs was reasonable to work through. Adhesive tape spaced by 0.1 m was placed on the beach slope and from this, vectors of time and shoreline location along the center line of the flume were determined based on the derived images as shown in Figure 3c. This approach minimized the influence of fluid boundary layers formed on the flume walls. Based on the outlined procedure, an accuracy for the manually processed shoreline location of ± 5 mm was estimated. For 50% of the experimental repetitions, PVC tracers with a diameter of 2 mm were used to increase the traceability of the wave front. The tracers' density is very close to the density of water which results in small settling velocities and similarly small inertial forces were required to accelerate the tracer particles close to the wave front. It was assured through preliminary tests that the shoreline dynamics were not affected by the presence of the tracers. In particular, this method proved useful for the run-up motion whereas inaccuracies might have occurred due to the fuzzyness of the withdrawing shoreline during the run-down process, which has to be looked into in future studies.

(a)

(b)

T = 30s, H = 2cm, Measurement 1, t = 0s

(c)

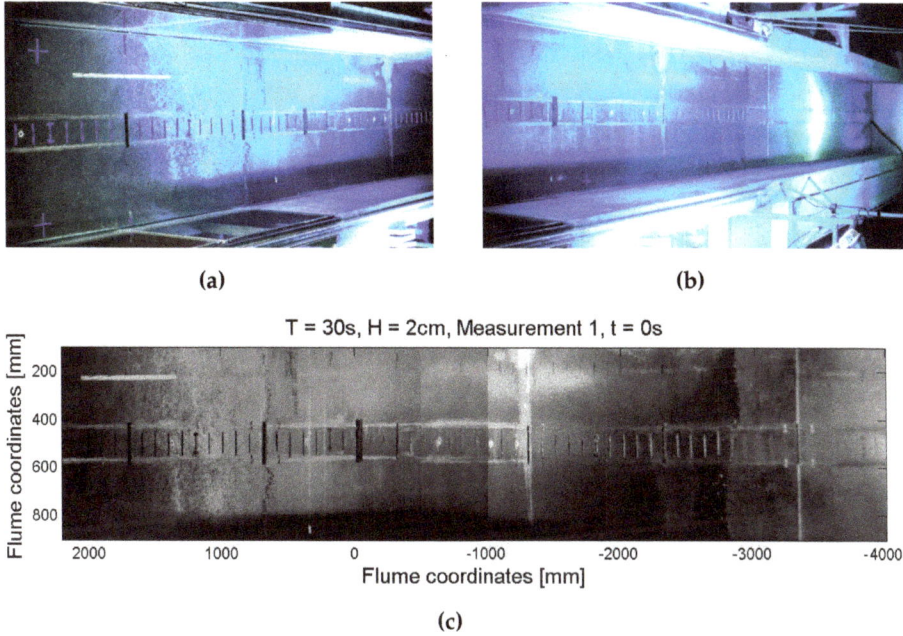

Figure 3. Stages of the image processing with raw data images and final result after image processing routines. (**a**) First camera scene; (**b**) Second camera scene; (**c**) Final stitched image for analysis.

Subsequently, the shoreline velocity was derived from the basic positional information along the center line of the flume based on

$$V_{i+1/2} = \pm \frac{a}{t_{i+1} - t_i} \tag{8}$$

Here, $V_{i+1/2}$ is conveniently assigned to time $t_{i+1/2} = (t_{i+1} + t_i)/2$ and positively defined in the offshore direction, $a = 0.1$ m denotes the distance between adjacent marker tapes, and t_i and t_{i+1} are the times where the shoreline crossed these tapes.

Due to irregular image acquisition measurement gaps occurred in the time series of shoreline location and shoreline velocity. These gaps were not considered in the analysis of the experimental data.

2.3. Scale Effects

The chosen length scale of the Froude-scaled physical model may cause scale effects at times when or at locations where the Reynolds and Weber numbers fall short of commonly accepted thresholds. For example, low Reynolds (Re) numbers in the physical model result in laminar boundary layers which in turn reduce the

75

effective roughness of the down-scaled model as shown in the context of landslide generated waves [39]. In the following, the possible influence of such scale effects in the present experimental setup is briefly discussed. A more general treatise on scale effects can be found in Heller [40].

The Reynolds number during the wave propagation over horizontal bottom, conveniently defined by wave celerity $c = \sqrt{gh}$, initial water depth h, and kinematic viscosity ν, yields

$$\mathrm{Re} = \frac{\sqrt{gh}\,h}{\nu} = \frac{1.72 \times 0.3}{10^{-6}} \approx 5.1 \times 10^5 \tag{9}$$

This value is non-critical with boundary layer turbulence fully developed and turbulence effects similar between prototype and scaled model. However, in the sequence of the wave propagation characteristic velocities and water depth decrease significantly during the wave run-up and run-down phase. Local water depth and run-up velocities are then characteristic quantities to define the Reynolds number on the sloping beach. These values eventually approach zero, and so does the corresponding Reynolds number. Schüttrumpf [41] defined a critical threshold of $\mathrm{Re_{crit}} = 10^3$ for overtopping experiments and this might be good guidance in the present case as well. As a direct result, viscous forces are likely to play a more dominant role compared with prototype conditions in the very shallow, phasing-out wave tongue region; Reynolds numbers fall short of the threshold for example for water depth smaller than 1 cm in combination with flow velocities smaller than $0.1 \ \mathrm{m \cdot s^{-1}}$. This finding aligns well with the analogue requirement for scaled coastal models proposed by Le Méhauté [42] who considered a minimal flow depth smaller than 2 cm and wave periods smaller than 0.35 s as critical. As a matter of fact, the scale effects related to very long wave run-up on gently sloped beaches have not been sufficiently addressed in the literature; experiments with varying model families at different length scales have not been compared until now. An occasional example of scale effect discussion in the context of impulsive wave run-up is a study by Fuchs and Hager [43].

In addition to viscous forces, surface tension has theoretically the potential to bias the experimental results. While in the prototype scale waves are rarely transformed under the forces carried to the water surface by surface tension, the effect might gain significant influence in small scale models. Short waves, which regularly have to be investigated in laboratory studies, can reach the region of capillary waves when the model length scale of a hydraulic scale model is designed too small. Dingemann [44] expressed the effectiveness of the surface tension in an increase of gravitational acceleration, but outlined that the influence of surface tension to wave action is dominant when, for example, capillary waves are investigated. The effect of surface tension is interpreted to be small for large Weber (We) numbers, indicating that driving forces dominate. The Weber number is defined as

$$We = \frac{\rho v^2 l}{\sigma} \tag{10}$$

where ρ is the fluid density, v and l are the characteristic velocity and length scales, and $\sigma = 0.073$ N·m^{-1} is the surface tension for 20 °C water. Using the equation above, the Weber number in the experiments reported herein was approximately 1.2×10^4 in the horizontal propagation section (applying the wave celerity c as characteristic velocity and the fluid depth h as length scale). This value, which is well above critical thresholds, also reflects the fact that the length of the long waves and capillary waves differ by orders of magnitude. In the wave tongue region earlier discussed in regard to the viscous effects, where fluid depth and velocity decrease to $l = 1$ cm and $v = 0.1$ m·s^{-1}, Weber numbers were however significantly reduced and yielded values as small as 1.4. A comprehensive review of small scale models and the influence of the Weber number has been presented in Peakall and Warburton [45]. They summarized various recommendations on the critical flow depth in small scale models reporting critical Weber numbers in the range of We $= 2.5 - 160$. It is thus likely, that surface tension has a major influence on the wave tip formation during the run-up process, as the Weber number found for the experiments fell below this given threshold. Surface tension does not properly scale in physical models governed by Froude similitude. Therefore, the wave run-up of long waves measured is likely be underestimated to a certain extent since tensile forces along the wave front with small surface radii counteract the inertia forces from the run-up flow.

Summarized, in the constant depth region the flow should be realistically modeled in the wave flume. In the run-up process, however, certainly some inaccuracies arise compared to realistic tsunami conditions. However, these are probably negligible compared to inaccuracies in modeling the topography and other parametrization.

2.4. Numerical Model

The numerical simulations presented in this study were executed using a one-dimensional shallow water model. In this model the equations are solved in conservation form with fluid depth h and discharge (momentum) hu being the primary variables. They are discretized using the explicit Runge-Kutta discontinuous Galerkin (RKDG) finite element method [46] with second-order accuracy. This scheme was chosen as a state-of-the-art discretization for hyperbolic conservation laws with source terms. The method is mass-conservative and preserves the steady state at rest (*i.e.*, it is well-balanced). A comprehensive presentation of the scheme including its validation with respect to the shallow water equations is given in Vater *et al.* [47].

An important part in the numerical scheme is the treatment of wetting and drying events. Here, the authors pursued a fixed grid approach where initially dry cells can be flooded and wet cells can dry out during the simulation. A cell can be either wet or dry, thus the wet/dry interface is only accurate up to one cell size. The wetting and drying treatment involves only one additional parameter, which is a wet/dry tolerance. Whenever the fluid depth h is below this tolerance, the velocity is set to zero. In [47] it was shown that the specific value of this parameter does not affect the stability of the scheme. It rather influences the accuracy of the wetting and drying computation. In the presented results, this tolerance was always set to 10^{-8} m.

The fixed grid approach implies that one does not get a continuous representation of the shoreline evolution. Instead, the scheme results in a discrete time series in which the shoreline jumps from one cell interface to the next, whenever one cell gets fully flooded and water penetrates into the next cell, or draining leads to the opposite process. Therefore, similarly to the data processing in the experiments these jumps were identified in the time series as points where the shoreline crossed a cell interface, and the space between the points was linearly interpolated. As the discrete solution led to some oscillations in the shoreline computation where the latter jumped back and forth in some situations only within a few time steps, all oscillations within a window of five time steps were removed in this time series. The shoreline velocity was then computed by a formula equivalent to Equation (8). In this time series, also some oscillations were filtered out, which occurred within a window of two time steps.

For the discretization in time, Heun's method was used, which is the second-order representative of a standard explicit Runge-Kutta total-variation diminishing (TVD) scheme [48,49]. Its stability is governed by a Courant–Friedrichs–Lewy (CFL) time step restriction [50] with $\Delta t \leq \text{cfl} \Delta x / c_{\max}$, where Δt and Δx are the time step and grid cell size, respectively, cfl is the CFL number and $c_{\max} = \max\{|u| + \sqrt{gh}\}$ is the maximum speed at which information propagates. To obtain linear stability and positivity of the water depth in the RKDG method, the validity of $\text{cfl} \leq 1/3$ has to be ensured [47,51].

The gravitational constant was set to $g = 9.81 \,\text{m}\,\text{s}^{-2}$ throughout the computations. No parametrization of bottom friction was included in the model, since the theoretical results of Madsen and Schäffer [14] also do not consider friction. However, some deviations compared to the experimental runs might be attributed to this fact. To obtain results, which are comparable to the experiments, the simulated domain was 33 m long and spanned from the dry area of the sloping beach at $x = -5.28$ m to the first wave gauge (WG1) at $x = 27.72$ m. For the discretization the domain was devided into 330 uniform cells with a cell size of $\Delta x = 0.1\,\mathrm{m}$. The time step size was fixed to $\Delta t = 0.01\,\mathrm{s}$, which results in a CFL number of cfl $= 0.18$. The (one-dimensional) bottom topography was given by $b(x) = \max\{0, h_0 - x/40\}$. The initial conditions for the numerical simulations were a fluid at rest, with

$$h(x,0) = \max\{0, h_0 - b(x)\}, \quad (hu)(x,0) \equiv 0 \tag{11}$$

2.5. Boundary Conditions

In this study, long wave conditions in the vicinity of the shore in shallow water were approximated by sinusoidal waves with leading depression. The chosen boundary conditions (BC) were used throughout this study for the experimental and the numerical method equally. Experimentally, those waves were generated by the pump-driven wave generator, as outlined in Goseberg *et al.* [26] and in Section 2.2 on the basis of ideal, analytical surface elevation time series of

$$\eta_b(t) = \begin{cases} \dfrac{H}{2} \cos\left(2\pi\dfrac{t - t_s}{T} + \dfrac{\pi}{2}\right) & \text{for } t_s \leq t \leq t_e \\ 0 & \text{otherwise} \end{cases} \tag{12}$$

where $H = 2A_0$ is the wave height, and T is the wave period. In the experiments only one period of a sinusoidal wave was modeled, which started after 10 s. The initial delay was necessary, because the pipe pumps needed some initialization time. This means that the starting time was set to $t_s = 10\,\mathrm{s}$ and the stopping time to $t_e = T + t_s$. It is to be noted that the experimentally generated waves exhibit imperfections (*cf.* Section 3.1) which are attributed to the tuning of the control loop as outlined in Section 2.2.

In the numerical simulations, the incident wave was modeled through the right boundary condition. For validation against analytical expressions the fluid depth was prescribed by Equation (12) with $h_b(t) = \eta_b(t) + h_0$. The velocity was then chosen such that the boundary data resulted in a single (simple) wave propagating to the left. In particular, it was set to

$$v_b(t) = 2\left(\sqrt{gh_0} - \sqrt{gh}\right) \tag{13}$$

The starting time in Equation (12) was set to $t_s = 0\,\mathrm{s}$, since no initialization phase is needed for the numerical model. For the stopping time two different simulations were conducted for each wave shape. In one simulation the stopping time was after one period as in the experimental setup, *i.e.*, $t_e = T$. In the other simulation the stopping time was set to infinity, which means that the incoming sinusoidal wave lasted until the end of the simulation. As will be shown in the results, the comparison between the respective two simulations revealed several deviations between the theory of periodic sinusoidal waves and the practical setup of a single period sinusoidal wave.

For comparison with the experiments the right boundary condition of fluid depth and velocity were set using the data from the time series at wave gauge WG1 of the experimentally measured data. The left boundary did not affect the numerical solution as it was in the dry part of the domain.

Table 1 summarizes the boundary conditions used in the current study. The wave characteristics were chosen to cover the significant range of surf similarity parameters ξ. To investigate the run-up of long waves on a plane beach six leading depression, non-breaking, single period sinusoidal waves were selected. The wave period varied from 20 s to 100 s which correspond to laboratory wave lengths from 34.31 m to 171.55 m. The wave height varied from 2 cm to 4 cm. For brevity, wave identifiers (Wave-ID) were used throughout the paper to label the waves used; waves are labeled with a naming scheme "Tx_Hy" where T denotes the wave period in seconds, H denotes the wave height in centimeter and x and y contain the actual quantity values. The long waves covered a range of surf similarity ξ between $4.42 \leq \xi \leq 15.62$. In total 36 experimental runs were conducted, since each wave was reproduced six times. This procedure allowed quantification of the repeatability of the experiments. The data acquisition time for each experiment was 120 s.

Table 1. Characteristics of sinusoidal waves used, surf similarity according to Equation (5), naming scheme indicates period and wave height, respectively.

Wave-ID (−)	Height $H(m)$	Period $T(s)$	Length $L(m)$	Non-linearity $\epsilon = \frac{A_0}{h}(-)$	Rel. Amplitude $\frac{A_0}{L}(-)$	Surf Similarity $\xi(-)$
T20_H2	0.02	20.00	34.31	0.033	2.91×10^{-4}	4.42
T30_H2	0.02	30.00	51.47	0.033	1.94×10^{-4}	6.63
T44_H3	0.03	44.00	75.48	0.050	1.99×10^{-4}	7.94
T58_H3	0.03	58.00	99.50	0.050	1.51×10^{-4}	10.46
T77_H4	0.04	77.00	132.09	0.067	1.51×10^{-4}	12.03
T100_H4	0.04	100.00	171.55	0.067	1.17×10^{-4}	15.62

From now on the term "experimental BC" will be used to indicate that the numerical model was initialized with surface elevation and velocity of a single cycle sinusoidal wave as measured at WG1 in the experiments. Similarly, "analytical BC" indicates numerical runs using a perfect periodic sinusoidal wave.

3. Results

In this section the numerical and experimental results are described and compared with the analytical expressions of Madsen and Schäffer [14]. First, the quality of the experimentally generated waves is discussed, and the numerical model is validated in terms of the explicit expressions of run-up height and velocity as given in [14]. Having confidence in the numerical model, the experimental and numerical results are compared and, where appropriate, matched against the analytical expressions. Finally, the maximum determined run-up/run-down shoreline position as well as velocity are investigated and compared with the analytical expressions from Section 2.1.

3.1. Quality of Experimentally Generated Waves

As the quality of the measured run-up heavily depends on the generated wave signal, the experimentally generated waves are first discussed prior to the presentation of the shoreline evolution in the subsequent sections. Therefore, the wave signal, which was measured by the pressure sensor at the water inlet, and the time series of the surface elevation at the three wave gauges will be investigated.

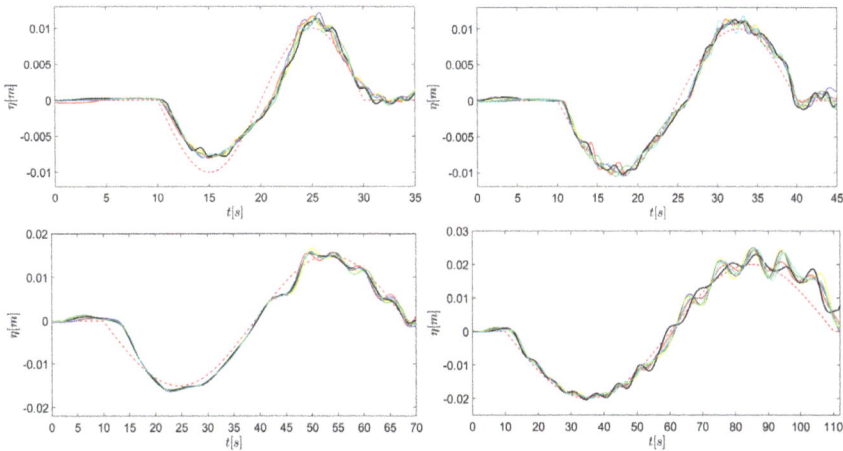

Figure 4. Time series of surface elevation for the reference curve (red dashed) and measured values (solid lines) for all six runs of T20_H2 (**top left**); T30_H2 (**top right**); T58_H3 (**bottom left**) and T100_H4 (**bottom right**) as measured by the pressure sensor at the water inlet.

Figure 4 shows the measured time series of the experimental surface elevation at the water inlet of the pump station for all six reproductions of the waves T20_H2, T30_H2, T58_H3 and T100_H4 compared to the corresponding set point, *i.e.*, reference values. In general, the sinusoidal long waves are generated sufficiently well. Deviations from the reference curve can mostly be attributed to the choice of tuning parameters for the wave generator (e.g., the smaller amplitude of T20_H2, top left); better tuning is principally feasible, but could not be achieved with acceptable effort at the time of the experimental data acquiring. The tuning parameters electronically steer the rate of rotation of the pumps and determine the behavior of the wave generator, therefore they strongly influence the quality of the generated waves. Especially in the second half of the wave period one can observe some spurious, parasitic waves riding on top of the main, long wave, which become more significant for wave shapes with longer wave length. As already discussed in Section 2.2, these riding waves are a result of the applied wave generation strategy.

The authors used the Brier score which is also known as the mean squared error as defined in [52] to further evaluate the quality of the generated experimental waves. Thereto, the Brier score and standard deviation were determined for each experimental run of each wave. Table 2 shows the minimum and maximum Brier score as well as the minimum and maximum standard deviation of the actual curve from the reference curve for all six sinusoidal waves from Table 1. Therewith, the values in Table 2 take into account the individual experimental runs. A Brier score of 0 would be a perfectly reproduced curve. The table shows that the waves T30_H2 and T58_H3 exhibit the best fits, although the latter already contains a significant amount of riding waves as can be seen in Figure 4. The wave T20_H2 has bad Brier scores which is probably due to a too small amplitude during the initial run-down, as opposed to the wave T100_H4 which has bad scores due to the high amount of riding waves present in the signal.

Table 2. Variation of Brier score [52] and standard deviation (STD) of all generated waves.

Wave-ID	Brier Score	STD (cm)
T20_H2	0.65–0.71	0.15–0.19
T30_H2	0.30–0.38	0.10–0.11
T44_H3	0.30–0.40	0.18–0.20
T58_H3	0.22–0.28	0.15–0.17
T77_H4	0.35–0.45	0.24–0.29
T100_H4	0.29–0.48	0.24–0.34

Figure 5 shows the temporal evolution of the surface elevation observed in the experiments and the numerical simulations at the three wave gauges for the same four wave shapes as in Figure 4. In addition to the data from the numerical simulation using experimental BC, also the time series using analytical BC is shown, but shifted by $t_{shift} = (4.2/\sqrt{gh_0} + 10)$s to account for the additional propagation section between the water inlet and WG1 in the experiments (which was not present in the numerical simulations) and the 10 s delay in the experimental wave generation (*cf.* Section 2.5). For all waves, the surface elevation obtained by the experiments and the numerical simulations using experimental BC agree well. Only at larger distances from the wave generation (e.g., at WG2 and WG3) the experimental time series show larger amplitudes, which are probably due to a small inconsistency in the velocity signal at WG1, where the numerical model obtains its boundary data from the experiments. Furthermore, the riding waves on top of the waves with larger period (here T58_H3 and T100_H4) evolve with the main wave towards the sloping beach.

In comparison to the surface elevation obtained from the numerical simulations using analytical BC, it can be seen that the deviation for the initial single cycle sinusoidal wave is mostly the same as in the signal at the wave inlet from Figure 4 at the different wave gauges. Furthermore, for the waves T20_H2 and T30_H2 there is a small hump visible after the sinusoidal wave has passed, which is not present in the "analytical" data. This hump is due to the shut down of the wave generation.

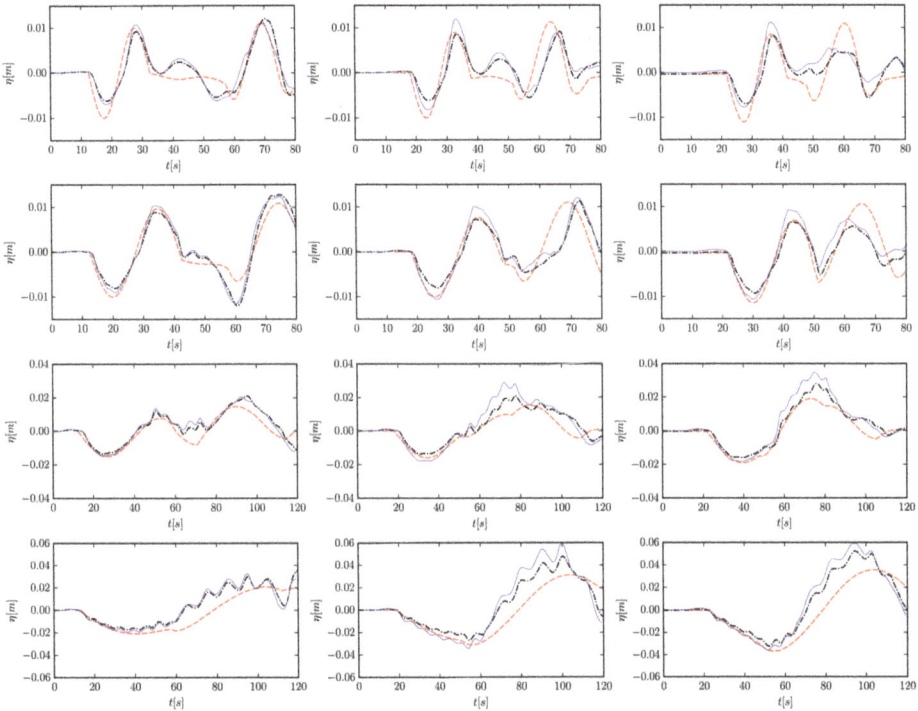

Figure 5. Time series of surface elevation η at wave gauges WG1 (left), WG2 (middle) and WG3 (right) for four different wave shapes. From top to bottom: T20_H2, T30_H2, T58_H3, T100_H4. Experimental data (blue), numerical results with experimental boundary data (black dash-dotted), numerical results with analytical boundary data shifted by $t_{\text{shift}} = (4.2/\sqrt{gh_0} + 10)$s to fit the experimental data (red dashed).

3.2. Validation of the Numerical Model

To have sufficient confidence in the numerical model, it was validated using the analytical BC as defined in Section 2.5. Similar to the experiments, the numerical model was always initialized with a steady state at rest, and the sinusoidal wave was only imprinted by the boundary condition. Note, that this is a deviation from the theory of periodic sinusoidal waves as presented by Madsen and Schäffer [14], whose expressions are derived for permanently incoming waves without any initialization. However, it was not possible to find the correct initial conditions for this setup. Furthermore, the two simulations with a single cycle and a periodic sinusoidal wave as described in Section 2.5 revealed the deviations to be expected between theory and the conducted experiments, which will be shown in the following. This shows the value of concurrent experimental and numerical studies, when it is not possible to design the experimental setup completely according to the theory.

Figure 6 shows a comparison of analytical (red dashed) and numerical shoreline position and shoreline velocity for the waves T20_H2, T44_H3 and T100_H4. For the numerical simulations the results are shown for both, the periodic sinusoidal (blue solid) and the single cycle sinusoidal wave (black dash-dotted). These waves were chosen as representatives for the validation, but the findings are also valid for the other wave shapes. Comparing the simulations using a periodic sinusoidal wave with the analytical expressions, one can see that the numerical model captures the theory well. Only in the beginning, when the still water solution passes on to the periodic one, a fast transition in the shoreline position with a high velocity peak is visible. Note that a positive shoreline velocity corresponds to an actual run-down. This is due to the choice of the coordinate system where the x-axis points positively offshore. One can further notice that the amplitude in the numerical simulations is often a bit too small, and there is a small phase shift, especially for the low frequency waves. These deviations can probably be attributed to the fact that the numerical solution does not oscillate everywhere from the beginning, but needs a spin-up.

The numerical results of the single cycle sinusoidal wave follow the one with the periodic sinusoidal wave for about a period. After this time the oscillating solution passes back on to the still water solution, which again comes along with a sharp transition in the shoreline position and an associated velocity peak. This behavior has also implications for the comparison of the experimental results with the analytical expressions of Madsen and Schäffer [14], especially concerning the extreme values in run-up elevation and velocity. In Figure 6, the occurrence of these extreme values in the first period is also indicated by red circles and diamonds, respectively. From this, one can conclude that the maximum run-down cannot be reliably determined in the experiments, since it is always biased by the transition between the still water and the oscillating state—during the initial run-down and after the maximum run-up was reached. This is also true for the maximum run-down velocity for long periods.

Therefore, these values will be omitted when the extreme values are compared to theory. On the other hand, the maximum run-up elevation and velocity should be well captured by a single cycle sinusoidal wave, provided there are no other disturbances in the experimental data.

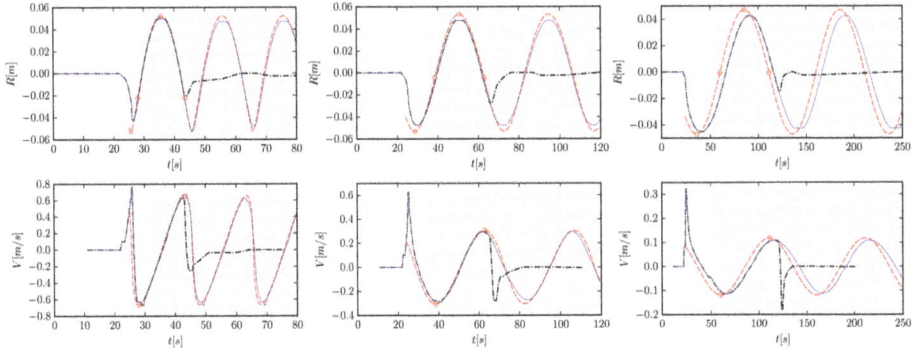

Figure 6. Time series of run-up elevation R (top) and velocity V (bottom) using analytical boundary data for three different wave shapes (**left**: T20_H2; **middle**: T44_H3; **right**: T100_H4. Depicted are the theoretical evolution for periodic sinusoidal waves (red dashed), the numerical simulation with a single cycle sinusoidal wave (black dash-dotted) and the simulation with periodic sinusoidal waves (blue solid). Also shown are the positions of the theoretical maximum/minimum run-up elevations (red circles) and velocities (red diamonds) in the first period. The shaded regions depict the intervals where the extreme values for the diagrams in Figures 9 and 10 were computed.

3.3. Shoreline Motion

After having discussed the quality of the experimental wave signal and the reliability of the numerical model concerning shoreline dynamics, the experimentally obtained shoreline motion data is finally presented. Shoreline location and shoreline velocity of a leading depression sinusoidal wave is compared to the numerical results using experimental BC and, where appropriate, to the analytical expressions of Madsen and Schäffer [14].

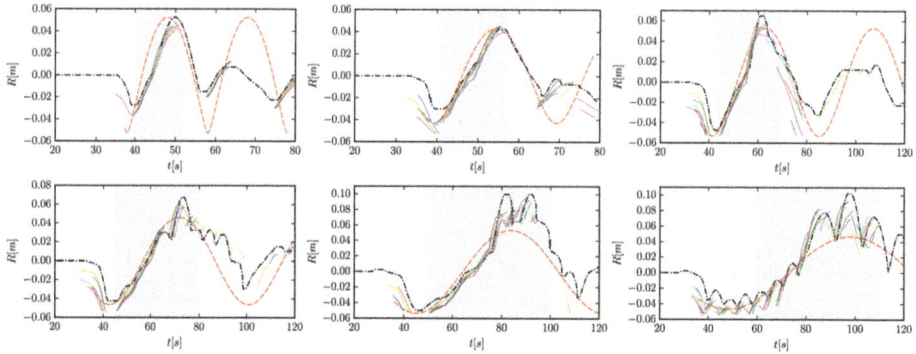

Figure 7. Time series of run-up elevation R from the experiments and the numerical simulation using experimental boundary data for all six wave shapes (left to right: T20_H2, T30_H2, T44_H3 **(top row)**; T58_H3 **(bottom left)**; T77_H4, T100_H4 **(bottom row)**). Depicted are the theoretical evolution for periodic sinusoidal waves (red dashed), the numerical simulation (black dash-dotted) and the results from the six experimental measurements (solid lines). The shaded regions depict the intervals where the extreme values for the diagrams in Figure 9 were computed.

3.3.1. Run-up and Run-down of Long Waves

Figure 7 shows the numerical, experimental and analytical shoreline location for all waves used in this study. The analytical expression (red dashed) represents the run-up of periodic sinusoidal waves while the numerical model was initialized with the surface elevation profiles of a single cycle sinusoidal wave as measured in the experiments. In case of the experimental data (solid lines) the results of all six measurements are shown for each wave as far as data was available. The gaps in the experimental data are consequences of a poor traceability of the shoreline. In particular, the run-down is a transient process, so that there was hardly a distinct evidence to trace the shoreline. In contrast, the shoreline could be traced well during run-up.

Good reproducibility is observed in the experiments. In general, the numerical, experimental and, where the wave has adjusted to the sinusoidal expression, also the analytical shoreline locations agree well for the waves T20_H2, T30_H2 and T44_H3 (top row). For the three longest waves T58_H3, T77_H4 and T100_H4 (bottom row) significant deviations of the experimental shoreline location from the analytical expressions are observed. These oscillations are caused by the over-riding waves, which are already present in the wave signal. One can clearly see that the riding waves overtake the main wave before the maximum run-up height is achieved and lead to false maxima in the measurements. In these intervals there are also some deviations between the experimental and the numerical measurements (e.g., around $t = 80\,\text{s}$ for the wave T77_H4). They could be due either to the breakdown of the shallow water theory in the vicinity of the shoreline, or to some incorrect measurements in the experimental data. Another deviation is present in the initial run-down, which is often underestimated by the numerical model. This is best visible in the waves T20_H2 and T30_H2, while in the other waves the maximum run-down could often not be determined since its location was located outside the field of observation covered by the cameras. This deviation is attributed to the smaller amplitude in the wave signal, already present in the time series at the wave gauges.

Disregarding the corrupting influence of the riding waves, the analytical shoreline location for periodic waves can be reproduced by laboratory single cycle sinusoidal waves with sufficient accuracy; accuracy will further increase in cases where a sufficient amount of time is spent to tune the wave generator to avoid spurious short waves during the wave generation or by following the improvements outlined in Bremm *et al.* [37].

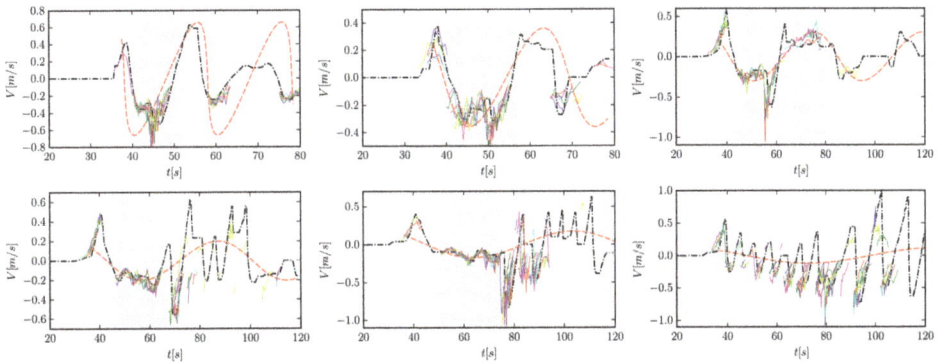

Figure 8. Same as Figure 7, but for run-up velocity V. The shaded regions depict the intervals where the extreme values for the diagrams in Figure 10 were computed.

3.3.2. Shoreline Velocity during Run-up and Run-down

Figure 8 shows the numerical, experimental and analytical shoreline velocity, again for all waves used in this study. As described in Section 2 the experimental shoreline velocity, and similarly the shoreline velocity for the numerical simulations, was derived from the shoreline location using Equation (8). Since the x-axis of the wave flume coordinate system points positively offshore, positive velocity values in Figure 8 indicate the wave run-down.

Similar to the shoreline motion the experimentally determined velocities agree well with the ones from the numerical simulation. In particular, this is true for the initial run-down and the first run-up phase (first maximum and minimum, respectively). As in the numerical runs with analytical BC the first run-down is not in accordance with the analytical shoreline velocity, but the run-up process essentially agrees well. An exception is the wave T20_H2 (top left in Figure 8), where the evolution of the analytical shoreline velocity exhibits a remarkable steep gradient that can neither be reproduced by the numerical model, driven by experimental initial conditions, nor by the physical experiments. The authors attribute this observation to the imperfect wave signal, which exhibits a too small amplitude in surface elevation during the initial depression and results in a shallower slope in the first edge of the sinusoidal cycle (*cf.* Figure 4). Note that this deviation is not present when computing the wave with (perfect) analytical BC as shown in Figure 6. Due to the the the lack of experimental data in the second run-down, a comparison for the velocity during this phase is only possible for the wave T44_H3, which agrees well with the analytical velocities. This is also true for the numerically determined values in these regions for the waves T20_H2, T30_H2 and T44_H3.

Similar to the shoreline location the riding waves also corrupt the experimental shoreline velocity. This becomes evident in high peaks in the time series, as seen around $t = 55\,$s for the wave T44_H3 or around $t = 70\,$s for the wave T58_H3. This effect is also evident for the two longest waves T77_H4 (bottom middle) and T100_H4 (bottom right). However, the numerical model simulates the riding waves well, since an overall good agreement between the numerical and experimental shoreline velocities is observed.

Table 3. Time intervals (in seconds) which were used for the computation of the extreme values in Figures 9 and 10.

Wave-ID	Analyt. BC, Periodic Wave All Extreme Values (s)	Experiments/exp. BC, Single Cycle Wave			
		R_{up} (s)	R_{down} (s)	V_{up} (s)	V_{down} (s)
T20_H2	30–80	40–52	–	40–52	–
T30_H2	30–80	40–57	–	40–57	–
T44_H3	35–120	43–70	–	43–54	60–80
T58_H3	40–150	45–80	–	45–67	–
T77_H4	40–200	50–100	–	50–74	–
T100_H4	50–250	60–110	–	–	–

3.4. Extremal Shoreline Dynamics

In this section the maximum run-up and run-down as well as the maximum shoreline velocity during run-up and run-down are addressed. If not stated otherwise, these extremal characteristics are investigated for the experimentally generated and the numerically simulated waves, with experimental as well as analytical boundary conditions, and compared with the theoretical results of Madsen and Schäffer [14]. Note, that the normalized maximum run-up shown in the following is defined as the maximum run-up normalized by the respective offshore amplitude of the long waves. Where experimental data was used, the values for maximum run-up and run-down and the corresponding maximum shoreline velocities were obtained through an average of all six runs of the waves. Note, that

88

the extreme values were not taken from the entire time period, but rather from carefully chosen intervals. The respective intervals are given in Table 3 and marked gray in Figure 6–8. The reason for this procedure was to detect the intervals in which the actual run-up took place and to disregard time spans where experiments and simulations deviated from the theory by design, or the data was corrupted by riding waves.

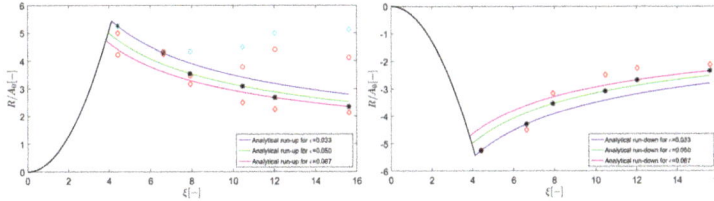

Figure 9. Maximum run-up (**left**) and maximum run-down (**right**) of long sinusoidal waves. Analytical (black stars), experimental (red circles) and numerical (diamonds)—with experimental (cyan) and analytical (red) boundary condition. Solid black line according to Equation (7). Colored solid lines are computed according to Equation (6).

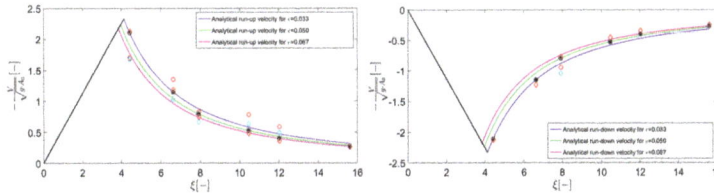

Figure 10. Maximum shoreline velocities during run-up (**left**) and run-down (**right**). Analytical (black stars), experimental (red circles) and numerical (diamonds)—with experimental (cyan) and analytical (red) boundary condition. Solid black line according to Equation (7). Colored solid lines are computed according to Equation (6).

3.4.1. Shoreline Location

Figure 9 (left) shows the experimental, numerical and analytical normalized maximum run-up for all long waves used in this study as a function of the surf similarity parameter ξ. The parameter ϵ defines the non-linearity of the wave, which is the ratio between the wave amplitude and the undisturbed water depth as defined in Section 2.1. As already stated in [53] the highest run-up occurs for surf similarity parameters $3 \lesssim \xi \lesssim 6$. This is the interval where the theoretical transition from breaking to non-breaking waves occurs. For increasing ξ the analytical expression from Madsen and Schäffer [14] predicts a decrease in the normalized run-up.

Using the analytical boundary conditions (red diamonds), the monotone dependence on ξ can be reproduced with the numerical model. However, the model under-predicts the maximum run-up for larger ξ (*i.e.*, longer wave lengths), which was already noted in Section 3.2 and is attributed to an inconsistency of the velocity signal at WG1. Note that for $\xi = 6.63$ (wave T30_H2) only one marker is visible for the numerical run-up. This is because the run-up is identical for both simulations (experimental and analytical BC).

The experimental normalized maximum run-up (circles) agrees well with the analytical expression (stars) for $\xi = 6.63$ and $\xi = 7.94$ (waves T30_H2 and T44_H3). However for $\xi = 4.42$ (wave T20_H2), it is lower than the analytical value, which the authors attribute to friction. For $\xi > 10$ the experimentally determined maximum run-up is significantly higher than the analytical run-up due to the riding waves that corrupt the shoreline evolution. Initializing the numerical model with the imperfect surface elevation profiles, *i.e.*, experimental BC, results in even higher run-up heights, which the authors also attribute to the neglected friction in the numerical model. However, the model is generally in good accordance with the experiments.

Figure 9 (right) shows a comparison between the analytical (stars) and numerical (analytical BC, red diamonds) normalized maximum run-down. Maximum run-down values for the experimental and numerical results using experimental BC were not computed, since they both result from a single cycle sinusoidal wave. As stated in Section 3.2, the maximum run-down is always biased by the transition between the still water and the oscillating state in this case and would have been misleading. Similar to the run-up case the maximum run-down is well reproduced by the numerical model using analytical BC, but it results in too small absolute values for large ξ.

3.4.2. Shoreline Velocity

Figure 10 (left) shows a comparison between analytical, experimental and numerical normalized maximum shoreline velocities during wave run-up. As stated in [53] and confirmed in this study, the highest maximum shoreline velocities occur for waves in the transition region with $3 \lesssim \xi \lesssim 6$, where the wave T20_H2 exhibiting the lowest surf similarity parameter of 4.42 has the highest shoreline velocity. Using analytical BC, the numerical maximum shoreline velocity agrees well with the analytical values. The maximum shoreline velocity could also be reproduced with the physical and numerical experiments, when the time intervals where riding waves corrupted the signal were excluded in the maximum computation. Deviations only occur for the wave T20_H2 ($\xi = 4.42$). This can be probably explained by the imperfect wave signal (*cf.* Section 3.1). The wave T100_H4 ($\xi = 15.62$) was completely corrupted by riding waves when it hit the beach, which made it impossible to determine an interval for a proper calculation of the maximum run-up velocity.

In analogy to the maximum shoreline velocity during run-up now the maximum shoreline velocity during run-down is addressed. The run-down velocity during the first run-down was not considered, since it was biased by the transition to the sinusoidal behavior of the wave. Furthermore, due to the lack of data from the experiments during the second run-down, no maximum values were computed from the experimental time series. One exception is the wave T44_H3, where data from several runs was available. Figure 10 (right) shows the comparison of maximum velocities during run-down.

Regarding the analytical expressions, the correlations between surf similarity ξ, magnitude of the shoreline velocity and wave non-linearity ϵ are the same as for the maximum run-up velocities. Initializing the numerical model with analytical BC yields again an overall good agreement of the numerical normalized maximum run-down velocities with the analytical expressions of Madsen and Schäffer [14]. Running the numerical model with experimental BC results in good agreement with the experimental data. However, since experimental data is available for one wave only, the results have to be interpreted with care.

4. Discussion and Conclusions

This study contributed to the answer of the basic research question, whether shallow water theory can explain and accurately model the run-up behavior of long waves on a plane beach. By conducting physical and numerical experiments and comparing them with analytical expressions derived from the underlying equations, the authors demonstrated the usefulness of such simultaneous experimentation.

A novel wave generator, based on hydraulic pumps that are capable of generating arbitrarily long waves (even exceeding the wave flume length) was used to generate wave periods of 20 to 100 s Combined with appropriate wave heights, surf similarity parameters between $4.4 \lesssim \xi \lesssim 15.6$ were realized. Sinusoidal wave shapes were adopted from [14], in order to obtain waves with corresponding analytical reference solutions. Due to the complex control problem for this type of wave generator in the current set-up, spurious over-riding small-scale waves were unavoidable. Scaling effects were also discussed and it could be concluded that the experimental scale imposes at most minor inaccuracies in the run-up area on the experimental results.

A Runge-Kutta discontinuous Galerkin method with a high fidelity wetting and drying scheme was applied to numerically solve the one-dimensional non-linear shallow water equations. Analytical (from [14]) and experimentally generated wave shapes were used as inflow boundary conditions for the numerical experiments.

In order to compare analytical, numerical and experimental data, the wave similarity measured by the Brier score, maximum run-up and run-down height, as well as run-up/run-down velocities were utilized as quantitative metrics. In a first analysis, periodic and non-periodic clean sinusoidal waves were compared to rule out differences due to the single sinusoidal wave generation in the wave flume. On further analysis, significant differences in experimental and analytically expected values are observed. However, with the combination of analytical, numerical, and experimental data it could be demonstrated that spurious over-riding small-scale waves lead to the observed deviations between analytical expression and experimental values. The numerical model serves as a linking element between theoretical and measured results, and can therefore explain rigorously the influence of small-scale spurious pollution.

Future investigations could be directed to solve the following problems: Better feedback control systems could minimize the spurious over-riding waves and allow for a better resemblance of experimental and analytical wave shapes. More appropriate wave shapes, like N-waves or wave shapes, reconstructed from tide gauges, could be generated in order to obtain a better representation of tsunami wave characteristics. Finally, a true two-dimensional simulation could investigate geometrical features of the experiment, taking into consideration small-scale reflections and inhomogeneous meridional velocity and wave height distributions.

Acknowledgments: The author Nils Goseberg acknowledges the support under a Marie Curie International Outgoing Fellowship for Career Development within the 7th European Community Framework Programme (*impLOADis*, Grant 622214). The authors Stefan Vater and Jörn Behrens gratefully acknowledge support through the ASCETE (Advanced Simulation of Coupled Earthquake and Tsunami Events) project sponsored by the Volkswagen foundation and the authors Nils Beisiegel and Jörn Behrens through ASTARTE—Assessment, STrategy And Risk Reduction for Tsunamis in Europe. Grant 603839, 7th FP (ENV.2013.6.4-3). The authors gratefully acknowledge the helpful reviews by C. Synolakis and four anonymous reviewers, who helped to improve the original manuscript significantly.

Author Contributions: This work is based on a master's thesis project of Ulrike Drähne, who conducted the experimental runs in the wave flume. Nils Goseberg designed the wave flume and supervised the tank experiments. Stefan Vater developed the one-dimensional discontinuous Galerkin numerical code and designed the set-up with data. Nils Beisiegel and Stefan Vater ran the numerical simulations. Jörn Behrens contributed to shaping the experimental set-up and the research questions. All authors contributed equally to the formulation of the manuscript.

Conflicts of Interest: The authors declare no conflict of interest.

Appendix

A supplementary file containing the experimental data presented in this study is available from the article's website or can be requested from the authors. It is a zip archive and provides the data files named *"TxxxHyExpz.txt"* and *"TxxxHyRunupExpz.txt"*. Both data sets are labeled with a naming scheme where T denotes the wave period in seconds, H denotes the wave height in centimeter, *Exp* means *Experiment* and xxx, y and z contain the actual quantity values. The files labeled *"TxxxHyExpz.txt"* contain seven columns of time series that are named and explained in the following:

- Time: experiment time in [s], range: 0 s to 120 s,
- Reference value: analytical surface elevation at the water inlet ($x = 31.92$ m) in [cm], ideal single cycle sinusoidal wave,
- Actual value: measured surface elevation at the water inlet ($x = 31.92$ m) in [cm] as generated by the wave generator,
- WG1, WG2, WG3: surface elevation in [cm] as measured at the three wave gauges,
- Velocity: measured wave velocity at WG1 in [cm/s] in propagation direction.

The sampling frequency for all measurements was 100 Hz. All data recorded with calibrated instruments was converted to SI units except for the velocity which is given in [cm/s]. The time series of wave gauge 1 (WG1) and the velocity appear to be noisier than other instrument data which might be due to electric field disturbances between neighboring instruments. Application of a suitable filter, (e.g., a moving average with 100 value window size as used in this study) might therefore be recommended before using the raw data presented here. The "Reference value" of the first experiment of wave T58_H3 (file T058H3Exp1.txt) exhibits a discontinuity at Time $t = 43$ s, which might be due to a temporary instrument failure.

The data in the files labeled "TxxxHyRunupExpz.txt" contains the experimental run-up data, *i.e.*, position and velocity of the shoreline as the wave climbs up the beach. Each text file contains three columns:

- t: time of measurement for shoreline position in [s]; duration depending on wave period,
- pos: horizontal position of shoreline at time t in [m] (not run-up height!),
- v: computed shoreline velocity in [cm/s].

The time in TxxxHyRunupExpz.txt is consistent with the one in TxxxHyExpz.txt; it is the overall time of the experiment. The position ("pos") in TxxxHyRunupExpz.txt is the horizontal distance of the shoreline to the zero water line. To calculate the run-up height (height above sea level), one has to use the relation: $R = pos \cdot \gamma$, where R is the run-up height, *pos* is the horizontal shoreline position (*i.e.*, "pos" in TxxxHyRunupExpz.txt) and γ is the constant beach slope (here 1/40). The velocity in TxxxHyRunupExpz.txt is calculated from two subsequent points of time and the corresponding shoreline positions. The NaN's in TxxxHyRunupExpz.txt indicate measurement periods where no data was available (e.g., lack of tracers or temporary camera failure).

References

1. Wang, X.; Liu, P.L. An analysis of 2004 Sumatra earthquake fault plane mechanisms and Indian Ocean tsunami. *J. Hydraul. Res.* **2006**, *44*, 147–154.
2. *Geophysical Loss Events Worldwide 1980–2014; NatCatSERVICE Munich Re*: Munich, Germany, 2015.
3. Synolakis, C.E.; Kong, L. Runup measurements of the December 2004 Indian Ocean tsunami. *Earthq. Spectra* **2006**, *22*, 67–91.
4. Synolakis, C.E.; Bernard, E.N. Tsunami science before and beyond Boxing Day 2004. *Philos. Trans. R. Soc. Lond. A Math. Phys. Eng. Sci.* **2006**, *364*, 2231–2265.
5. Briggs, M.J.; Synolakis, C.E.; Hughes, S.A. Laboratory measurements of tsunami run-up, 1993. In Proceedings of the Tsunami, Wakayama, Japan, 23–27 August 1993.
6. Gedik, N.; Irtem, E.; Kabdasli, S. Laboratory investigation of tsunami run-up. *Ocean Eng.* **2005**, *32*, 513–528.
7. Jensen, A.; Pedersen, G.; Wood, D. An experimental study of wave run-up at a steep beach. *J. Fluid Mech.* **2003**, *486*, 161–188.
8. Behrens, J. Numerical Methods in Support of Advanced Tsunami Early Warning. In *Handbook of Geomathematics*; Freeden, W., Nashed, M.Z., Sonar, T., Eds.; Springer Verlag: Heidelberg, Berlin, Germany, 2010; pp. 399–416.
9. LeVeque, R.J.; George, D.L.; Berger, M.J. Tsunami modelling with adaptively refined finite volume methods. *Acta Numer.* **2011**, *20*, 211–289.
10. Liu, P.L.F.; Yeh, H.H.J.; Synolakis, C. Advanced numerical models for simulating tsunami waves and run-up. *Advances in Coastal and Ocean Engineering*; World Scientific Publishing Company Incorporated: Ithaca, NY, USA, 2008; Volume 10.
11. Madsen, P.A.; Bingham, H.B.; Liu, H. A new Boussinesq method for fully nonlinear waves from shallow to deep water. *J. Fluid Mech.* **2002**, *462*, 1–30.
12. Rakowsky, N.; Androsov, A.; Fuchs, A.; Harig, S.; Immerz, A.; Danilov, S.; Hiller, W.; Schröter, J. Operational tsunami modelling with TsunAWI—Recent developments and applications. *Nat. Hazards Earth Syst. Sci.* **2013**, *13*, 1629–1642.
13. Carrier, G.F.; Greenspan, H.P. Water waves of finite amplitude on a sloping beaching. *J. Fluid Mech.* **1958**, *4*, 97–109.
14. Madsen, P.A.; Schäffer, H.A. Analytical solutions for tsunami runup on a plane beach: Single waves, N-waves and transient waves. *J. Fluid Mech.* **2010**, *645*, 27–57.
15. Goring, D.G. *Tsunamis—The Propagation of Long Waves onto a Shelf*, Technical Report Caltech KHR: KH-R-38; California Institute of Technology: Pasadena, California, 1978.
16. Synolakis, C.E.; Deb, M.K.; Skjelbreia, J.E. The anomalous behaviour of the runup of cnoidal waves. *Phys. Fluids* **1988**, *31*, 3–5.
17. Liu, P.L.F.; Cho, Y.S.; Briggs, M.J.; Kanoglu, U.; Synolakis, C.E. Runup of solitary waves on a circular island. *J. Fluid Mech.* **1995**, *302*, 259–285.
18. Madsen, P.A.; Fuhrman, D.R.; Schäffer, H.A. On the solitary wave paradigm for tsunamis. *J. Geophys. Res.* **2008**, *113*, C12012.
19. Monaghan, J.J.; Kos, A. Scott Russel's wave generator. *Phys. Fluids* **2000**, *12*, 622–630.

20. Synolakis, C.E. The Runup of Long Waves. PhD thesis, California Institute of Technology, Pasadena, CA, USA, 1986.

21. Synolakis, C.E. The runup of solitary waves. *J. Fluid Mech.* **1987**, *185*, 523–545.

22. Titov, V.; Synolakis, C.E. Numerical Modeling of Tidal Wave Runup. *J. Waterway Port Coastal Ocean Eng.* **1998**, *124*, 157–171.

23. Hammack, J.L. A note on tsunamis: Their generation and propagation in an ocean of uniform depth. *J. Fluid Mech.* **1973**, *60*, 769–800.

24. Goseberg, N. A laboratory perspective of long wave generation. In Proceedings of the International Offshore and Polar Engineering Conference, Rhodes, Greece, 17-23 June 2012; pp. 54–60.

25. Chanson, H.; Aoki, S.I.; Maruyama, M. An experimental study of tsunami runup on dry and wet horizontal coastlines. *Sci. Tsunami Hazards* **2003**, *20*, 278–293.

26. Goseberg, N.; Wurpts, A.; Schlurmann, T. Laboratory-scale generation of tsunami and long waves. *Coastal Eng.* **2013**, *79*, 57–74.

27. Titov, V.; Gonzalez, F.J. *Implementation and Testing of the Method of Splitting Tsunami (MOST) Model*; NOAA Technical Memorandum ERL PMEL-112 1927; NOAA: Seattle, WA, USA, 1997.

28. Hesthaven, J.S.; Warburton, T. *Nodal Discontinuous Galerkin methods: Algorithms, Analysis, and Applications*; Springer: New York, NY, USA, 2008.

29. Giraldo, F.X.; Hesthaven, J.S.; Warburton, T. Nodal high-order discontinuous Galerkin methods for the spherical shallow water equations. *J. Comput. Phys.* **2002**, *181*, 499–525.

30. Kelly, J.; Giraldo, F. Continuous and Discontinuous Galerkin Methods for a Scalable 3D Nonhydrostatic Atmospheric Model: limited-area mode. *J. Comput. Phys.* **2012**, *231*, 7988–8008.

31. Synolakis, C.E.; Bernard, E.N.; Titov, V.V.; Kanoglu, U.; Gonzalez, F.I. Validation and Verification of Tsunami Numerical Models. *Pure Appl. Geophys.* **2008**, *165*, 2197–2228.

32. Battjes, J.A. Surf similarity, 1974. In Proceedings of the 14th International Coastal Engineering Conference (ASCE), Copenhagen, Denmark, 24–28 June 1974,

33. Goseberg, N. The Run-up of Long Waves—Laboratory-Scaled Geophysical Reproduction and Onshore Interaction with Macro-Roughness Elements. PhD Thesis, The Leibniz University Hannover, Hannover, Germany, 2011.

34. Goseberg, N. Reduction of maximum tsunami run-up due to the interaction with beachfront development—Application of single sinusoidal waves. *Nat. Hazards Earth Syst. Sci.* **2013**, *13*, 2991–3010.

35. Goseberg, N.; Schlurmann, T. Non-stationary flow around buildings during run-up of tsunami waves on a plain beach. In *Coastal Engineering Proceedings*; Lynett, P., Ed.; World Scientific Publishing Company Incorporated: Seoul, South Korea, 2014; Volume 1.

36. Goseberg, N.; Bremm, G.C.; Schlurmann, T.; Nistor, I. A transient approach flow acting on a square cylinder—Flow pattern and horizontal forces. In Proceedings of the 36th IAHR World Congress, Hague, The Netherlands, 28 June–3 July 2015; pp. 1–12.

37. Bremm, G.C.; Goseberg, N.; Schlurmann, T.; Nistor, I. Long Wave Flow Interaction with a Single Square Structure on a Sloping Beach. *J. Mar. Sci. Eng.* **2015**, *3*, 821–844.

38. Goseberg, N.; Schlurmann, T. Interaction of idealized urban infrastructure and long waves during run-up and on-land flow process in coastal regions. *Proceedings of the International Conference on Coastal Engineering*; Lynett, P., Smith, J.M., Eds.; World Scientific Publishing Company Incorporated: Santander, Spain, 2012.

39. Müller, D.R. Auflaufen und Überschwappen von Impulswellen an Talsperren. PhD Thesis, Mitteilungen des Instituts, Versuchsanstalt für Wasserbau, Hydrologie und Glaziologie der Eidgenössischen Hochschule Zürich, Zürich, Germany, 1995. (In German)

40. Heller, V. Scale effects in physical hydraulic engineering models. *J. Hydraul. Res.* **2011**, *49*, 293–306.

41. Schüttrumpf, H. Wellenüberlaufströmung bei Seedeichen—Experimentelle und theoretische Untersuchungen. PhD Thesis, Technische Universität Carolo-Wilhelmina, Braunschweig, Germany, 2001. (In German)

42. Le Méhauté, B. *An Introduction to Hydrodynamics & Water Waves*; Springer: New York, NY, USA; Heidelberg, Berlin, Germany, 1976.

43. Fuchs, H.; Hager, W.H. Scale effects of impulse wave run-up and run-over. *J. Waterway Port Coastal Ocean Eng.* **2012**, *138*, 303–311.

44. Dingemann, M.W. Water wave propagation over uneven bottom. Part 1. Linear wave propagation. In *Advanced Series on Ocean Engineering*; World Scientific: Ithaca, NY, USA, 1997; Volume 13.

45. Peakall, J.; Warburton, J. Surface tension in small hydraulic river models—The significance of the Weber number. *J. Hydrology* **1996**, *53*, 199–212.

46. Cockburn, B.; Lin, S.Y.; Shu, C.W. TVB Runge-Kutta local projection discontinuous Galerkin finite element method for conservation laws III: One-dimensional systems. *J. Comput. Phys.* **1989**, *84*, 90–113.

47. Vater, S.; Beisiegel, N.; Behrens, J. A Limiter-Based Well-Balanced Discontinuous Galerkin Method for Shallow-Water Flows with Wetting and Drying: One-Dimensional Case. *Adv. Water Resour.* **2015**, *85*, 1–13.

48. Shu, C.W.; Osher, S. Efficient Implementation of Essentially Non-oscillatory Shock-Capturing Schemes. *J. Comput. Phys.* **1988**, *77*, 439–471.

49. Gottlieb, S.; Shu, C.W.; Tadmor, E. Strong Stability-Preserving High-Order Time Discretization Methods. *SIAM Rev.* **2001**, *43*, 89–112.

50. Courant, R.; Friedrichs, K.O.; Lewy, H. Über die partiellen Differenzengleichungen der mathematischen Physik. *Math. Ann.* **1928**, *100*, 32–74.

51. Cockburn, B.; Shu, C.W. The Runge-Kutta Local Projection P^1-Discontinuous-Galerkin Finite Element Method for Scalar Conservation Laws. *RAIRO Modél. Math. Anal. Numér.* **1991**, *25*, 337–361.

52. Brier, G.W. Verification of forecasts expressed in terms of probability. *Mon. Weather Rev.* **1950**, *78*, 1–3.

53. Madsen, P.A.; Fuhrman, D.R. Run-up of tsunamis and long waves in terms of surf-similarity. *Coastal Eng.* **2008**, *55*, 209–223.

Long Wave Flow Interaction with a Single Square Structure on a Sloping Beach

Gian C. Bremm, Nils Goseberg, Torsten Schlurmann and Ioan Nistor

Abstract: In the context of dam breaks, tsunami, and flash floods, it is paramount to quantify the time-history of forces by the rapidly transient flow to vertical structures and the characteristics of the induced flow patterns. To resemble on-land tsunami-induced flow, a free-surface-piercing structure is exposed to long leading depression waves in a tsunami flume where long waves run up and down a 1:40 smooth and impermeable sloping beach after its generation by a volume-driven wave maker. The structure and its surrounding were monitored with force transducers, pressure gauges and cameras. Preparatory steady-state experiments were accomplished to determine the drag force coefficient of the square cylinder at various water depths. The flow during wave run-up and draw-down acting on the structure resulted in distinct flow pattern which were characteristic for the type of flow-structure interaction. Besides bow wave propagating upstream, a standing or partially-standing wave was observed in front of the structure together with a wake formation downstream, while a von Kármán vortex street developed during the deceleration phase of the flow motion and during draw-down. Force measurements indicated a sudden increase in the stream-wise total force starting with the arrival of the flow front during initial run-up. Lateral velocities showed significant oscillations in correlation with the von Kármán vortex street development. A comparison of the total measured base force with the analytically-calculated share of the drag force revealed that forces were prevailingly drag-dominated.

Reprinted from *J. Mar. Sci. Eng.* Cite as: Bremm, G.C.; Goseberg, N.; Schlurmann, T.; Nistor, I. Long Wave Flow Interaction with a Single Square Structure on a Sloping Beach. *J. Mar. Sci. Eng.* **2015**, *3*, 821–844.

1. Introduction

1.1. Background

This paper addresses the interaction of a transient flow with a surface-piercing height-finite structure and presents new detailed laboratory data. It includes the description of the spatio-temporal evolution of the flow fields around the structure and the resulting time series of the induced stream-wise forces. After the 2011 Great East Japan Earthquake and Tsunami which claimed the lives of thousands and caused unprecedented damage to coastal infrastructure, it is of vital importance to deepen the understanding of the interaction of rapidly advancing flow and the induced forces. The evaluation of effects caused by the build environment is pivotal for the simulation-based risk assessment and reduction for tsunami-prone areas [1,2]. Commonly, the approaching flow caused by water-related phenomena such as dam-breaks or tsunami are transient in nature.

Current or wave induced forces on circular cylinders are well-researched in disciplines with applications in aerodynamics [3], hydrodynamics [4], coastal engineering [5,6], and geophysics [7]. Less attention was paid to square or rectangular obstacles. Several studies were conducted to assess the effect of side-by-side arranged structures in 2D (two-dimensions) under stationary currents [8]. The propagation of pure fluid bores on dry beds classifying bores on the basis of relative wave height into strong turbulent and undular flow was reported likewise [9]. Other investigations involved the impact of bores generated from dam breaks on various freestanding structures [10] or researched the process of attenuation to very long waves on onshore house placed in various geometric patterns [11]. Other researchers investigated the effect of mangrove forests [12]. Park et al. [13] addressed how flow characteristics change when solitary waves interact with coastal infrastructure. Goseberg and Schlurmann [14] investigated the time-history of the wake angle of single and side-by-side arranged square cylinders for tsunami-like onland flow.

In the case of onshore propagating tsunami-induced inundation, experimental studies are often limited to solitary wave or dam break wave condition (WC). However, time- and length-scales for such experiments have to be chosen carefully as this strategic attempt is crucial to yield meaningful estimates of induced forces similar to those exerted at prototype scale. The investigated flow characteristics usually depend on the flow characteristics, the nearshore bathymetry or inland topography as well as the wave transformations which can modify features of the transient fluid motion. Tsunami periods commonly range from 5 to 30 min [15]. However, the vast majority of the studies looking into tsunami near-shore action model the WC by means of solitary waves [16,17] or by dam-break waves [10,18,19]. This study however follows a more recent approach to simulate prototype-like tsunami with physically-sound length and time scales under experimental conditions. Following this approach, tsunami generation is achieved by means of a volume-driven wave maker consisting of high-capacity pipe pumps, as described in Goseberg et al. [20], to generate leading depression sinusoidal waves.

Morison et al. [21] applied a superposition of inertia and drag forces to predict forces exerted by regular waves onto a vertical, cylindrical pile. Linear inertia forces were derived from potential theory in order to account for forces exerted to the pile in the presence of horizontal accelerations from a regular (or irregular) oscillatory fluid motion. Following Journée and Massie [22] and Chakrabarti [23], the inertia force consists of the Froude-Krilov force attributed to the pressure gradient acting on the vertical pile surface and a virtual force formulated in order to correct for the fact that a cylinder in a velocity field modifies the proximity velocity field and thus flow accelerations around the perimeter (locally). The effect of the pressure gradient on the flow and disturbance induced by the pile, which theoretically takes values between 1 and 2 , are combined and expressed through the inertia or added mass coefficient C_M (see Equation (2)) which is derived experimentally. Drag forces, which in an oscillatory flow occur with a phase shift of $90°$, are additionally exerted onto a vertical, cylindrical pile; the experiment proved that the drag force is proportional to the horizontal fluid velocity squared. A drag force coefficient, C_M, displayed in Equation (1), has yet to be determined experimentally for each particular case. A vast amount of literature is dedicated to the computation of forces and the appropriate determination of C_D and C_M values in connection with circular piles; i.e., for short-crested waves [24], for internal wave loads [25], for additional loads attributed to wave run-up and impact [26] or to account for the reciprocal influence of piles in pile groups [27].

Forces exerted on obstacles in a transient flow include buoyancy and hydrodynamic forces. Stream-wise forces encompass surge/slamming/impulsive forces, forces due to debris impact as well as hydrodynamic drag and inertia forces [28]. This paper focuses on an accurate assessment of drag force which occur during the interaction with a transient flow which can thus be described by the following equations

$$F_D = 0.5 C_D \rho A \vec{v}^2 \tag{1}$$

with F_D – drag force, C_D – drag force coefficient, ρ – fluid density, A – projected flow-facing area and $\vec{v}$ – flow velocity. Typical drag force coefficients were reported by Yen and Liu [29] and by Hashimoto and Park [30] to be $C_D = 2.05$. While the drag force is proportional to the dynamic head $0.5\rho\vec{v}^2$ and the flow-facing area A, the inertial force which are estimated here as well is proportional to the acceleration and deceleration of the fluid interacting with the structure. It reads:

$$F_I = C_M \rho V \frac{d\vec{v}}{dt} \tag{2}$$

with F_I inertial force, $C_M = 1 + k_m$ mass coefficient, k_M added mass, V volume and $d\vec{v}/dt$ acceleration or deceleration of the fluid. C_M has typical values in the range $C_M = [1..2]$ depending on the setting and boundary conditions.

1.2. Objectives

This paper attempts to enhance the existing knowledge of the development of drag and inertial forces of a single surface-piercing square structure. The forces resulting from the impact of a transient approach flow with prototype tsunami characteristics have been studied for the first time. In a first step, the drag force coefficients are determined for the structure in a stationary current at different Reynolds numbers as a basis for the analytical determination of the drag forces. Secondly, the study aims at deriving stream-wise forces exerted onto a structure by a transient flow. This study investigates whether the maximum force on the structure—A key factor for design—Is dominated by either the drag or by the inertial forces. The authors also investigated how the relation between the two forces develops. An optimized control strategy for the volume-driven wave maker is also presented herein.

The paper is organized as follows: Section 2 describes the experimental set-up, the applied instrumentation and the used lab facilities are outlined. Section 3 addresses the experimental results and depicts a characterization of the transient approaching flow, the resulting flow pattern and the horizontal forces. Results and main findings are presented in Section 4.

2. Experimental Setup

2.1. Stationary Force Measurement Tests

In order to determine drag force coefficient for single structures subjected to a stationary current, preliminary tests were conducted in a 1 m wide and 20 m long current flume. The maximum flow capacity of the flume is of $0.25\,\mathrm{m^3\,s^{-1}}$. The flow entered at one end through a flow straightener and then propagated over a horizontal bottom and finally flowed out into a bottom outlet. The structure was an aluminum square pipe of 0.20 m in length and with a cross section of 0.1 m × 0.1 m. The bottom side was sealed by welding and water-proofed. On the inside, the bottom also held a bend-proof connection to accommodate the force transducer. The structure was mounted onto a rigid frame and placed at the center axis of the flume, 7.74 m downstream of the flow straightener. A two-axis radial force sensor with a maximum capacity of 100 N (Honigmann RFS 150 XY, accuracy class 0.25) was used to record the time-history of the current-induced stream-wise forces exerted onto structure. Prior to the tests, the force sensor was calibrated with a lab-certified scale. In addition, the time-history of flow velocities as well as the water levels were measured 1.6 m upstream of the structure. Each force measurements was recorded over 3 min with sampling frequency of 200 Hz and time-averaged during the subsequent data processing analysis.

2.2. Transient Flow Tests

Following the stationary tests conducted to estimate the drag force coefficients, transient flow conditions were used with the structure. The experimental setting closely follows those of Goseberg [11] and Goseberg and Schlurmann [14]. The generation of the WC utilized the procedure outlined in Goseberg et al. [20] and involved the same high-capacity pipe pumps to accelerate or decelerate a specific volume of water. The method offers the advantage of precisely controlling the desired wave period and wave height by applying a proportional-integral-derivative controller (PID controller) as a control loop feedback mechanism. The method allows for small differences between the desired WC and the actually generated WC. In order to decrease the time needed to adjust the control variables, an improved step was introduced to the previous generation routine. While the generation of any WC was constantly controlled in the original wave generation scheme, during the tests reported herein, the controller was merely applied to record the time series of the rotational rate of the pipe pumps once per WC. The additional step consisted of bypassing the continuous PID-controller and feeding the previously recorded rotational rate time series in smoothed form to the pumps directly.

Figure 1 presents a comparison between the original and the improved wave generation method. Shown is the process value η of the repeated WC. Differences between the two methods are minor during the generation of the wave through. In order to avoid prominent artefact oscillations during the generation of the wave, the novel attempts to directly control the wave generation clearly demonstrates a major improvement during peak and descending phase of the flow. Contrary, the original method introduced higher frequency surface waves caused by an insufficiently tuned PID-controller. These disturbances would have been further dampened with additional effort and more tuning. However, the improved wave generation method allowed for a faster generation of the desired flow and reduces the existence of parasitic surface waves of higher frequencies which alter the wave height and force measurements in the flume. The investigated WCs are listed in Table 1. All three WCs are leading-depression sinusoidal waves. The base case has a wave height of $H = 8\,\mathrm{cm}$ and a period of $T = 60\,\mathrm{s}$. Comparatively to the base case a low amplitude wave and a long period wave are investigated.

Table 1. Summary of tested wave conditions (WCs). All three WCs are leading-depression sinusoidal waves.

WC-ID [-]	Wave Condition [-]	Wave Height H [m]	Wave Period T [s]	Wave Length L [m]	a_0/h_0 [-]	a_0/L [-]	Iribarren Number ξ [-]
01	Base case	0.08	60	102.9	0.133	3.9×10^{-4}	6.63
02	Low amplitude	0.04	60	102.9	0.067	1.9×10^{-4}	9.37
03	Long period	0.08	90	154.4	0.133	2.6×10^{-4}	9.94

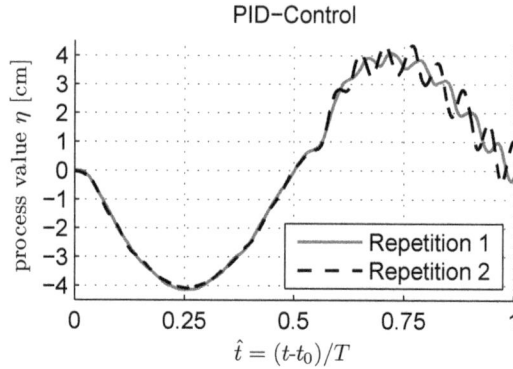

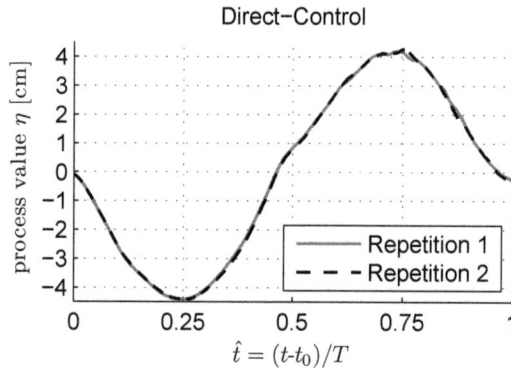

Figure 1. Comparison of the repeatability of the generation of the WC by the volume-driven wavemaker. With the original method of PID-control (**a**) and with the improved method which involved direct control (**b**). Zero dimensionless time refers to beginning of the wave generation.

The time-history of the surface elevations in the proximity of the wave generator are plotted for the different WCs (Figure 2). Additionally, a solitary wave is plotted in the figure in order to exemplify how length and time scales differ between the used sinusoidal and idealized solitary wave. Madsen *et al.* [31] clearly pointed out that the use of solitary waves results in major discrepancies between down-scaled and real-world tsunami waves when analyzing the tsunami propagation, its nearshore transformation and the subsequent on-land flow. These discrepancies result from the fact that, when up-scaled, the experimental solitary wave periods and their wave lengths, respectively, remain orders of magnitude too short when compared to observed tsunami waves. Table 1 also contains ratios of the relative amplitude and the relative wave length. In addition, the Iribarren number, following the definition of [32], is added to provide guidance about the respective wave conditions used. The Iribarren number, ξ in its recast form presented by [32] reads:

$$\xi = \sqrt{\pi} \left(\frac{a_0}{h_0} \right)^{\left(-\frac{1}{2} \right)} \left(\frac{\Omega^2 h_0}{g \tan \alpha^2} \right)^{\left(-\frac{1}{2} \right)} \tag{3}$$

where $\Omega = \frac{2\pi}{T}$ denotes the wave frequency, a_0 is the wave amplitude, h_0 is the water depth, while α is the beach slope.

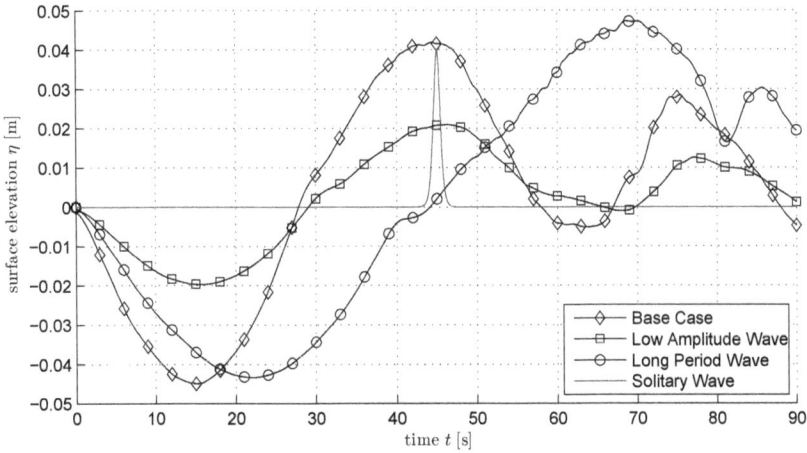

Figure 2. Time-history of the water surface elevation for the three tested wave cases (WCs) measured in the proximity of the wave generator compared to a solitary wave.

A closed-circuit current flume was used for the experiments and was divided into a propagation section and reservoir. The propagation section included a 1:40 sloping beach with aluminium board surface. Length scale were set to 1:100. More experimental details are presented in Goseberg [11]. The structure was used for the transient flow tests too. Water depth in the flume was kept constant at 0.32 m throughout the experiments and the structure was placed onshore at a horizontal distance of 0.20 m from the still water line. A gap of 2 mm was maintained between the structure and the beach surface in order to prevent small grains to hinder correct force measurements. Besides measuring forces in x- and y-directions, additional informations were recorded using pressure sensors positioned inside the beach, wave gauges (with a sampling rate of 100 Hz), an electro-magnetic velocimeter (with a sampling rate of 100 Hz) and two high-definition cameras sampling at 30 Hz. Data recording included the time-history of the water level in front of the structure and, the surface velocity vector field by means of particle image velocimetry (PIV). Graphic details of the experimental set-up are depicted in Figure 3.

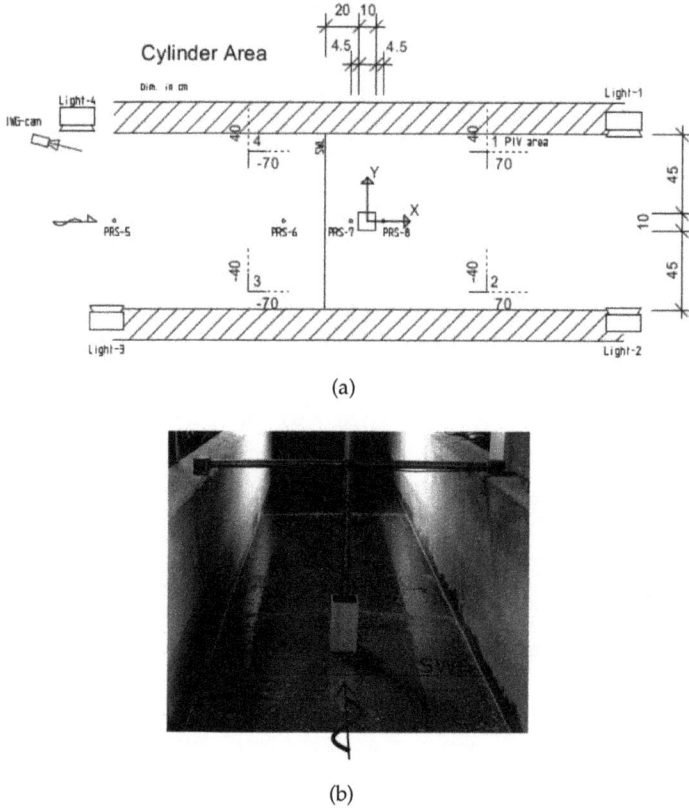

(a)

(b)

Figure 3. Plan view sketch of the structure located 0.20 m from the still water line with PIV field of interest (**a**) and support frame of the surface-piercing structure (**b**).

3. Experiments

3.1. Time-Variant Transient Flow

The water stage development and velocity depth profiles at the beach toe ($x = -12.8$ m) are presented in Figure 4 for the case with no structure. While the temporal dimension was normalized by the respective period and related to the first water level change at this position, wave heights were normalized by means of the still water depth while velocities were normalized by the wave propagation speed derived for still water condition. Figure 4 shows the time-histories of the non-dimensional stream-wise velocity $\hat{u}$ of the incoming wave train at the location of the beach toe in 3 different submergence depth of $h/h_0 = 0.25, 0.44$, and 0.63, beneath the still water elevation in a color-coded manner. Despite the varying submergence depth, three velocity time-histories show an almost identical development in terms of velocity magnitude and phase. This indicates fairly uniform distribution of horizontal velocities over the water depth. It can however be speculated that horizontal velocities beneath the recorded level, in particular beneath $h/h_0 < 0.2$, resemble more closely the velocities predicted by the wall logarithmic law [33], especially since the bottom surface of the flume and the beach were not hydraulically smooth. In the subsequent computations of the drag and inertial forces it is however hypothesized that the horizontal velocity profile was vertically uniform and thus in agreement with the shallow water approximation. However, a perceptible phase-shift between the long wave elevation and the corresponding velocities beneath the water surface occurs, indicating that smallest velocities continually occur prior to the minimum water depth during the wave trough. Highest onshore-directed flow velocities occur during flow run-up between the leading depression wave through and the maximum run-up. The base case had twice the amplitude of the low amplitude, wile sharing the same period. The Minimum and maximum wave-induced velocities of the base case were therefore roughly doubled compared with the low amplitude wave. Flow velocities of the long period wave resided in between the minimum and maximum flow velocities of the other two WCs. Slack time for the three considered WCs was approximately $\hat{t} = 0.3$.

In order to compute the drag and inertia forces for comparison with the measured horizontal forces at the test specimen position, it was important to obtain reliable data on flow depths and velocity field in sufficient spatio-temporal dimension. Water depth information was yielded through analysis of the pressure sensor recordings positioned around the structure position from the undisturbed model tests. Under the assumption of a shallow water wave and its hydrostatic characteristic, assuming that the turbulent shear stresses and the vertical acceleration are small compared to the gravitational acceleration, g. This is underlined by the constant horizontal velocity profiles underneath each generated long wave. The pressure sensorreadings were converted into time series of water depth by a hydrostatic calibration factor in front of and behind the structure. A linear interpolation between data from PRS-7 and PRS-8 (cp. Figure 3) allowed to approximately determine the water level at the structure position. Surface velocities were derived from PIV analysis [34] using the MatPIV toolbox [35] on the basis of the images recorded by the over-head mounted camera. PIV-derived velocities were averaged in spatial domain. Before using the water depth and velocity time series to compute drag and inertial forces, a smoothing filter was used to average the data temporally. For the remainder of this paper, dimensionless times were related (and zeroed) to the particular instant in time when the flow hit the upstream facing front of the structure.

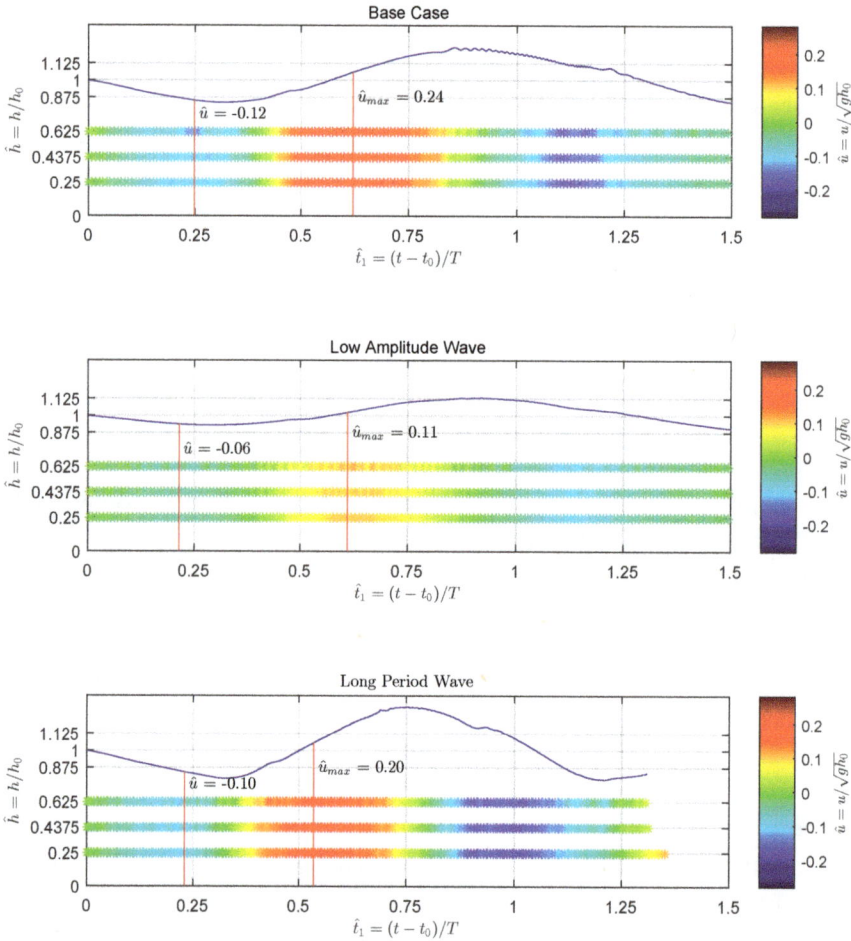

Figure 4. Non-dimensional water depth and vertical velocity profiles for the three investigated WCs. Water depth are normalized by still water depth h_0, time is normalized by the wave period T. Data were recorded at the toe of the beach slope.

Figure 5 presents the time history of the water level and the stream-wise velocity at the structure position for the three investigated experimental configurations. The maximum water level at the structure location reached $h_{max} = 0.11\,\text{m}$ for the base case, while its maximum for WC-ID 02 was almost half that of that of the base case with $h_{max} = 0.06\,\text{m}$. An increased wave period and identical wave height of the WC-ID 03 compared with the base case resulted in a higher maximum flow height at the structure position ($h_{max} = 0.13\,\text{m}$). An increase of the period also resulted in a run-down, which in the dimensionless consideration appeared to be finished earlier than the two cases with the shorter period (WC-ID 01 and WC-ID 02).

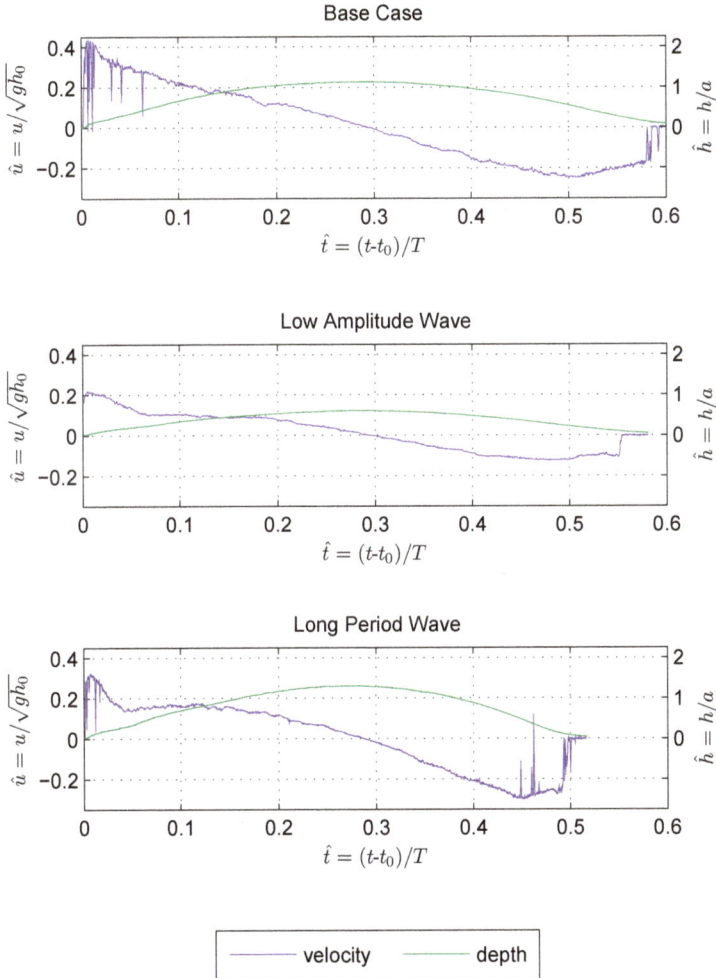

Figure 5. Time-history of dimensionless stream-wise velocities based on PIV analysis and water depth based on interpolated pressure sensor readings at the position of the structure for the three experimental configurations ($x = 13.05\,\text{m}$ from beach toe to structure center) in the undisturbed wave run-up and draw-down.

Stream-wise velocities derived from PIV analysis followed an identical pattern for all three WCs. Highest velocities were measured during the run-up phase and slightly smaller maximum velocities during the draw-down of the wave. In contrary to findings from the beach toe position, zero velocities occurred simultaneous to the highest water levels. Velocities rapidly increased as the flow front hit the structure and then gradually decreased, and turned negative as the draw-down of the flow started taking place. Spikiness of the presented velocity data resulted from the PIV and not from real processes during the flow. An increase of the period obviously reduced the maximum positive velocity for WC-ID 03 at the beginning of the run-up.

3.2. Flow Pattern

Figure 6a–f shows plan view images of the 2D hydrodynamic vector field induced by the presence of the structure for the base case. Dimensionless times again were referred to the instant in time when the flow hit the upstream front of the structure. The black vertical line results from blocking the camera sight on the flow surface. The flow approached structure and a subsequent hydraulic jump was generated. The later slowly radiated in the upstream direction while the downstream side of the structure remained shadowed from the incoming flow. Between $\hat{t} = [0.05..0.15]$, the protected downstream area was subsequently filled with water rushing up on the sloping beach and a turbulent wake started developing (Figure 6b). Flow velocities at the front of the structure generally decreased until they approached zero at the structure's surface.

On both the upstream and downstream sides of the structure the bow wave is advected due to the run-up velocity of the approaching flow. As its velocity decreases, a von Kármán vortex street develops (Figure 6c). Although the ambient currents approaches zero, vortices from the decelerated, but transient von Kármán vortex street are still traceable, as shown in Figure 6d. Draw-down phase flow features are also recorded (Figure 6e,f). Run-down velocities around $\hat{V} = 0.2$ appear and this time a von Kármán vortex street develops after velocity inversion with local velocities reach as high as $\hat{V} = 0.3$.

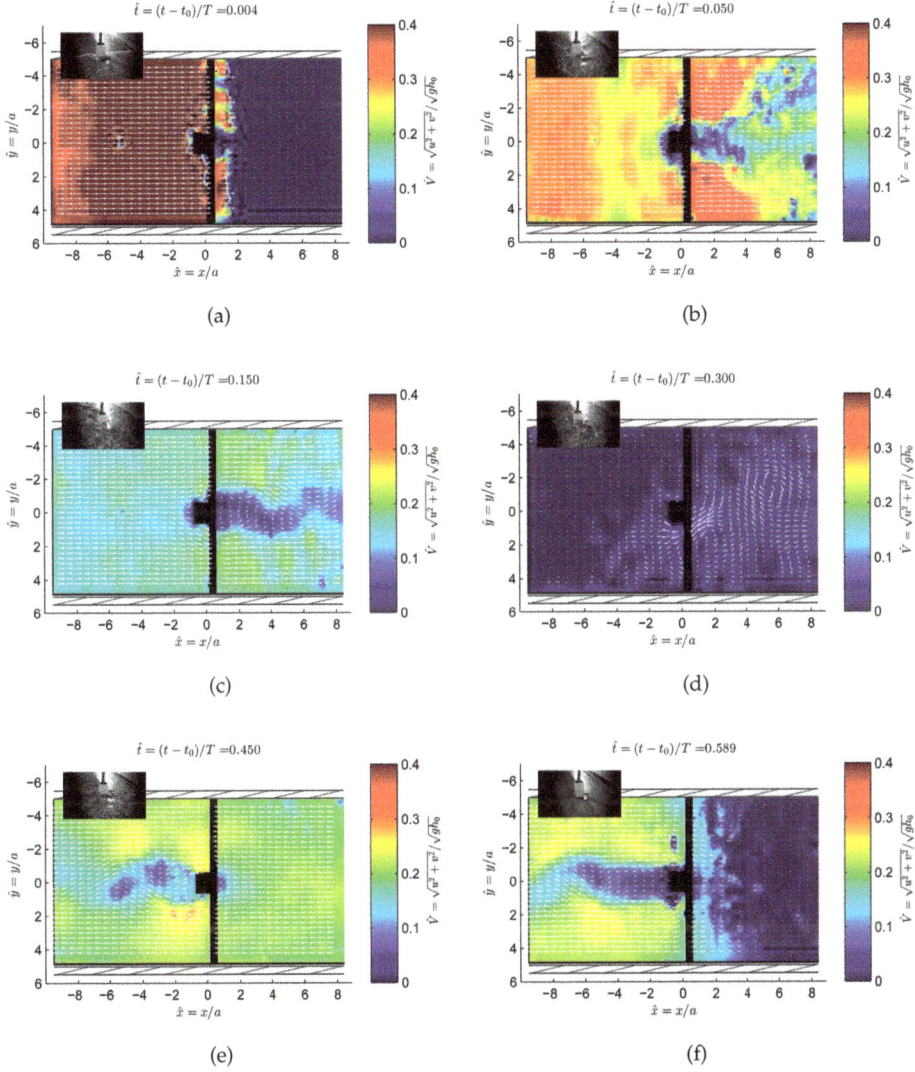

Figure 6. Snapshots of the base case flow interaction around the rectangular structure. Shown is the non-dimensional velocity field $\hat{V}$ with directions indicated by white arrows. The grayscale image is a perspective view of the flow-column interaction.

3.3. Horizontal Forces

Horizontal forces recorded by the two-axial force transducer are presented for the three investigated configurations in Figure 7 and are plotted against the dimensionless time. Time development of the flow run-up at the upstream front of

109

the structure as derived from image processing of the second high-definition camera were additionally plotted. The flow run-up, p, was normalized by the width of the structure, $a = 0.1$ m. The time development of the flow run-up was depicted in Figure 7 for the three hydraulic configurations and was phase with the pressure-based water level measurements for the undisturbed case presented in Figure 5. Highest run-up was found for the longest flow with WC-ID 03.

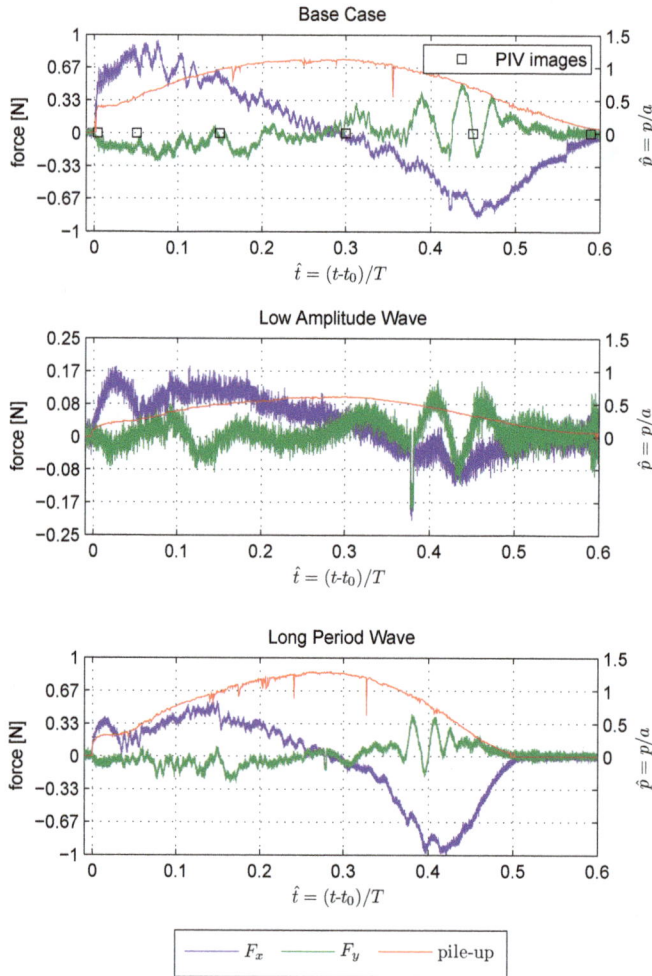

Figure 7. Time history of horizontal forces in x- and y-directions and local run-up (pile-up) at the upstream front of the structure for the three investigated WCs. The square boxes indicate some relevant instants in time with PIV results (cp. Figure 6).

All flows had a sudden increase of the water level and, thus, the flow run-up at the structure front was observed for all hydraulic WCs investigated. The largest increase in water level in front of the structure was yet found for the flow with $T = 60$ s. Though peak water levels were higher for the long period wave ($T = 90$ s). Stream-wise total horizontal forces exhibited interesting features. While stream-wise forces for all cases increased rapidly once the flow front reaches the structure, it appeared that rates of change of the force were significantly higher for the flows with twice the amplitude (WC-ID 01 and 03). The horizontal stream-wise force gradient for the short flow with small amplitude was lower than the case using the larger amplitudes and dropped sooner after reaching the peak. Generally, horizontal stream-wise forces were much smaller compared to those occurring when the larger amplitudes were used. Based on the cases investigated, it was not yet possible to define a distinct correlation between the minimal or maximal forces and amplitudes or periods respectively. The base case showed most similar maximum and minimum forces of $F_{x,max} = 0.94$ N and $F_{x,min} = -0.86$ N. The force peak occurred approximately $\hat{t} = 0.07$ after the flow front reaches the location of the structure. The change from positive to negative horizontal stream-wise force coincided with the highest flow run-upon the structure front. A plateau phase similarly reported by Nouri *et al.* [10] for dam break flow was, however, not found for this case. Although the magnitude of the horizontal stream-wise force increased and then decreased with time until a minimum is reached in the two variations, some distinct differences were found. Both variations exhibit a first peak which might be due to an impulsive transfer of momentum which was followed by a decrease and another increase of the force before they decreased and eventually started exhibiting negative values. Absolute values of the maximum forces for WC-ID 02 were $F_x = 0.18$ N and $F_x = -0.08$ N. Negative forces exerted onto the structure during the draw-down phase seemed to increase significantly for WC-ID 03 where the flow period was 50% longer than for the other two cases. Maximum value is $F_x = 0.56$ N while the minimum value yielded $F_x = -1.00$ N. Significantly smaller forces were recorded in the transverse direction to the flow. Maximum and minimum lateral forces were generally 50% smaller than stream-wise forces. Significant force in the lateral direction were of oscillatory nature and they always occurred during the draw-down phase and were likely to occur in parallel to the formation of the transient von Kármán vortex street. Visual inspection of the force time series allow the assumption that distinct vortex shedding frequencies exist. Tough at present, this assumption can neither be verified nor disproved. However, due to space restrictions a more thorough analysis cannot be presented. For a summary of the dimensionless forces observed see Table 2 , which also details the maximum water levels, h_{max}, measured during the undisturbed experiments and the corresponding hydrostatic force. The hydrostatic force was subsequently used to normalize the recorded forces to allow for generalization and further use of the results. The hydrostatic force was calculated as follows:

$$F_{HS} = \frac{1}{2}\rho g a h_{max}^2 \qquad (4)$$

where density $\rho = 1000\,\text{kg}\,\text{m}^{-3}$, g is the gravitational acceleration, while a is the width of the square cylinder face.

Table 2. Summary of the maximum loads on the structure by the different wave conditions (WCs). Plus and minus refers to the direction in the coordinate system (see Figure 3).

WC-ID	h_{max} [m]	F_{HS} [N]	F_x/F_{HS} [-]		F_y/F_{HS} [-]	
			up	down	left	right
01	0.12	6.60	0.14	−0.13	0.08	−0.05
02	0.06	1.89	0.10	−0.11	0.07	−0.10
03	0.13	7.79	0.07	−0.13	0.06	−0.03

Forces induced by the flow around the square cylinder were normalized by the hydrostatic force on the front face of the square cylinder as expressed by Equation (4). The minimum or maximum recorded base shear forces never exceeded 15 % of the respective normalization force for the investigated wave conditions. As the used wave periods were similar to prototype tsunami periods, it is worthwhile to compare measured base shear forces from sinusoidal waves with base shear forces exerted to structures under the attack of much shorter solitary waves. Although comparable research is scarce, Mo et al. [36] investigated the base shear forces exerted on a circular cylinder subjected to a shoaling and breaking solitary wave running up a 1:50 sloping beach. The authors presented the measured and computed force time-history normalized by $\rho g D^3$, with D being the cylinder diameter. Based on their findings (Figure 18a,b in Mo et al. [36]), it was however possible to reconstruct the original dimensional force time-history. By using the maximum wave run-up on the pile and the maximum force calculated, one could then calculate the ratio of the maximum base shear force to the hydrostatic unit force times the structure's diameter; this yielded $(\frac{F_D}{F_{HS}})_{\text{Mo et al. (2014)}} = 0.80$ which is roughly six times larger than the highest value found for the present case wave conditions. A two-dimensional attempt to determine forces and overturning moments exerted onto a beach front house at locations near the still water shoreline was presented by Xioa and Huang [37] who used a numerical model to predict the base shear force time-history as a result of a solitary wave impact. They showed that non-dimensional forces are a function of the distance between the structure and the still water shoreline. Unfortunately, no time-histories of the water level were reported in order to reconstruct the original force time-histories; thus, it can only be conjectured that the forces were also higher than in the long sinusoidal wave condition case reported herein. The authors hypothesized that base shear forces incurred by solitary wave attack are greatly increased compared to those generated by longer, prototype-like sinusoidal waves; yet it remains unsolved until further research is conducted which type of wave to apply for the design of infrastructure under the threat of tsunami waves.

In order to highlight how wave periods and amplitudes other than those investigated might change the forces exerted to the square cylinder, the maximum, base shear forces were plotted against the surf similarity parameter as provided by Equation (3) [32]. Figure 8 presents stream-wise and lateral force ratios against the surf similarity parameter.

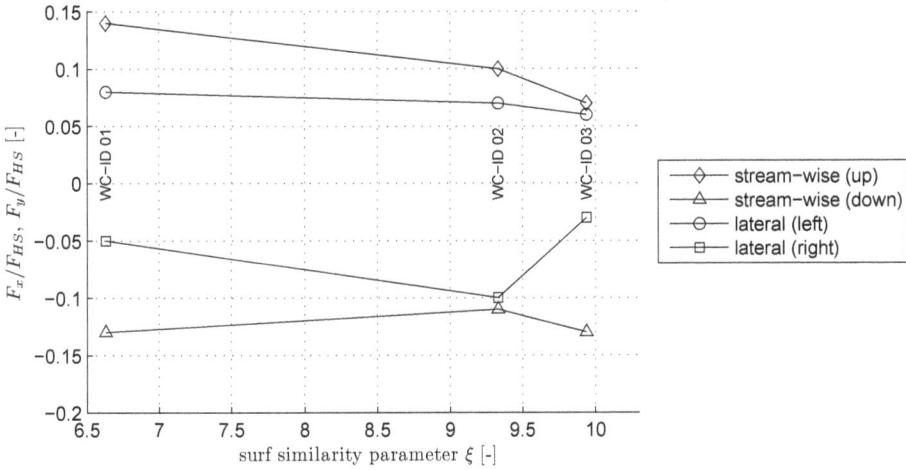

Figure 8. Non-dimensionalized base shear forces for the lateral and the stream-wise direction as a function of the surf similarity parameter ζ.

The measured base shear forces are firmly affected by the distribution of the turbulence in the water column and in particular influenced in the near-field around the square cylinder. Once the run-up and draw-down flow is fully developed, the bottom boundary layer is fully developed and the Reynolds number exceeds the threshold for turbulent flow conditions which affects the base shear force measurement similarly. At the same time, the boundary layer changes along the lateral sides of the square cylinder: the boundary layer separation induced by the sharp structure edges results in a lateral pressure drop which in turn effects the lateral base shear forces. However, the investigation of the temporal variation and separation of the boundary layer is beyond the scope of this study.

4. Discussion

4.1. Drag Force Coefficient, Reynolds, Froude, Keulegan-Carpenter Number

Resulting drag force coefficients C_D derived from preliminary tests are shown in Figure 9. The C_D values presented herein were derived from stream-wise force measurements of the total base force exerted onto the structure by a stationary flow in the flume setting as described in Section 2.1 and using the rearranged Equation (1) with the stream-wise velocity u to yield C_D. The measurements were conducted in the range of Reynolds numbers $Re = [10^3 .. 2 \times 10^4]$. For this measurement range, a power-law was fitted to easily determine the C_D values corresponding to the Reynolds number which were further used for the computation of the analytical drag forces during the transient flow process. For larger Reynolds numbers beyond the valid measurement limit of these experiments, drag force coefficients were suitably supplemented with constant values reported in literature.

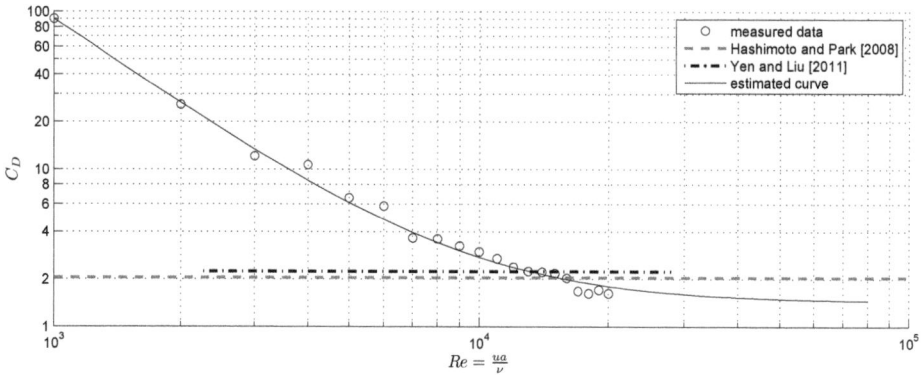

Figure 9. Resulting drag force coefficients C_D derived from stationary force measurements preliminary studies.

In order to compute drag forces from Equation (1), it is necessary to determine the Reynolds numbers for selecting the correct drag force coefficients from the relationship $C_D(Re)$ obtained from experimental preliminary tests presented in Section 4.1. Figure 10a illustrates the time-history of the Reynolds number based on two different characteristic length scales. The top panel presents Re based on the structure width, a, while Re_h is computed based on the hydraulic diameter, R_h. Reynolds numbers are defined as:

$$Re = \frac{\vec{v}a}{\nu} \tag{5}$$

$$Re_h = \frac{4\vec{v}R_h}{\nu} \tag{6}$$

Since this paper predominantly focuses on flow-structure interaction processes, it seems more reasonable to relate the flow processes to the structure-based Reynolds number as it resided well above $Re > 2 \times 10^4$ most of the time. In hydraulic model tests with Froude similitude it is important to maintain Reynolds numbers high enough to assure hydraulically rough conditions. Based on overtopping experiments conducted by Schüttrumpf [38], Reynolds numbers are recommended to exceed $Re_{crit} = 10^3$.

This requirement was fulfilled during both the flow run-up and draw-down. However, present experiments were still smoother in terms of the Reynolds numbers compared with the tests conducted by Arnason *et al.* [39] who reported experimental research on bore impingement on a vertical column involving Reynolds numbers in the range of 0.86×10^5 to 3.8×10^5. Thus, prototype Reynolds numbers usually range from 10^6 to 10^7 depending on the characteristic length scale and the velocities involved. This implies that force contributions stemming from the turbulent characteristics of the flow might not be fully represented in the experiments reported herein since the Reynolds numbers used are smaller than desired. Based on the Reynolds numbers used in these experiments, it was thus possible to determine the correct drag force coefficients and subsequently determine analytically the drag force (cp. Figure 10b). The computation of the drag forces is further discussed in Section 4.2.

Froude numbers are also presented for the base case for the flow run-up and draw-down. Figure 11 presents the time history of the Froude number at the location of the structure for the base case. The values of Fr are based on the following equation:

$$Fr = \frac{\vec{v}}{\sqrt{gh}} \tag{7}$$

In these experiments, the Froude number rarely exceeded $Fr = 1$ during the flow run-up and draw-down. Supercritical flow occurred directly at the beginning of the flow-structure interaction from $\hat{t} = 0.00$ to $\hat{t} = 0.05$ and transitions afterwards to subcritical flow. As the flow was smoothly decelerated during this transition through the run-up on the mildly-sloping shore, no obvious additional features such as standing waves were found. It thus can be ruled out that effects from flow transition could influence the flow-structure interaction at the position of the structure.

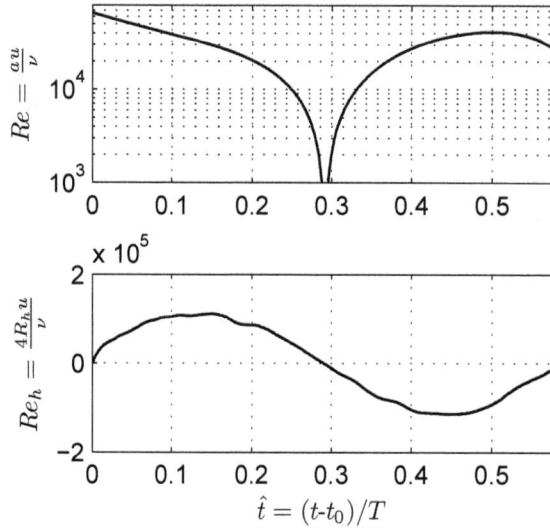

(a) Reynolds number

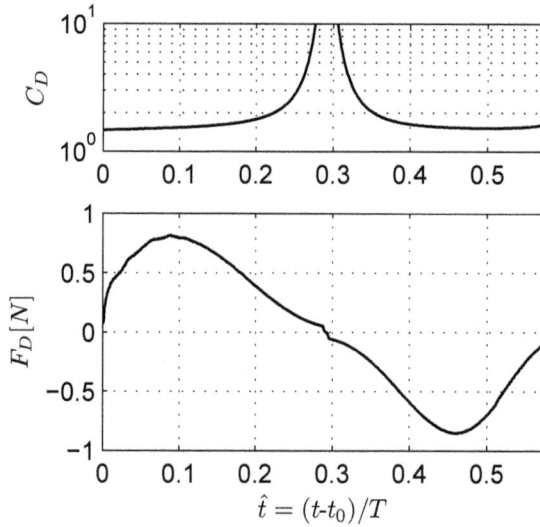

(b) Drag coefficient

Figure 10. The time-history of (**a**) Reynolds numbers and (**b**) drag coefficient and the subsequent time-history of the horizontal stream-wise velocity for the base case (Time is normalized by the period T).

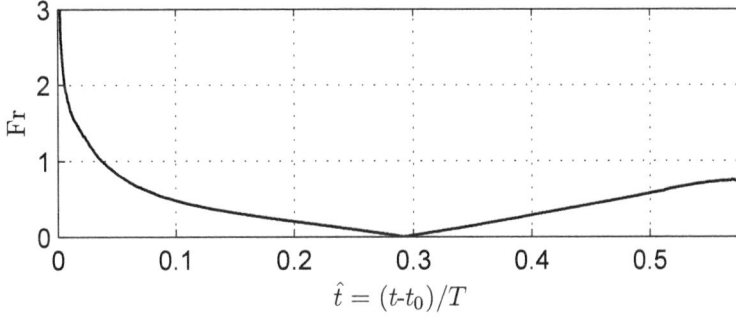

Figure 11. Time-history of the Froude number (time is normalized by the wave period T).

According to Chakrabarti [23], the drag force and added mass coefficients were tested in detail, but for cylindrical piles under wave and current loading over a horizontal bottom; the coefficients were found to be dependent on the Keulegan-Carpenter (KC) number and the Reynolds number. The KC number, which indicates the relative importance of the drag forces over the inertial forces, is expressed as [23]

$$KC = \frac{(u_{amp}T)}{a} \tag{8}$$

with u_{amp} - velocity amplitude, T - wave period, and a - width of the structure. Assuming an approximate velocity amplitude u_{amp} = 0.4 m/s, the KC number yields 240, and 360 for periods of T = 60 s and 90 s, respectively. This KC range underlines the fact that, in the present case, the force regime is clearly drag-dominated and that the inertial forces are negligible in the design of coastal beachfront under tsunami conditions.

4.2. Comparison with Prototype-Scale Cases

Fritz *et al.* [40] derived flow depth and flow velocities for Kesennuma Bay during the 2011 Great East Japan Tsunami from recorde videos. Highest flow depth was determined to be 9.0 m and flow velocities varied roughly from $4\,\mathrm{m\,s^{-1}}$ to $9\,\mathrm{m\,s^{-1}}$. Hayashi and Koshimura [41] analyzed aerial images and found tsunami on-land velocities to vary between $6\,\mathrm{m\,s^{-1}}$ to $8\,\mathrm{m\,s^{-1}}$, decreasing in magnitude as they propagated inland. Based on the scaling applied to the experiments reported herein, flow run-up heights at the structure are in the range of 6 m to 18 m for the different configurations given a length scale of $\lambda_L = 100$. Time scale under Froude similitude is thus $\lambda_T = 10$. Largest stream-wise velocity for the base case WC without disturbance by the structure reaches $3.17\,\mathrm{m\,s^{-1}}$ while the largest draw-down velocity is $4.40\,\mathrm{m\,s^{-1}}$. Compared with these values reported in literature, the chosen hydraulic configuration appears to resemble sufficiently to prototype conditions.

4.3. Comparison of Experimental and Computed Drag Forces

For the range of hydraulic configurations tested, the most significant contribution towards the total stream-wise force was estimated to precisely differentiate the drag and inertial forces. The experimentally-derived forces presented in Section 3.3 showed little evidence of short-duration flow slamming effects once the wave front hits the structure. It thus was reasoned that no impulsive forces had to be taken into account for the tested conditions. However, impulsive forces may occur under some hydrodynamic conditions and for structures located close or within the wave breaking area.

In order to verify the above assumption, the drag and inertial forces were also calculated analytically based on Equations (1) and (2) and by using the time-history of smoothed water depth and velocities at the structure position, shown in Figure 12. Drag coefficients were selected from the derived C_D function shown in Figure 9 in conjunction with the relevant structure Reynolds number. Structure's area exposed to hydrodynamic loading was computed based on the water depth and the width of the structure, a, as $A = h \times a$. Velocities were taken from the reference configuration with no structure. The resulting drag force development plotted over the relevant run-up and draw-down phase along with the absolute differences between the measured and computed drag forces are shown in Figure 12 for the three hydraulic configurations. The resulting computed drag forces on the basis of the analytical formulation appear to be in good agreement with the experimentally derived ones during both the run-up and the draw-down stages. Some larger deviations can be observed at the very beginning of the flow-structure interaction and this may be also related to inaccuracies in the velocity calculations which were derived from surface PIV measurements. In particular, the initial phase of the flow front interacting with the structure was characterized by a depth-uniform flow. The flow at this initial phase and again around $\hat{t} = [0.1 .. 0.2]$, when the strongly turbulent wake behind the structure forms, is rather three dimensional in nature and the assumption of uniform depth did not hold. Oscillations found in the measured force were not reproduced by applying Equation (1). A potential reason may be due to the fact that the velocity data used for the drag force computation was taken from the undisturbed case whereas contributions towards the oscillating force are more likely to stem from the vortex shedding which occurred in the back of the structure during run-up. However, it has to be stressed that maximum and minimum computed drag forces compare well in both magnitude and phase with the experimentally-derived ones. For the design engineer, it is thus possible to use Equation (1) with reasonable confidence in cases where detailed studies are not available or feasible.

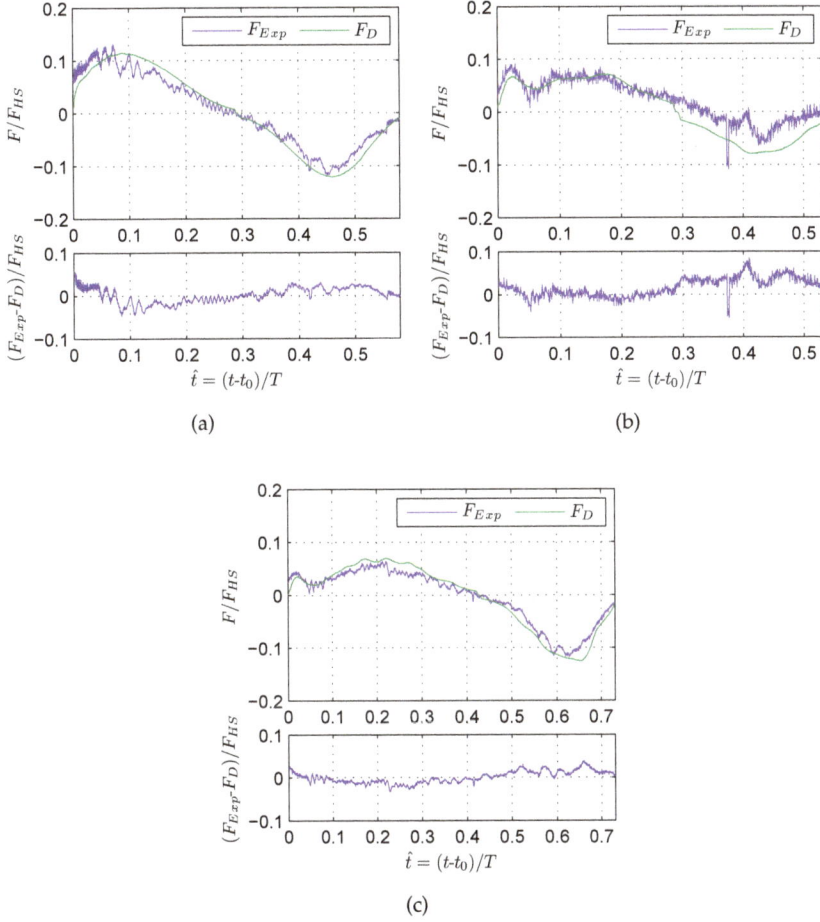

Figure 12. Comparison of the time-histories of the measured horizontal stream-wise forces and computed one (using Equation (1)) for the three WCs. Lower panels of the Figures show the residual forces. All forces were normalized by the hydrostatic force acting on the square cylinder face at maximum inundation depth. (**a**) Base case (WC-ID 01); (**b**) Low amplitude case (WC-ID 02); (**c**) Long period case (WC-ID 03).

4.4. Analytically-Derived Inertial Forces

The well-known first study of Morison *et al.* [21] introduced the concept of dividing forces induced by oscillatory fluid motion into drag and inertia forces. In this regard, inertial forces have been attributed to the acceleration of the fluid involved in the flow-structure interaction. It was found that inertial forces increases with increasing ratio of the width of a structure to the wave length of the flow. In this study, the authors attempted to determine how large the inertial forces for the investigated hydraulic configurations were in comparison with the drag forces. Time-history of the acceleration needed to determine the inertial force at the location of the structure was calculated based on the smoothed stream-wise velocity which was in its turn derived from the PIV analysis and differentiated numerically with Matlab. Figure 13a presents the accelerations du/dt derived for the base case. After the slack phase, water was slowly accelerated in the upstream direction which was also expressed by a negative acceleration (positive acceleration is defined in the direction of the flow).

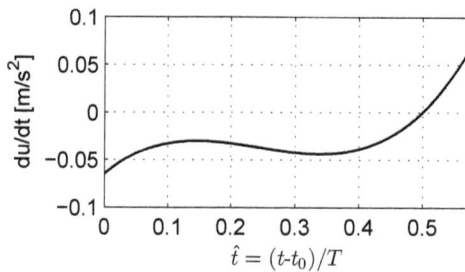

(a) Stream-wise acceleration du/dt

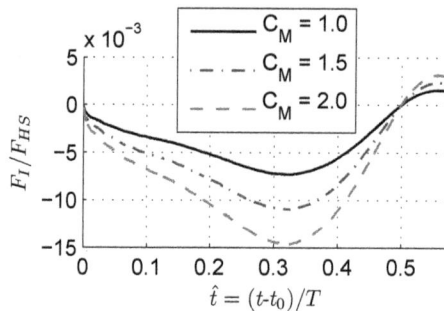

(b) Inertia force under various C_M

Figure 13. Time-history of the acceleration and inertia force for the base case.

The mass coefficient of $C_M = [1.0..2.0]$ has been estimated depicting the ranges given in literature. Figure 13b thus presents the resulting inertial force for the base case computed with Equation (2). The displaced water volume has been calculated on the basis of the water depth derived from pressure sensor readings as described in Section 3.1. Apparently, inertial forces which were found analytically are an order of magnitude smaller than the drag forces. Minimum inertia force which occurs around slack time is as small as $F_I = -0.072\,\text{N}$ and the maximum inertia force is found at the end of the draw-down with a value of $F_I = 0.018\,\text{N}$.

5. Conclusions

This research work concerns flow-structure interaction processes between a single structure mounted on a 1:40 sloping beach with leading depression long waves having different periods and amplitudes. To the author's knowledge, this type of experimental configuration combined with the recording of the time-dependent drag and inertia forces as well as of the transient von Kármán vortex shedding phenomena have not been previously investigated. The transient hydraulic configurations have been chosen in order to represent time-varying tsunami-induced inland flow as occurred during recent catastrophic tsunami events such as the 2004 Indian Ocean Tsunami or the 2011 Great East Japan tsunami. These experiments are also of relevance for studies involving dam-break flows which are also transient. The following conclusions can be drawn from the findings elaborated in this paper:

- Drag force coefficients which are a function of Reynolds numbers vary over the course of the flow-structure interaction. For the analytical computation of drag forces for a structure impacted by a transient flow it is thus important to incorporate the time-history of drag force coefficients rather than constant values. It was found that the calculated drag coefficients are well within the range of reported values found in the literature.
- Flow patterns around the structure consist of a series of complex, but meaningful physical processes and incorporate bow wave propagation upstream, hydraulic jump condition, turbulent wake development behind the structure and vortex shedding during the flow run-up and the draw-down over the beach which led to a considerably contribution towards the lateral total forces in all three cases.
- Using the water depth, the stream-wise velocity and drag force coefficients available at the location of a structure involved in flow-structure interaction it was possible to accurately predict the time-history of the drag forces generated from a given hydraulic configuration. However, using this method, it was not possible to confirm if oscillation-induced force contributions occur as a result of vortex shedding. To investigate this in more detail, experimental tests remain a valuable resource of information.
- As the hydraulic configurations applied are transient in nature, inertial forces might contribute to the total force exerted to the structure under investigation. However, inertia forces which were also found to occur during the flow run-up and draw-down are of an order of magnitude smaller than those drag forces exerted onto the structure. For the range of flow conditions investigated in this experimental program, it was found that one may neglect inertial forces for flows exhibiting the investigated range of flow periods.

Acknowledgments: The Franzius-Institute for Hydraulic, Estuarine and Coastal Engineering, University of Hannover, Germany provided the hydraulic research laboratory. Further, the manuscript preparation was partially supported by a Marie Curie International Outgoing Fellowship within the 7th European Community Framework Programme granted to NG.

Author Contributions: GB undertook the experimental research, prepared the figures and tables, and provided the content in report form. NG wrote the main manuscript based on the report and supervised the research work. TS supervised the research work and the writing of the manuscript. IN supervised the writing of the manuscript and substantially contributed to revisions. All authors reviewed the manuscript.

Conflicts of Interest: The authors declare no conflict of interest.

References

1. Taubenböck, H.; Goseberg, N.; Setiadi, N.; Lämmel, G.; Moder, F.; Oczipka, M.; Klüpfel, H.; Wahl, R.; Schlurmann, T.; Strunz, G.; *et al.* Last-Mile preparation to a potential disaster—Interdisciplinary approach towards tsunami early warning and an evacuation information system for the coastal city of Padang, Indonesia. *Nat. Hazards Earth Syst. Sci.* **2009**, *9*, 1509–1528.

2. Taubenböck, H.; Goseberg, N.; Lämmel, G.; Setiadi, N.; Schlurmann, T.; Nagel, K.; Siegert, F.; Birkmann, J.; Traub, K.P.; Dech, S.; *et al.* Risk reduction at the "Last-Mile": An attempt to turn science into action by the example of Padang, Indonesia. *Nat. Hazards* **2013**, *65*, 915–945.

3. Cheung, J.; Melbourne, W. Turbulence effects on some aerodynamic parameters of a circular cylinder at supercritical numbers. *J. Wind Eng. Ind. Aerodyn.* **1983**, *14*, 399–410.

4. Sumer, B.; Christiansen, N.; Fredsoe, J. The horseshoe vortex and vortex shedding around a vertical wall-mounted cylinder exposed to waves. *J. Fluid Mech.* **1997**, *332*, 41–70.

5. Li, C.W.; Lin, P. A numerical study of three-dimensional wave interaction with a square cylinder. *Ocean Eng.* **2001**, *28*, 1545–1555.

6. Sundar, V.; Vengatesan, V.; Anandkumar, G.; Schlenkhoff, A. Hydrodynamic coefficients for inclined cylinders. *Ocean Eng.* **1998**, *25*, 277–294.

7. Cui, X.; Gray, J. Gravity-driven granular free-surface flow around a circular cylinder. *J. Fluid Mech.* **2013**, *720*, 314–337.

8. Alam, M.; Zhou, Y.; Wang, X. The wake of two side-by-side square cylinders. *J. Fluid Mech.* **2011**, *669*, 432–471.

9. Ramsden, J. Forces on a vertical wall due to long waves, bores, and dry-bed surges. *J. Waterw. Port Coast. Ocean Eng.* **1996**, *122*, 134–141.

10. Nouri, Y.; Nistor, I.; Palermo, D.; Cornett, A. Experimental Investigation of Tsunami Impact on Free Standing Structures. *Coast Eng. J.* **2010**, *52*, 43–70.

11. Goseberg, N. Reduction of maximum tsunami run-up due to the interaction with beachfront development—Application of single sinusoidal waves. *Nat. Hazards Earth Syst. Sci.* **2013**, *13*, 2991–3010.

12. Strusinska-Correia, A.; Husrin, S.; Oumeraci, H. Tsunami damping by mangrove forest: A laboratory study using parameterized trees. *Nat. Hazards Earth Syst. Sci.* **2013**, *13*, 483–503.

13. Park, H.; Cox, D.T.; Lynett, P.J.; Wiebe, D.M.; Shin, S. Tsunami inundation modeling in constructed environments: A physical and numerical comparison of free-surface elevation, velocity, and momentum flux. *Coast. Eng.* **2013**, *79*, 9–21.

14. Goseberg, N.; Schlurmann, T. Non-stationary flow around buildings during run-up of tsunami waves on a plain beach. In Proceedings of the 34th Conference on Coastal Engineering, Seoul, Korea, 15–20 June 2014.

15. Kinsman, B. *Wind Waves: Their Generation and Propagation on the Ocean Surface*; Prentice-Hall: Englewood Cliffs, NY, USA, 1965.

16. Liu, P.L.F.; Cho, Y.S.; Briggs, M.J.; Kanoglu, U.; Synolakis, C.E. Runup of solitary waves on a circular island. *J. Fluid Mech.* **1995**, *302*, 259–285.

17. Seiffert, B.; Hayatdavoodi, M.; Ertekin, R.C. Experiments and computations of solitary-wave forces on a coastal-bridge deck. Part I: Flat Plate. *Coast. Eng.* **2014**, *88*, 194–209.

18. Aguiniga, F.; Jaiswal, M.; Sai, J.; Cox, D.; Gupta, R.; Van De Lindt, J. Experimental study of tsunami forces on structures. In Proceedings of the ASCE Engineering Mechanics Conference, Los Angeles, CA, United States, 8–11 August 2010.

19. Chanson, H. Tsunami surges on dry coastal plains: Application of dam break wave equations. *Coast. Eng. J.* **2006**, *48*, 355–370.

20. Goseberg, N.; Wurpts, A.; Schlurmann, T. Laboratory-scale generation of tsunami and long waves. *Coast. Eng.* **2013**, *79*, 57–74.

21. Morison, J.; Johnson, J.; Schaaf, S. The force exerted by surface waves on piles. *J. Pet. Technol.* **1950**, *2*, 149–154.

22. Journée, J.; Massie, W. *Offshore Hydromechanics*, 1st ed.; Delft University of Technology: Delft, The Netherlands, 2001.

23. Chakrabarti, S. *Handbook of Offshore Engineering*; Elsevier: Amsterdam, The Netherlands, 2005; Volume 1.

24. Zhu, S.; Moule, G. Numerical calculation of forces induced by short-crested waves on a vertical cylinder of arbitrary cross-section. *Ocean Eng.* **1994**, *21*, 645–662.

25. Cai, S.; Long, X.; Gan, Z. A method to estimate the forces exerted by internal solitons on cylindrical piles. *Ocean Eng.* **2003**, *30*, 673–689.

26. De Vos, L.; Frigaard, P.; de Rouck, J. Wave run-up on cylindrical and cone shaped foundations for offshore wind turbines. *Coast. Eng.* **2007**, *54*, 17–29.

27. Hildebrandt, A.; Sparboom, U.; Oumeraci, H. *Wave Forces on Groups of Slender Cylinders in Comparison to an Isolated Cylinder due to Non-Breaking Waves*; World Scientific: Hamburg, Germany, 2008.

28. Yeh, H. Maximum Fluid Forces in the Tsunami Runup Zone. *J. Waterw. Port Coast. Ocean Eng.* **2006**, *132*, 496–500.

29. Yen, S.; Liu, J. Wake flow behind two side-by-side square cylinders. *Int. J. Heat Fluid Flow* **2011**, *32*, 41–51.

30. Hashimoto, H.; Park, K. Two-dimensional urban flood simulation: Fukuoka flood disaster in 1999. *WIT Trans. Ecol. Environ.* **2008**, *118*, 59–67.

31. Madsen, P.A.; Fuhrman, D.R.; Schäffer, H.A. On the solitary wave paradigm for tsunamis. *J. Geophys. Res.* **2008**, *113*, C12012.

32. Madsen, P.A.; Schäffer, H.A. Analytical solutions for tsunami runup on a plane beach: single waves, N-waves and transient waves. *J. Fluid Mech.* **2010**, *645*, 27–57.

33. Nezu, I.; Nakagawa, H. *Turbulence in Open-Channel Flows*; A.A. Balkema: Rotterdam, The Netherlands, 1993.

34. Raffel, M.; Willert, C.E.; Kompenhans, J. *Particle Image Velocimetry: A Practical Guide*; Springer: Berlin, Germany; New York, NY, USA, 1998.

35. Sveen, J.K. An introduction to MatPIV v. 1.6. 1. *Preprint series. Mechanics and Applied Mathematics*; the Free Software Foundation, Inc.: Boston, MA, USA, 2004.

36. Mo, W.; Jensen, A.; Liu, P.L.F. Plunging solitary wave and its interaction with a slender cylinder on a sloping beach. *Ocean Eng.* **2014**, *74*, 48–60.

37. Xioa, H.; Huang, W. Numerical modeling of wave runup and forces on an idealized beachfront house. *Ocean Eng.* **2008**, *35*, 106–116.

38. Schüttrumpf, H. Wellenüberlauf bei Seedeichen—Experimentelle und theoretische Untersuchungen. Ph.D. Thesis, Technical University Carolo-Wilhelmina, Braunschweig, Germany, 2001.

39. Arnason, H.; Petroff, C.; Yeh, H. Tsunami bore impingement onto a vertical column. *J. Disaster Res.* **2009**, *4*, 391–403.

40. Fritz, H.M.; Phillips, D.A.; Okayasu, A.; Shimozono, T.; Liu, H.; Mohammed, F.; Skanavis, V.; Synolakis, C.E.; Takahashi, T. The 2011 Japan tsunami current velocity measurements from survivor videos at Kesennuma Bay using LiDAR. *Geophys. Res. Lett.* **2012**, *39*, doi:10.1029/2011GL050686.

41. Hayashi, S.; Koshimura, S. The 2011 Tohoku tsunami flow velocity estimation by the aerial video analysis and numerical modeling. *J. Disaster Res.* **2013**, *8*, 561–572.

A Methodology for a Comprehensive Probabilistic Tsunami Hazard Assessment: Multiple Sources and Short-Term Interactions

Grezio Anita, Roberto Tonini, Laura Sandri, Simona Pierdominici and Jacopo Selva

Abstract: We propose a methodological approach for a comprehensive and total probabilistic tsunami hazard assessment (*TotPTHA*), in which many different possible source types concur to the definition of the total tsunami hazard at given target sites. In a *multi*-hazard and *multi*-risk perspective, the approach allows us to consider all possible tsunamigenic sources (seismic events, slides, volcanic eruptions, asteroids, *etc.*). In this respect, we also formally introduce and discuss the treatment of interaction/cascade effects in the *TotPTHA* analysis and we demonstrate how the triggering events may induce significant temporary variations in short-term analysis of the tsunami hazard. In two target sites (the city of *Naples* and the island of *Ischia* in Italy) we prove the feasibility of the *TotPTHA* methodology in the *multi*−source case considering near submarine seismic sources and submarine mass failures in the study area. The *TotPTHA* indicated that the tsunami hazard increases significantly by considering both the potential submarine mass failures and the submarine seismic events. Finally, the importance of the source interactions is evaluated by applying a triggering seismic event that causes relevant changes in the short-term *TotPTHA*.

Reprinted from *J. Mar. Sci. Eng.* Cite as: Anita, G.; Tonini, R.; Sandri, L.; Pierdominici, S.; Selva, J. A Methodology for a Comprehensive Probabilistic Tsunami Hazard Assessment: Multiple Sources and Short-Term Interactions. *J. Mar. Sci. Eng.* **2015**, *3*, 23–51.

1. Introduction

In recent years the *Probabilistic Tsunami Hazard Assessment* (*PTHA*) has been proposed in different areas of the globe using different methodologies that can be grouped in:

(*a*) scenarios ([1–4]);
(*b*) statistical evaluations of the historical earthquakes ([5–8]);
(*c*) statistical analysis of tsunami sources ([9–11]);
(*d*) Bayesian inferences ([12–14].

In (a), the sources are generally identified *a priori*, neglecting the aleatory uncertainties on tsunami sources. In (b), the probability strongly depends on the catalogue completeness and data availability. In (c, d), the aleatory uncertainty of the sources is considered predominant so that the study is focused on the random nature of the tsunami generation due to the intrinsic unpredictability of the event occurrence. Also, the epistemic uncertainty is often treated explicitly merging many information into a single statistical model through Logic Trees or Bayesian inference, as recently proposed for the cases $(b–d)$.

Most of the *PTHA*s focus on one single type of tsunami source, usually the seismic sources ([9,10,15,16]). However, it has been demonstrated that in many areas, major causes of tsunami are the *non*-seismic sources, such as mass failures (whether or not triggered by seismic events) ([17,18]), volcanic activity ([19–25]). In particular, volcanic flows ([26,27]), landslides and rock slides on coastal areas ([28–32]) and submarine mass failures ([33–36]) may lead to the catastrophic tsunamis, but very few analyses treated such *non*-seismic tsunamigenic sources in a probabilistic frame ([13,37]). Meteorite impacts can also trigger tsunamis, even though they are extremely rare events ([38]).

In order to provide an unbiased and a complete *PTHA*, all possible tsunami sources should be considered and combined together and their relative importance properly quantified.

Also, in *PTHA* a bias may be induced by neglecting an external event that is not significant for tsunami generation but is able to activate another type of tsunamigenic source. This external event is commonly defined a triggering event. Formally including all possible interaction chains, triggers and cascade effects would make the number of conditional hazards to be assessed too large and neither large databases nor theoretical background are available in those cases. As a result, the underestimation of source interactions could result in an underestimation of the hazards. Thus, several simplifications are needed, such as: (i) reducing the analysis to few significant event types; or (ii) considering several single specific cascade scenarios. In case (i) a triggering event may be the volcanic unrest (deformations may produce tsunamigenic landslides), or coastal/submarine earthquakes (that may cause tsunamigenic earthquakes, or landslides); In case (ii) the occurrence of one event possibly affects another hazard.

Multi-hazard and *multi*-risk assessments are new approaches ([39,40]) that allow the comparison of different hazards and risks and their interactions, accounting for multiple events that may exponentially increase the impact of disasters on society. Both the comparison of different sources and the interaction/cascade effects are among the most important factors of *multi*-hazard and *multi*-risk assessments, as treated in several recent analyses ([39,41]). In this perspective, the *PTHA* represents a *multi*-source problem because of the different type of sources (earthquakes, landslides and volcanic eruptions) to be considered in a *multi*-risk framework (natural, industrial, nuclear, coastal infrastructure risk) since consequences are highly impacting.

Indeed, a *multi*-source approach for $PTHA$ is fundamental for an unbiased and a complete analysis. Firstly, in the present paper we indicate how to simultaneously deal with different types of tsunamigenic sources in a coherent probabilistic framework by proposing a comprehensive method to provide a *Total Probabilistic Tsunami Hazard Assessment* (*TotPTHA*) and evaluating the representativeness of the used tsunamigenic source dataset/catalogue.

Secondly, we pose the problem of interaction/cascade effects considering when they may play a major role in *TotPTHA* and how to estimate them in short- and long term assessments.

Finally, the applicability of the *TotPTHA* methodology is exemplified for the city of *Naples* and the Island of *Ischia* (Italy), accounting for (1) different types of sources (*multi*-source approach); and (2) specific triggering event (source interaction approach).

2. A Comprehensive TotPTHA Methodology

The *TotPTHA* indicates the comprehensive $PTHA$ in a given target location. Here, the term "comprehensive" means an exhaustive tsunami hazard assessment based on datasets including all sources. In the present methodology the *TotPTHA* formally merges the $PTHA$s due to all different types of tsunamigenic sources. The *TotPTHA* is defined as the probability that in a given exposure time Δt a selected value z of a tsunami parameter Z is overcome by any possible tsunamigenic source

$$TotPTHA \Rightarrow p(Z \geq z, \Delta t) \tag{1}$$

In Equation (1), the tsunami parameter Z can be referred to any parameter describing the tsunami characteristics and intensity. It can be run-up, velocity, energy, moment flux, wave height, or any other variable required by the tsunami hazard/vulnerability/risk assessment.

Previous studies treated the different possible types of tsunamigenic sources separately, producing a $PTHA$ in a specific region, with special emphasis on one type of source, generally the seismic one. If only one type of source is taken into account then only a portion of the total number of tsunamis is considered, leading to a bias in the $PTHA$ analysis due to the neglected types of tsunamigenic sources. This relates to the issue on whether *non*-seismic sources (or the "*non*-prevalent" tsunamigenic source) can be ignored.

For example, in a global analysis, about 75% of the historical tsunami are generated by seismic sources and the remaining 25% by *non*-seismic ones (mass failure, volcanic, meteorological, cosmogenetic, and anthropogenic sources) ([42]). Such proportion may drastically change in a particular area where the hazard posed by *non*-seismic sources is significant. Also, tsunami events generated by *non*-seismic causes often have a more devastating impact in near-field areas compared to the tsunami caused by seismic sources.

2.1. TotPTHA: The Effects of Multiple Source Types of dIfferent Potentially Tsunamigenic Source Events

In the present study, the term Source Type (ST) indicates a selected set of potentially tsunamigenic sources (e.g., source types refer to submarine seismicity, mass failures, volcanic activity, meteorite impacts, *etc.*), and the term Source Event (SE) indicates a potentially tsunamigenic event of a specific source type (e.g., a given submarine earthquake with specific fault parameters, magnitude and location).

Given a specific i-th tsunamigenic source type ST_i the relative $PTHA_i$ indicates the contribution to the *TotPTHA*

$$PTHA_i \Rightarrow p(Z \geq z, ST_i; \Delta t) = p(Z \geq z | ST_i) p(ST_i, \Delta t) \tag{2}$$

The specific contribution $PTHA_i$ can be isolated by conditioning the *TotPTHA* with respect to the occurrence of the specific ST_i.

For example, in the case of seismic sources it is

$$\begin{aligned} p(Z \geq z, \Delta t) &= p(Z \geq z, E_q s; \Delta t) + p(Z \geq z, \overline{E}_q s; \Delta t) \\ &= p(Z \geq z | E_q s) p(E_q s, \Delta t) + p(Z \geq z | \overline{E}_q s) p(\overline{E}_q s, \Delta t) \end{aligned} \tag{3}$$

where $E_q s$ means earthquakes and $ST_i \equiv E_q s$. The first addend (denoted by $E_q s$) takes into account the tsunamis caused only by earthquakes and the second one (denoted by $\overline{E}_q s$) the tsunamis occurred even though earthquakes do not occur (*i.e.*, tsunamis generated by different STs).

Locally, the $PTHA_i$ due to earthquakes can be considered equivalent to the *TotPTHA* if, and only if, in Equation (3) the *non*-earthquake $p(Z \geq z, \overline{E}_q s; \Delta t)$ is negligible respect to the earthquake $p(Z \geq z, E_q s; \Delta t)$. This may happen if we assume negligible the probability of occurrence of the other STs (the term $p(\overline{E}_q s, \Delta t) \simeq 0$) and/or the probability of overcoming the selected parameter z due the other STs (the term $p(Z \geq z | \overline{E}_q s) \simeq 0$).

In general, the *TotPTHA* may be expressed by

$$p(Z \geq z, \Delta t) = 1 - \prod_{i=1}^{N_{TOT}} [1 - p(Z \geq z, ST_i; \Delta t)] \tag{4}$$

where the product over the N_{TOT} types of tsunamigenic ST_i ($i = 1, ..., N_{TOT}$) represents the probability that none of the STs produces a value of the tsunami parameter Z larger than z in Δt. This formulation assumes that the occurrences of the Z values generated by different STs are mutually independent. This assumption of independence is valid also considering that:

(a) the propagations of different tsunami waves are independent in the case of almost simultaneous events,
(b) the sizes of different STs are independent in the case of triggering events (see paragraph 2.2), and
(c) the $PTHA$s are relative to the next main tsunami event and possible subsequent tsunamis are not included (similarly to other hazard assessments: aftershocks are not examined in $PSHA$s, [43], or subsequent eruptions are not encompassed in $PVHA$s, [44]).

In fact, when one significant tsunami event occurs many factors characterizing the system (morphology of the coasts, bathymetry, exposure, etc.) might change and the previous hazard/vulnerability/risk assessments should be updated. For this reason, sequences of tsunami waves immediately after a major tsunami event should be excluded by the $PTHA$s and eventually removed from the catalogues ([45]). In contrast, the cascade effects should be examined for each ST under the assumption that a triggering event may not produce any tsunami (even if potentially tsunamigenic) but its tsunamigenic potential may be assessed on the basis of the triggered event only.

We calculate the $TotPTHA$ through Equation (4) by evaluating separately each $PTHA_i$ from the different ST_i. As a special case, the Equation (4) can represent a single tsunamigenic ST if we set $N_{TOT} = 1$. Then, the $TotPTHA$ is equivalent to the $PTHA_i$ of Equation (2). In the example of Equation (3) for the seismic sources, a bias would be introduced by neglecting the second addend of the equarion.

Each i-th ST_i can be represented by a sufficient large number N_i of independent physical sources. The j-th SE relative to the i-th ST is indicated by SE_{ij} and the correspondent $PTHA_i$ is given by

$$p(Z \geq z, ST_i; \Delta t) = 1 - \prod_{j=1}^{N_i}[1 - p(Z \geq z, SE_{ij}; \Delta t)], i = 1, ..., N_{TOT} \qquad (5)$$

In the seismic example, SE_{ij} is the j-th specific earthquake event of given location, size and magnitude relative to the i-th type, that is the seismic one.

The underlying assumptions of Equation (5) are essentially the same of Equation (4) and the final $TotPTHA$ is obtained by substituting Equation (5) in Equation (4)

$$p(Z \geq z, \Delta t) = 1 - \prod_{i=1}^{N_{TOT}} \prod_{j=1}^{N_i} [1 - p(Z \geq z, SE_{ij}; \Delta t)] \qquad (6)$$

The total number of sources is $\sum_{i=1}^{N_{TOT}} N_i$, that is the sum of N_{TOT} sets of tsunamigenic source types ST_i (seismic, mass failure, volcanic, meteorologic, cosmogenetic, anthropogenic) and each set is defined by N_i tsunamigenic source events SE_{ij}. In Equation (6) each $p(Z \geq z, \Delta t; SE_{ij})$ represents the $PTHA$ relative to a single independent tsunamigenic physical source and is expressed by

$$p(Z \geq z, SE_{ij}; \Delta t) = p(Z \geq z, |SE_{ij}; \Delta t)p(SE_{ij}, \Delta t) \qquad (7)$$

The factors $p(SE_{ij}, \Delta t)$ and $p(Z \geq z|SE_{ij})$ may be related respectively to the "source phase" and the "inundation phase" of a tsunami event. This formulation is useful to separate the temporal occurrence of tsunami sources (evaluated often by *Poisson* models, [46]) from the tsunami impacts (simulated usually by scenario models, [47]).

2.2. Source Interactions and Triggering Events

Equations (6) and (7) are useful to consider possible triggers on sources and their interactions. The term Source Interaction SI is referred to the interaction process that may produce an extended set of potential events either tsunamigenic or *non*-tsunamigenic. The triggering event TE is a specific event of the extended set and it belongs to the same ST or a different ST.

For example, cases of SIs occur when slides are generated by earthquakes or eruptions. A case of TE is referred to: a specific volcanic eruption, that triggers an earthquake or landslide; an earthquake, that triggers another earthquake or landslide; a volcanic unrest episode that triggers an earthquake or landslide.

Given the hypotheses of Equations (4) and (6), even potentially tsunamigenic events may not generate tsunami. Thus, we can state that a generic event (tsunamigenic or not) may trigger another tsunamigenic event by only acting on the source phase in Equation (7). Also, all STs must be assumed independent in Equations (4) and (7), when one triggering event TE (or a set of independent TEs) occurs. In practice, each single source probability $p(SE_{ij}, \Delta t)$ $\forall i, \forall j$ can be updated in Equation (7) accounting for the occurrence of the TE. Then, the updated source probabilities can be used in Equation (6) to assess the $TotPTHA$.

Given a specific k-th triggering event TE the probability of occurrence of the specific SE_{ij} due to the occurrence of the TE_k is

$$p(SE_{ij}, TE_k; \Delta t) = p(SE_{ij}|TE_k)p(TE_k, \Delta t) \tag{8}$$

Assuming that the SIs may produce different and independent TEs, the cumulated effect of all N_{TE} possible TEs is

$$\begin{aligned} p(SE_{ij}, TE_1, ..., TE_{N_{TE}}; \Delta t) &= 1 - \prod_{k=1}^{N_{TE}} [1 - p(SE_{ij}|TE_k)p(TE_k, \Delta t)] \\ &\approx \sum_{k=1}^{N_{TE}} p(SE_{ij}|TE_k)p(TE_k, \Delta t) \end{aligned} \tag{9}$$

where the approximation holds if $p(SE_{ij}|TE_k) \ll 1$ and the terms of order higher than 2 are neglected. The interaction matrix $p(SE_{ij}|TE_k)$ represents the interactions among the different events and the terms of order higher than 2 in this matrix are related to a chain of interactions and triggering events. For example, the probability of occurrence is neglected in the case of a volcanic eruption that triggers an earthquake that triggers a slide that causes a tsunami (cascade of events).

We define $p^*(SE_{ij}, \Delta t)$ the total probability of occurrence of the SE_{ij} in the time interval Δt due to its intrinsic probability of occurrence and to a possible SI

$$\begin{aligned} p^*(SE_{ij}, \Delta t) &= p(SE_{ij}, \Delta t) + [1 - p(SE_{ij}, \Delta t)]p(SE_{ij}, TE_1, ..., TE_{N_{TE}}; \Delta t) \\ &\approx p(SE_{ij}, \Delta t) + p(SE_{ij}, TE_1, ..., TE_{N_{TE}}; \Delta t) \end{aligned} \tag{10}$$

The approximation is possible since *non*-interactive source probabilities are usually $\ll 1$, then $(1 - p(SE_{ij}, \Delta t)) \approx 1$. We underline that Equation (10) is based on two fundamental assumptions:

(1) the triggering mechanisms are independent, meaning that two or more TEs sum their effects in term of probability, but the combined effects do not result in a further amplification of the tsunami parameters at the coast;
(2) only first order interactions are modeled and second order (or higher) are assumed negligible.

We substitute the source phase probability $p^*(SE_{ij}, \Delta t)$ of Equation (10) for the *non*-interactive probability $p(SE_{ij}, \Delta t)$ of Equation (6). Finally, using the updated probability of Equation (7) we evaluate the SI effects due to all STs and assess the $TotPTHA$.

2.3. Catalogue Completeness and Source Representativeness

In general, the completeness analysis consists of a screening of the data collected in a database/catalogue in order to evaluate the degree of representativeness of the real system and the influence of the missing information on the final assessment. In many cases, the tsunami catalogues are much less complete than other catalogues (for example the seismic catalogues) and this plays a fundamental role in hazard assessment. In fact, (*i*) small datasets imply a high level of uncertainties and (*ii*) *non*-representative datasets may introduce important biases in the results (to be compensated by other means). Indeed, those aspects pose an intrinsic limitation when a database/catalogue of rare source events is used for probabilistic hazard assessments because it is not adequate to explore the whole potential aleatoric variability.

Now we consider the problem of the representativeness of the subset of the innumerable possible sources (in the cases of *SE*s, *ST*s, *TE*s) and we discuss the unbiased *PTHA*, assuming a certain completeness of the available catalogues of the past events.

- *Source representativeness (SE$_{repres}$)*

  It is related to the fact that some *SE*s of a given *ST* may be not considered in the *PTHA*. Given a set of SE_{ij}, we may evaluate the source representativeness for the target site computing M_1/T_1, where M_1 is the number of tsunamis generated by the SE_{ij} and T_1 is the total number of tsunamis due to that *ST*. For example, in the case of the seismic *ST* let us assume that the sources SE_{ij} (*i* = earthquakes, and *j* = 1, ..., N_i numbers of events) are located in a delimited study area at given distance from a target site. In this case, the most distant seismic sources are neglected and only near sources are assumed representative of all seismic sources, meaning that the *PTHA* cannot be influenced by far earthquakes. However, tsunami are known to travel long distances and excluding far-field sources may lead to underestimations of the tsunami hazard. M_1 is the number of tsunamis from seismic sources belonging to the study area and T_1 the total number of tsunamis due to the earthquakes. If the ratio M_1/T_1 is close to 1, we might have chosen a good subset of all possible *SE*s of the ST_i to evaluate the *PTHA* at the target site. For simplicity in the example we assume that all sources are well represented with respect to one source parameter, that is the location. In a complete discussion, the source representativeness should include all other parameters describing the source (magnitude, depth, focal mechanism, fault geometry).

- *Source-type representativeness (ST$_{repres}$)*

  In this case, several *ST*s may be neglected in the *PTHA*. A source-type representativeness can be estimated by M_2/T_2 where M_2 is the number of tsunami events at the target sites as reported by catalogues and and T_2 is the total number of tsunami events from any possible *ST*.

For example, in the case of different STs like earthquakes, submarine mass failures and eruptions we need to evaluate the bias introduced by neglecting sub-aerial slides and meteorites. If the number M_2/T_2 is close to 1, it is reasonable to neglect other STs.

- *Interaction-source representativeness* (TE_{repres})

The interactions of the sources may significantly modify the $PTHA$, especially in short-term applications. This implies that if one (among many) interaction-source is not considered, the resulting $PTHA$ is biased.

An interaction-source representativeness can be formulated by M_3/T_3, where M_3 is the number of tsunami due to cascade effects known by experimental or modeled interaction-source cases, and T_3 the total number of all possible tsunami. However, this ratio is subject to underestimation, since the identification of all the interactions is difficult and in many cases the evaluation of T_3 may be subjective ([39,40]).

2.4. Source Interactions in Long- and Short-Term TotPTHA

In the long-term hazard assessments it is common to consider the Δt of order of years/decades. In the short-term hazard assessments the Δt is of order of hours/days/weeks/months.

The effects of the SIs are automatically included into the estimation of the source probability in long-term hazard assessments based on catalogues of the past SE occurrence. For example, this is the case of the undifferentiated catalogues of the seismic sources. Also, in the tsunami catalogues the interaction episodes are usually included, and thus $p^*(SE_{ij}, \Delta t)$ is directly estimated using the past occurrence of the TEs. However, such interaction effects may be not present in long-term applications if the analysis is based on a single type of source. This analysis can be supposed unbiased only if the probability $p(SE_{ij}, \Delta t; TE_1, ..., TE_{N_{TE}})$ is assumed negligible with respect to the *non*-interactive source probability in Equation (9)

$$\sum_{k=1}^{N_{TE}} p(SE_{ij}|TE_k)p(TE_k, \Delta t) \ll p(SE_{ij}, \Delta t) \tag{11}$$

If $p(TE_k, \Delta t) \approx p(SE_{ij}, \Delta t)$ and $p(SE_{ij}|TE_k) \ll 1 \; \forall i$, for each specific TE_k in Equation (11), then the standard *non*-interactive $p(SE_{ij}, \Delta t)$ can be used for the long-term tsunami hazard assessment in Equation (7)

$$p(SE_{ij}, \Delta t) \Rightarrow TotPTHA_{long-term} \tag{12}$$

If the specific TE_k occurs, then the probability $p(TE_k, \Delta t)$ is ≈ 1 and the approximation in Equation (11) is not valid. Indeed, $p^*(SE_{ij}, \Delta t)$ should replace $p(SE_{ij}, \Delta t)$ in Equation (7) because $p^*(SE_{ij}, \Delta t)$ varies remarkably from its background long-term value. In this case, we explicitly account for the SI effects and the occurrence of TE_k may strongly affect the tsunami hazard in the short-term assessment

$$p^*(SE_{ij}, \Delta t) \Rightarrow TotPTHA_{short-term} \qquad (13)$$

For example, during an off-shore seismic sequence the probability of strong earthquakes may increase of several orders of magnitude with respect to long-term assessments ([48]). During the seismic sequence the probability of all seismic events increases significantly and in the same area also the probability of tsunamigenic earthquakes may increase. As a consequence, $p(TE_k, \Delta t)$ (≈ 1) is not comparable to $p(SE_{ij}, \Delta t)$ ($\ll 1$). Similarly, during a volcanic unrest episode the probability of tsunamigenic eruptions (and also earthquakes, and landslides) may increase significantly. Then, "seismic sequence" and/or "volcanic unrest" should be part of the set of possible TEs and their effects should be evaluated to assess the short-term $TotPTHA$. Those cases pose the issue of the time delay dt passed from the triggering event to the subsequent induced source event. An event is supposed to be a trigger if $dt \ll \Delta t$.

3. The *TotPTHA* Application: Naples and Ischia (Italy)

The scope of the present study is the evaluation of feasibility of the comprehensive *TotPTHA* methodology in the *multi*−hazard/*multi*−risk context for the *ByMuR* Italian project: *Bayesian Multi-Risk Assessment: A case study for natural risks in the city of Naples (http : //bymur.bo.ingv.it/)*. The *ByMuR* region was chosen since it may be threatened by several natural hazards and risks of different origin (e.g., volcanic, seismic and tsunamigenic). In the application, the tsunami *multi*-source assessment is focused on two different types of source (near-field seismic events and mass failures) considering a sufficiently large number of sources. This raises a comment on the other types of tsunamigenic sources. According to the Italian Tsunami Catalogue ([49]), tsunami waves due to volcanic activities occurred in 79, 1631 and 1906. The events are reported with tsunami intensity 2 in Ambraseys-Sieberg Scale (generally, an event of tsunami intensity ≥ 3 produces run-up of approximately 1 m while for tsunami intensity equal 2 there is not a clear estimation) and with different reliability. The volcanic sources identified in the region are the *Vesuvius* and the *Campi Flegrei* caldera. For a complete evaluation of the hazard, these volcanic sources should be examined in a separate volcanic tsunami hazard assessment to evaluate how the volcanic activity produces tsunami waves in the target area. Once the impact of the tsunamis generated by those sources is evaluated, it would be possible to include them in a *multi*-source context.

Considering the illustrative purposes of this methodological study, volcanic sources are not included and the selected sets of *STs* for the *TotPTHA* in the region are

- *Submarine Seismic Sources (SSSs),*
- *Submarine Mass Failures (SMFs).*

Figure 1 reports the spatial framework of the potentially tsunamigenic sources and the target sites. The target sites for the tsunami hazard assessment are located in the port of the city of *Naples* and in the island of *Ischia*.

For the *TotPTHA* at the target sites the chosen tsunami parameter Z is the run-up and is calculated for each tsunami caused by each *SE* of each *ST*, similarly to [12] and relative application, [13]. Main characteristics and parameters of the sources are described in the next subsections explaining how they are sampled to construct the statistical datasets. In general, each *SE* is assumed independent from the others. The *SEs* may be different in magnitudes and size, even if they belong to the same specific *ST*. Also, each *SE* is defined by a set of parameters that simplify the geometric shape of the correspondent source.

The potentially tsunamigenic *SEs* of the two *STs* determine the tsunami generation phase. The generated initial tsunami waves propagate up to the coast where their impacts are evaluated in terms of the overcoming threshold values $z \in [0.5\,\text{m}, 1\,\text{m}, 2\,\text{m}, \dots, 10\,\text{m}]$.

Then, we estimate the *TotPTHA* by Equation (6) computing each $PTHA_i$ $(i = SSS, SMF)$. For simplicity, in the $PTHA_i$ factorization we indicate the $p(Z \geq z, SE_{ij}; \Delta t)$ by P_i. This is the probability that the tsunami parameter Z overcomes the threshold values z in the time interval Δt considering N_j $(j = 1, ..., N_i)$ source events *SEs* of N_i $(i = SSS, SMF)$ source types *STs*. In order to determine the $PTHA_i$, the probabilities of occurrence of the *SEs* of the different *STs* are assessed through the following independent factors: spatial probability P_i^{spat}, frequency-size probability P_i^{size}, and temporal probability P_i^{temp}. Different *STs* have different likelihoods and the relative return periods can be longer than the historical records. So that, we express the probability by $P_i \times year^{-1}$ for the comparisons.

Finally, we present a discussion on how the *TotPTHA* changes in the short-term after a possible seismic triggering event in the region.

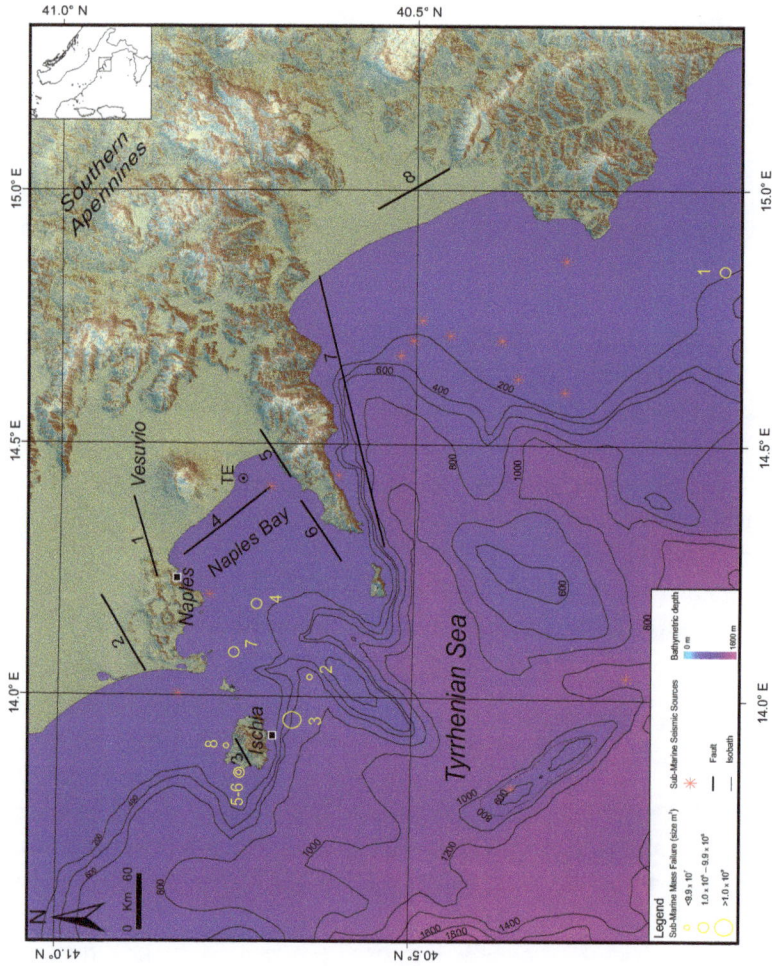

Figure 1. Topography-Bathymetry of the application region: red stars are the submarine seismic sources (*SSSs*), the white circles are the past sub-marine mass failures (*SMFs*) of different sizes, and the black circle is the triggering event (*TE*) location. The square black boxes correspond to the target sites *Ischia* and *Naples*.

3.1. SSSs

- *Spatial Identification.* The seismic sources are spatially identified on the active seismic areas and we consider that even small seismicity indicates the presence of major faults ([50]). So that, epicenters are defined considering the recorded seismic events. We extracted the locations of the seismic events in the application region from the Italian Seismological Instrumental and Parametric Database (*ISIDe, http : //iside.rm.ingv.it/*) available for the period 1983–2009. From the database we selected the latitudes and longitudes of those seismic events occurred with magnitude above the completeness magnitude ($M_w = 2.3$) in the upper ocean crust at depth (<15 km). In the exemplified study 14 submarine earthquakes occurred in that period and their locations are the *SSSs* locations (Figure 1).

- *Magnitudes and sizes.* The moment magnitudes M_w associated to each seismic source location are hypothetic in the interval [5.5–7.5] sampled with 0.1 magnitude increment. The introduction of hypothetic magnitudes was necessary because in the *ISIDe* database the maximum M_w recorded in the area was 5.4 and did not caused tsunami. The lower limit was chosen considering the Mediterranean instrumental tsunami observations: a tsunami occurred in the *Gulf of Corinth* in 11-02-1987 ([51]) after an earthquake of 5.5 magnitude and the concomitant submarine slide. The upper limit was chosen on the basis of the background knowledge of the Tyrrhenian sismo-tectonics indicating that there are no evidence of earthquakes larger than 7.5 in this area in the last two millennia ([52]).

- *Geometric Parameters.* The geometric parameters of the faults are associated to each *SSS* location. According to the hypothetic M_w, the lengths L, widths W and slips S of the fault are scaled by *Wells and Coppersmith*'s formulas ([53]). To introduce a certain variability the fault parameters were sampled twice considering different values (randomly chosen) in the relative interval determined by the standard deviation of the fault parameters provided by *Wells and Coppersmith*. Orientation angles (strike, dip, and rake) are predefined for each location by associating the angles of the nearest submarine fault coherently with the geological/tectonic setting of the area considering also inland faults (Figure 1 and Table 1). The choice of faults is based on the current knowledge of Holocenic active faults even in terms of kinematics and geometry in the vicinity of the study area. Indeed, we considered existing catalogs of Holocenic active faults (DISS) and relevant published papers ([54,55]).

- *Tsunami Initial Wave and Tsunami Wave Propagation.* Two widely accepted assumptions in tsunami modeling are: *(i)* setting the tsunami initial condition to be equal to the vertical sea floor deformation and *(ii)* using the linear shallow water theory to describe tsunami propagation in deep waters, where *non*-linear effects are still negligible. The former is computed by *Okada*'s analytical formulas ([56]) using the source parameters described in the previous section. The latter solves numerically, the conservation of mass and momentum equations for an incompressible fluid expressed by

$$\frac{\partial \eta}{\partial t} + \frac{\partial M}{\partial x} + \frac{\partial N}{\partial y} = 0 \tag{14}$$

$$\frac{\partial M}{\partial t} + gD\frac{\partial \eta}{\partial x} = 0 \tag{15}$$

$$\frac{\partial N}{\partial t} + gD\frac{\partial \eta}{\partial y} = 0 \tag{16}$$

where t is the time variable, x and y are the horizontal space coordinates, g is the gravity, D is the total water column ($D = \eta + H$, being H the average water fluid height at the undisturbed level), η is the surface water elevation and M and N are the discharged fluxes, defined as the product of the total water column D and the corresponding horizontal components of the velocity u and v directions, respectively in the x and y directions, *i.e.*, $M = Du$ and $N = Dv$. Equations (15)–(17) are used to propagate the tsunami waves in waters deeper than the 50 m isoline, since in shallower water depths the contribution of the *non*-linear terms cannot be neglected. Similarly, the frictional terms are not included in this formulation, since their importance become relevant as the waves reach the shallower areas close to the coast.

The computational grid is a regular mesh (composed by squared cells having side length = 50 m) and it has been set up by matching the *GEBCO* (*http : //www.gebco.net/*) 30 arc-second bathymetry and the *SRTM* (*http : //srtm.csi.cgiar.org/*) 90 meters topography datasets.

The two assumptions defined above are still a powerful compromise that allows us to represent the most relevant tsunami features (*i.e.*, the main front height and its directivity) with a relatively low computational time cost for each simulation.

- *Coastal Effects and Tsunami Impact Evaluation.* The final number of initial tsunami waves is $N_{SSS} = 588$ according to the described *SSSs*. The tsunami amplification at the coast is estimated for each tsunami wave using the *Green*'s law based on the energy conservation law. By considering that (*i*) the wave front is almost parallel to the coast because wave refraction is towards

138

the directions of the down-gradient depth and (*ii*) the convergence and/or divergence of the rays can be neglected we apply the formula:

$$H_1 = \sqrt[4]{\frac{d_{50}}{d_1}} H_{50} \tag{17}$$

where $H_{1,50}$ and $d_{1,50}$ are respectively the tsunami amplitude and the sea depth and the suffixes 50 and 1 refer respectively to the 50 m and 1 m depth isolines. Here, run-up estimation is found by setting $H_1 = 1$ m as the last wet point, in order to avoid singularity in Equation (17). The use of the *Green*'s law close to the coasts is a common approach for rapid estimation of the maximum tsunami wave height for both probabilistic hazard ([11]) and warning purposes ([57]) and its good agreement with computational results has been demonstrated by [57]. Despite the limit of neglecting the wave reflection that somehow the coast may introduce, an overestimation of the run-ups resulted not relevant. As a consequence, we consider the H_1 values equivalent to the run-ups Z at the coastal points of the target sites.

3.2. P_{SSS}

- *Spatial probability P_{SSS}^{spat}.* The locations of the submarine epicenters associated to the instrumental records are considered equiprobable and the relative depths of the hypocenters are assigned randomly between 0 and 15 km of the crust. If necessary, the depths of the hypocenters are conveniently lowered for consistency with the *Wells and Coppersmith*'s formulas.
- *Frequency-Size probability P_{SSS}^{size}.* Large earthquakes are less likely to occur compared to events of smaller magnitude. The SSSs frequency-size relation, known as the *Gutenberg-Richter*'s law ([58]) is computed using a large set of instrumental data in the Tyrrhenian Sea. The events are extracted from the ISIDe database (with location at sea and epicenter depths $\leq$ 15 km). The completeness magnitude is $M_w = 2.3$ for this database and the resulting b-value is equal 1.059. Each 0.1 interval of magnitude in the range [5.5–7.5] identifies a class of magnitude which is weighted using the SSSs frequency-size relation and is associated to each seismic source.
- *Temporal probability P_{SSS}^{time}.* The annual probability is calculated by the *Poisson* occurrence that is $1 - e^{-\lambda \Delta t}$, where λ is the annual rate of occurrence of the potential tsunamigenic SSSs and Δt is the exposure window. The λ is computed in the Tyrrhenian Sea using the *ISIDe* instrumental data and is 0.01059.

Table 1. *SSS* geometric fault parameters by: [1] [59], [2] [60].

Id	Fault Name	Length	Width	Min Depth, Max Depth	Strike	Dip	Rake
1	*Posillipo* [1]	30.0	13.0	1.0, 9.8	74	60	270
2	*Neaples*[1]	30.0	13.0	1.0, 9.8	74	60	270
3	*Nord Campi Flegrei* [1]	13.0	13.0	1.0, 9.8	58	60	270
4	*Casamicciola Terme* [2]	5.0	4.0	0.0, 3.5	235	85	270
5	*Castellamare* [1]	5.0	13.0	1.0, 9.8	58	60	270
6	*Vico Equense* [1]	15.0	13.0	1.0, 9.8	58	60	270
7	*Golfo di Salerno* [1]	48.0	13.0	1.0, 9.8	78	60	270
8	*Ponte Barizzo* [1]	10.0	13.0	1.0, 9.8	150	60	270

3.3. SMFs

- *Spatial Identification.* The spatial failure probability of a *SMF* to occur is identified on the basis of some predominant controlling factors deduced by previous background studies: (1) statistical analysis indicated that the slope angle and the depth of the centre of mass are principal controlling factors ([61]); (2) scars observations indicated that the past mass failures left unstable margins ([62]); (3) earthquake occurrences increase the slide instability ([63]). In order to compute the spatial probability, the Tyrrhenian Sea was divided in square cells of $1'$. The length of the cells is of the same order of magnitude of the major slide indicated by the past events referred in Table 1. Each cell is designed by a weight that is the sum of scores that quantify the informative features. On the basis of points (1), (2) and (3) the proposed scores in the Tyrrhenian Sea are:

(a) 1 in the deep basin or in the coastal zones,

(b) 10 where the average depth of the cell is in the range 1000–1300 m,

(c) 10 where the mean slope of the cell is between 3°–5°,

(d) 10 where a factor of safety indicates slide instability in the case of earthquakes,

(e) 20 where the mean slope is $>5°$,

as discussed in another application study [13]. The factor of safety F_s is computed following [64]:

$$F_s = \frac{[1 - (\gamma/\gamma')a_y - (\gamma/\gamma')a_y \tan\phi - r_u/\cos^2\phi]}{[1 - (\gamma/\gamma')a_y + (\gamma/\gamma')a_x \tan\phi]} \tan\rho'/\tan\phi \quad (18)$$

where ϕ is the slope angle, γ and γ' are the total and buoyant unit weights of the sediment (typical values are chosen considering the sediment distributions in the Tyrrhenian Sea by [65]), a_x and a_y are the horizontal and vertical ground acceleration due to the earthquakes expressed in terms of % of gravity, r_u ($=u_e/\gamma'\zeta$) is related to the exceeded pore pressure u_e and the sediment

140

thickness ζ, $\tan \rho'$ is the coefficient of friction (ρ' generally ranges between $20°$ and $35°$).

We simplify by eliminating the terms containing a_y and r_u because the vertical acceleration may be neglected. Also, the u_e excess pore pressure is considered equal *nil*. This simplification does not take into account the excess pore pressure development for the slides during a seismic event ([66]). The use of the u_e value requires a specific knowledge of the accumulated materials at sea that is not available. The case $F_s < 1$ indicates instability, $F_s = 1$ is the limit equilibrium and $F_s > 1$ indicates stability. The horizontal peak ground acceleration (PGA) is set from the best estimate values of the *Italian Hazard Map* (2004 - $http : //zonesismiche.mi.ingv.it/$). In each cell we consider the PGA values with the probability of excedence of 16% and 84% in 50 years. The correspondent median value is used in Equation (14) with the underlying assumption that such a value represents a reliable indication of the earthquake load that can increase the slide instability and the subsequent probability of mass failure. The highest scores are used for the steep slopes because often they belong to flanks of volcanic structures in the Tyrrhenian Sea. The cell weights range from 1 (the case of cells presenting low mass failure probability to occur where the slope is $<3°$, the depths are either <1000 m or >1300 m and the seismicity is low) up to 40 (the case of cells presenting high mass failure probability to occur where the slope gradient is sharp and the depths are between 1000 m and 1300 m in an area of high seismicity). The geological records indicating past failures are reckoned by additional weights. The increment corresponds to the number of the past SMFs (Table 2). Finally, based on the cell weights we compute by normalization the SMF probability of occurrence, that is predominant: (1) on open continental margins; (2) on the flanks of volcanic islands; and (3) in the areas of high earthquakes occurrence.

- *Magnitudes and sizes.* The mass failure volumes V are chosen in the interval $[5 \times 10^5 - 5 \times 10^{10}]$ m^3 which was divided in 5 classes of reasonable sizes considering that the tsunami heights and run-ups are related to the volume of the slides ([67]). The classes are chosen on the basis of the past regional events. The lower class is set according to the mass failure of the Stromboli volcano in December 2002 estimated about 5×10^5 m^3 by [68]. The upper class is derived from historical slide sizes mapped using marine geological technique by [69]. A wide set of potential mass failures is produced associating each SMF class to each square cell of $1'$ in the Tyrrhenian Sea. For practical reasons, we consider the SMF geometrical parameters in two separate dataset: *slides* (generally defined as thin, translational failures traveling long distances) and *slumps* (generally defined as thick, rotational failures occurring with minimal displacement). The slump failures span in the first 4 classes for a total number of events $N_{slump} = 10,000$, whereas the slides span in the full range of the 5 volume classes for an equivalent number of $N_{slide} = 10,000$. The total number of SMF events are $N_{SMF} = 20,000$.

- *Geometric Parameters*. On the basis of the rigid body approximation the length l, thickness s and width w deduced by the hypothetic volumes of the mass failures are the basic parameters describing the mass failures. Their regular shape is further simplified as a function of the SMF length, following [70,71]. In other words, it is $s \approx 0.01\ l$ and $w \approx 0.25\ l$ in the case of submarine slides, and $s \approx 0.1\ l$ and $w \approx l$ in the case of submarine slumps. As a consequence, having set the mass failure volumes we simply computed the related parameters l.

- *Tsunami Initial Wave and Tsunami Wave Propagation*. It is calculated by empirical laws that represent approximations at the first order of the initial tsunami wave amplitude η' ([71]) in the case of slides

$$\eta' \approx 1.7410^{-5}l(1 - 0.750sin\phi)[(lsin\phi)/d]^{1.75} \tag{19}$$

and in the case of slump

$$\eta' \approx coef_{\Delta\Phi}l(sin\phi)^{0.25}(1 + 2.06\sqrt{d/l})^{-1}(l/d)^{1.25} \tag{20}$$

where ϕ is the incline angle, d is the depth of the centre of mass of the submarine slide or slump, $coef_{\Delta\Phi}$ is the difference between the initial and the final angles of the centre of mass. For specific values we refer to previous application in [13].

- *Coastal Effects and Tsunami Impact Evaluation*. The initial wave is propagated by the empirical run-up law by [72]

$$Z = 2.381(cot\phi')^{\frac{1}{2}}\eta^{\frac{5}{4}}d^{-\frac{1}{4}} \tag{21}$$

where ϕ' is the average coastal sea-bottom slope angle in front of the target sites, η (=$2\eta'$) is the tsunami wave height, and d is the water depth at the source.

This empirical law does not consider: (*i*) wave propagation effects (refraction, diffraction, *etc.*) and (*ii*) nonlinear process (breaking, dispersion, *etc.*). Here, we produce a first order tsunami hazard assessment of mass failures that has to be compared with the tsunami hazard assessment of the seismic type in order to apply the methodology in the region.

142

3.4. P_{SMF}

- *Spatial probability P_{SMF}^{spat}*. The spatial domain is divided in cells where the spatial probability is calculated. The score parameters associated to each cell (see previous paragraph) are converted in spatial probability and normalized to 1, following [73].
- *Frequency-Size probability P_{SMF}^{size}*. A power law is calculated for the *SMFs* similarly to the *SSSs* case by considering known submarine mass failures in the Tyrrhenian Sea ([13,74–76]). The corresponding *b*-value (= 1.3114) is computed in order to weight the mass failure classes. Considering the wide range of events the volumes of the *SMFs* were grouped in 4 (slump cases) or 5 (slide cases) classes weighted by the frequency-size relation.
- *Temporal probability P_{SMF}^{time}*. Also in this case the annual probability is calculated by the *Poisson* occurrence $1 - e^{-\lambda \Delta t}$, where λ is the annual rate of occurrence of potential tsunamigenic *SMFs* in the Δt exposure window. The λ is 0.013, computed using the available geological background knowledge.

Table 2. *SMF* parameters by: [1] [13], [2] [74], [3] [75], [4] [76].

Id	*SMF* Name	Size m^3	*Time*, Years Before Present
1	*Capo Licosa* [1]	3.2×10^8	14,000
2	*Baia Napoli (Gaia Bank)* [2]	3.8×10^6	15,000–6000
3	*Baia Napoli (Dohrn Canion)* [2]	100×10^6	15,000–6000
4	*Baia Napoli (Miseno Bank)* [2]	200×10^6	15,000–6000
5	*Ischia Nord* [3]	$15–20 \times 10^6$	3000–2400
6	*Ischia Sud* [3]	1.5×10^9	23,000
7	*Ischia West* [3,4]	$80–150 \times 10^6$	5500
8	*Ischia West* [3,4]	1×10^6	5500

4. The $TotPTHA_{long-term}$

Finally, the $PTHA_i$ is assessed by each factor P_i through Equation (7)

$$p(Z \geq z, SE_{ij}; \Delta t) = \mathcal{H}(Z \geq z) \cdot (P_i^{spat} \cdot P_i^{size} \cdot P_i^{time}) \tag{22}$$

where $\mathcal{H}(Z \geq z)$ is the Heaviside step function (that is 1 if the run-ups Z are larger than the threshold levels of $z \in [0.5\ m, 1\ m, 2\ m, \ldots, 10\ m]$ and 0 in the other cases) and the factors P_i represent the *non*-interactive source probabilities.

If P_i defines the probability of occurrence of at least one tsunami event and $(1 - P_i)$ the generic probability that no tsunami occurs, then the final probability $P_{long-term}$ is

$$P_{long-term} = 1 - [(1 - P_{SSS})(1 - P_{SMF})] \tag{23}$$

that combines both factors P_i ($i= SSS$ and SMF). The $TotPTHA_{long-term}$ is shown in Figure 2 and the relative probabilities $\times$ $year^{-1}$ of $z \geq 0.5$ m and $z \geq 1$ m at the target sites *Naples* and *Ischia* are listed in Table 3. Equation (23) takes into account also tsunami events produced by $SMFs$ triggered by $SSSs$ assuming that the sources are independent.

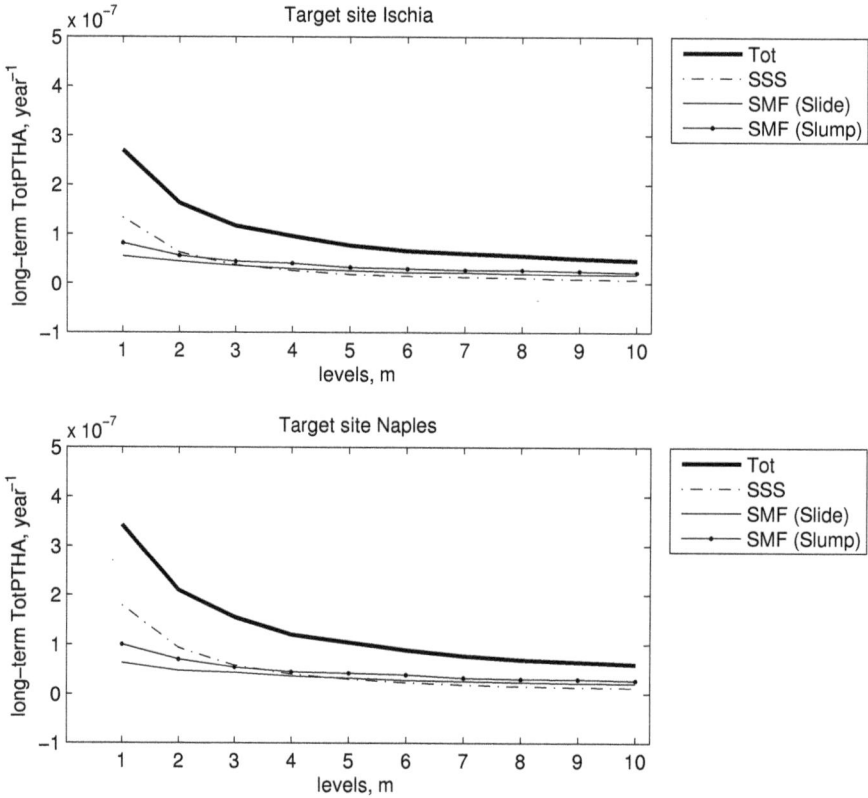

Figure 2. $TotPTHA_{long-term}$ at the target sites. The $TotPTHA$ takes into account the $PTHA_i$ where i indicates $SSSs$, $SMFs$ (Slides and Slumps).

Table 3. $TotPTHA_{long-term} \times year^{-1}$ to overcome 0.5 m and 1 m run-ups at the island of *Ischia* and the port of the city of *Naples*.

	Ischia: 0.5 m	1 m	Naples: 0.5 m	1 m
SSS	1.353×10^{-6}	3.915×10^{-7}	2.400×10^{-6}	1.230×10^{-6}
SMF (slide)	1.121×10^{-7}	8.222×10^{-8}	1.342×10^{-7}	9.344×10^{-8}
SMF (slump)	1.625×10^{-7}	1.274×10^{-7}	1.837×10^{-7}	1.422×10^{-7}
$TotPTHA_{long-term}$	1.627×10^{-6}	6.011×10^{-7}	2.718×10^{-6}	1.466×10^{-6}

This assumption is realistic by considering the time lag between a SSS event and the triggered SMF event. In other words the two events are well separated in time: the triggering event occurs in a time duration of an order of seconds, the triggered event may occur after minutes or hours.

5. The $TotPTHA_{short-term}$ in the Case of the TEs

The $TotPTHA$ can be modified (and eventually increased) by the occurrence of an earthquake. Now we compute the $TotPTHA_{short-term}$ through time at both target sites in case a triggering event TE occurs, that is one submarine earthquake of magnitude M_{TE} at time t_{TE}. To model the influence of TE on the other potential seismic sources we apply the epidemic-type aftershock sequence ($ETAS$) model ([77–80]) to the seismicity of $SSSs$ in the region.

The fundamental concept of the $ETAS$ model is that each earthquake can trigger subsequent events ([50,81,82]) and seismicity can be generated by the aftershocks. Also the model assumes that the aftershock activity cannot always be predicted by a single modified $Omori$ function but each aftershock is able to perturb the earthquake rate and generate its own $Omori$ aftershock decay dependent on the magnitude ([83,84]). In general, the temporal seismicity rate λ' of a specific area is the sum of two terms ([85]:

$$\lambda' = \lambda_b + \lambda_{ETAS} \tag{24}$$

the first term (λ_b) is the rate of background that is not triggered by the previous earthquakes in the catalogue and the second one (λ_{ETAS}) is the rate of the triggered event that is associated to the stress perturbations generated by the previous earthquakes in the catalogue. In this practical application, the background seismicity rate is considered equal to the one computed in the long-term analysis. According to the $ETAS$ model assumptions the total space-time conditional intensity is the probability of an earthquake occurring in the infinitesimal space-time volume conditioned to the past history [85]. For simplicity, we consider only one main shock TE and we neglect the sequence of aftershocks. It is worth to note that this results in an underestimation of the effect of a large TE on the final tsunami hazard. The λ_{ETAS} represents the contribution to the annual seismicity rate in the region in each location and it is expressed according to [85] by

$$\lambda_{ETAS} = \frac{K}{(t_{ob} - t_{TE} + c)^\Pi} e^{\alpha(M_{TE} - M_c)} \frac{\frac{(q-1)}{\pi} G^{2(q-1)}}{(r_i^2 + G^2)^q} \beta e^{-\beta(M_w - M_c)} \tag{25}$$

where $t_{TE}(=0)$ is the time when the TE of moment magnitude $M_{TE} = 5.4$ occurs, t_{ob} is the observation time after the TE, M_c is the completeness magnitude of the instrumental catalogue, r_i is the distance between the potential tsunamigenic $SSSs$ and the epicenter of the TE and $\beta = b\ln(10)$ is the parameter of the $Gutenberg$-

145

Richter's law ($b = 1.059$). The values of the other parameters in Equation (25) are: $K = 0.011, c = 0.00004, \Pi = 1.16, \alpha = 1.3, G = 1.1$ and $q = 1.5$ according to [85].

As stated in Equation (13), the $TotPTHA_{short-term}$ is performed again through Equation (6) where the source phase probability in Equation (7) is assessed accounting for SI through Equation (10)

$$p^*(SE_{ij}, \Delta t) = p(SE_{ij}, \Delta t) + 1 - e^{-\lambda_{ETAS}\Delta t} \tag{26}$$

where the first term is assessed by Equation (22) and λ_{ETAS} through Equation (25). We consider an exposure time window Δt of 7 days. The results are shown in Figure 3 at the target sites. Immediately after the TE event (after 1 day, 2 days and 7 days) the probabilities increase. For example in the case of 0.5 m and 1 m run-up the $TotPTHA_{short-term}$ is three order of magnitudes higher than the $TotPTHA_{long-term}$ (Table 4). On the other hand, 30 days after the TE there are no more significant effects on the $TotPTHA_{short-term}$ and the probability values are back to the $TotPTHA_{long-term}$ assessment.

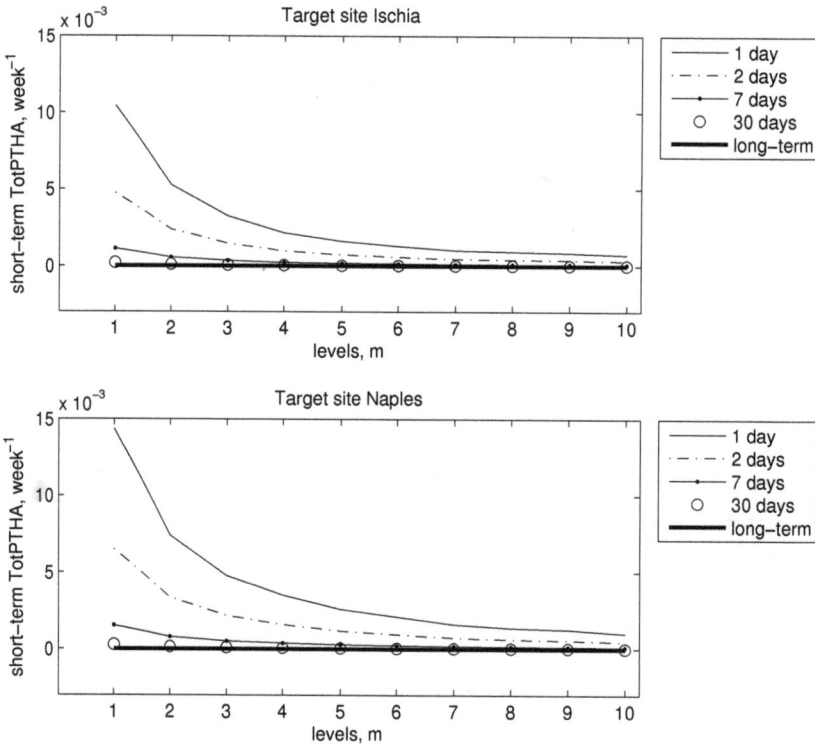

Figure 3. Probability at the target sites, applying the $ETAS$ model after the triggering event TE, a submarine earthquake of magnitude $M_w = 6.8$.

In the case of Equation (26) the independence between the TE and the SSS is assumed. Also, the choice of $M_{TE} = 5.4$ makes the TE an external event to the dataset used to evaluate the $p(SE_{ij}, \Delta t)$.

Table 4. $TotPTHA_{short-term}$ to overcome 0.5 m and 1 m run-ups after 1 day, 2 days, 7 days and 30 days due to the TE considering the exposure time window of 7 days at the island of $Ischia$ and the port of the city of $Naples$.

$TotPTHA_{short-term}$	Ischia: 0.5 m	1 m	Naples: 0.5 m	1 m
Day 1	1.624×10^{-2}	5.786×10^{-3}	7.333×10^{-2}	2.412×10^{-2}
Day 2	7.363×10^{-3}	2.615×10^{-3}	3.378×10^{-2}	1.096×10^{-2}
Day 7	1.737×10^{-3}	6.158×10^{-4}	8.052×10^{-3}	2.588×10^{-3}
Day 30	1.155×10^{-3}	8.556×10^{-4}	1.496×10^{-3}	4.799×10^{-4}

6. Discussion, Limitations and Further Work

In this paper we describe a general methodology for a comprehensive and total probabilistic tsunami hazard assessment ($TotPTHA$) that coherently merges the probabilistic hazards computed on the basis of a number of tsunamigenic events due to different source types. In practice, we evaluate the methodology and its potential by selecting two target sites in Italy as illustrative cases, the city of $Naples$ and the Island of $Ischia$. In order to show the applicability of the $TotPTHA$ methodology many important simplifications are required and their limitations in the application are discussed. However, we would like to remark that this case study means to be illustrative of the proposed methodology for the $TotPTHA$ without meaning to be complete for the final hazard assessment.

On the basis of historical information from the Italian Tsunami Catalogue ([49]) three definite (maximum reliability) tsunami events occurred in $Naples$ (in 1631, 1805, 1906) with tsunami intensity 2 in Ambraseys-Sieberg Scale. They were associated to earthquakes or volcanic activity. There is no explicit reference to tsunami caused by submarine mass failures in the historical period. On the other hand, extensive long-range side scan sonar survey ([86]) showed that a catastrophic collapse occurred in pre-historical time (3000–2400 years before present) at the $Ischia$ Island driven by the volcano-tectonic uplift of Mt. Epomeo. This major collapse was likely followed by less catastrophic terrestrial and submarine failures. Simulation of such event showed that giant waves could impact on the Tyrrhenian coasts ([87]).

Firstly, to compute a fully comprehensive $TotPTHA$ in the region, multiple tsunamigenic sources must be taken into account. In fact, the estimation of the $PTHA$ considering only a single type of source (the seismic one) may generate biases and underestimations. In the application at the two target sites we evaluate the tsunami hazard produced by the submarine seismic events and the submarine mass

147

failures. Other sources, related to the volcanic sources like the pyroclastic flows and the caldera activities should be considered. Even if we recognize their importance in the area in terms of ST_{repres} and the possible implications on the $TotPTHA$ they are not explored in this application. Major information are missing on the tsunami waves associated to volcanic events in the area. Therefore, in the present application we do not include the volcanic sources in order to avoid additional biases.

Secondly, the analysis is focused on the near sources, both SSSs and SMFs, and the choices of the source datasets pose the problem of the SE_{repres}. Source information on seismic faults decrease with the distance from the Gulf of *Naples* and are often unknown/incomplete. In the past two millennia there is no reported seismic event far in the Tyrrhenian Sea that produced tsunami waves at the target sites. The use of the instrumental database is an additional way to identify the active faults offshore. In the application case, the database *ISIDe* indicates a region of interest wider than the existing faults but covers a very short time period. The missing information in the database have to be compensated by the general background knowledge of the tectonic/geological setting. In order to overcome the afore-mentioned limitations of the databases/catalogues we should consider alternative strategies aimed to enlarge the number of events considered in the dataset. A solution could be the introduction of a gridded domain to systematically explore the potentially tsunamigenic SSSs in unknown locations. In this way, the source dataset is expanded by identifying the grid centers and/or corners as locations of the epicenters in the binned domain. Also, we can use a larger set of faults by adding more randomness in the other parameters. Other relevant far sources may be identified in the Southern Tyrrhenian Sea and the Northern African coasts to complete the dataset of the SSSs. The limitations posed by the completeness and the length of the databases/catalogues indicate the need of further studies in the area of the application.

Beside the problem of a missing complete catalogue also for the SMF case, the description of the mass failures and the computation of the relative run-ups undergo to simplifications and approximations that do not take into account distance, orientation and bathymetry between sources and target sites [88]. A slide modeling would improve the determination of the run-ups by using more realistic slope and water depth in the shallowest part of the domain. As a consequence, uncertainties related to the Z parameter would be reduced.

Despite the discussed limitations, important achievements have been obtained by applying the $TotPTHA$ methodology for the *multi*−source case. In both target sites the multiple-source effect on the $TotPTHA$ is the increasing of the hazard in the long-term analysis compared to the hazard posed by a single type of source. In fact, the $TotPTHA$ in the long-term case remarkably increases after considering submarine mass failures along with the potential submarine seismic events. Even in not definitive analysis (like the present application), once the $multi-$hazards are formally assessed and the relative importances of the $PTHA_{SSS}$ and the $PTHA_{SMF}$ terms are computed in the $TotPTHA_{long-term}$ then it is possible to consider the source interactions. In the case of a triggering seismic event there are relevant changes in the short-term $TotPTHA$.

7. Conclusions

The scope of this study is to present a methodology for the total probabilistic tsunami hazard assessment ($TotPTHA$) produced by multiple tsunamigenic sources in the framework of the $ByMuR$ Italian project: *Bayesian Multi-Risk Assessment: A case study for natural risks in the city of Naples* ($http : //bymur.bo.ingv.it/$).

In an exemplified application we illustrate the methodology at two target sites (the port of *Naples* and the island of *Ischia*) focusing the analysis on: (1) the *multi*-hazard approach and (2) the *long*- and *short*-term hazard assessments.

1. We demonstrate the applicability of the proposed comprehensive $TotPTHA$ to treat the interactions among different tsunami hazards. Indeed, in the selected region the tsunami waves may be generated by different types of sources (near submarine seismic events $SSSs$ and submarine mass failures $SMFs$) at the target sites. We computed the $TotPTHA_{long-term} \times year^{-1}$ and we show that the hazard curve increases when *non*-seismic STs are included in the analysis and formally considered together rather than separately;

2. We compute the $TotPTHA_{short-term}$ by considering a submarine earthquake as triggering event TE. In other words, a seismic event (external to the seismic dataset used to produce the hazard) acts as a perturbation of the regional seismicity. We show that the $TotPTHA_{short-term}$ changes remarkably compared to the $TotPTHA_{long-term}$ and increases for a limited time window (few days).

In order to perform a more realistic and complete $TotPTHA$ for the selected target sites, further improvements are needed by considering, for example, the tsunamigenic volcanic sources and the far field seismic sources in the Southern Tyrrhenian Sea and Northern Africa coasts. However, we have illustrated how the proposed $TotPTHA$ methodology can be applied in order to potentially consider any kind of tsunami sources and their interactions for both short- and long-term hazard purposes.

Acknowledgments: We thank the Editor Valentin Heller and we are particularly grateful to three anonymous reviewers/referees for the constructive suggestions and comments that strongly improved the paper. This study was supported by the Italian project *ByMuR http : //bymur.bo.ingv.it*.

Author Contributions: Conceived and designed the methodology: AG, LS, JS. Identified and provided the seismic sources: SP. Performed the numerical simulations: RT. Wrote the paper: AG. Commented on and improved the review: AG, RT, LS, SP, JS.

Conflicts of Interest: The authors declare no conflict of interest.

Abbreviations

Abbreviations/Nomenclature

α: parameter to determine the triggering capability depending on magnitude

a_x: horizontal ground acceleration

a_y: vertical ground acceleration

β: parameter of the the *Gutenberg — Rchter*'s law

b: coefficient of the *Gutenberg — Rchter*'s law

c: parameter of the modified *Omori* Law

d: depth

D: height of the total water column

ϕ: slope and/or incline angle

ϕ': coastal sea-bottom slope angle

F_s: factor of safety

γ: total unit weight of the sediment

γ': buoyant unit weight of the sediment

g: gravity

G: parameter characterizing the spatial probability density function of the triggered event

H: average water fluid height

K: parameter of the modified *Omori* Law

λ: annual rate of occurrence

λ': rate of background

λ_{ETAS}: rate of the triggering event

l: slide and/or slump length

L: fault length

M: discharged water fluxes in the eastward direction

$M_{1,2,3}$: number of generated tsunami events

M_c: magnitude threshold for Italian territory in *ETAS* model

M_{TE}: magnitude of the triggering event

M_w: Moment Magnitude

η: tsunami surface elevation

η': tsunami wave amplitude

N: discharged water fluxes in the northward direction

N_{slide}: number of slides

N_{slump}: number of slumps

N_{SMF} : number of *Submarine Mass Failures*

N_{SSS} : number of *Submarine Seismic Sources*

N_{TE} : number of *Triggering Events*

N_{TOT}: total number of events

Π: parameter of the modified *Omori* Law

P_i: probability

PSHA : *Probabilistic Seismic Hazard Assessment*

PTHA : *Probabilistic Tsunami Hazard Assessment*

PVHA : *Probabilistic Volcanic Hazard Assessment*

q: parameter characterizing the spatial probability density function

ρ': coefficient of friction

r_i: distance between the potential tsunamigenic seismic sources and the epicenter of the triggering event

r_u: parameter of the exceeded pore pressure

s: slide and/or slump thickness

S: slip fault

SI : *Source Interaction*

SE : *Source Event*

SE_{repres} : *Source Event* representativeness

SMF: *Submarine Mass Failure*

SSS: *Submarine Seismic Source*

ST : *Source Type*

ST_{repres} : *Source Type* representativeness

t : time

t_{ob}: observation time after the triggering event

t_{TE}: time of the triggering event

Δt: exposure time window

$T_{1,2,3}$: number of total tsunamis

TE : *Triggering Event*

TotPTHA : *Total Probabilistic Tsunami Hazard Assessment*

u: x-component of the horizontal velocity

u_e: exceeded pore pressure

v: y-component of the horizontal velocity

V: slide and/or slump volume

w: slide and/or slump width

W: fault width

ζ: sediment thickness

z: thresholds of the level run-ups

Z: run-ups

References

1. Liu, Y.; Santos, A.; Wang, S.M.; Shi, Y.; Liu, H.; Yuen, D.A. Tsunami hazards along Chinese Coast from potential earthquakes in South China Sea. *Phys. Earth Plan. Int.* **2007**, *163*, 233–244.

2. Power, W.; Downes, G.; Stirling, M. Estimation of tsunami hazard in New Zealand due to South American earthquakes. *Pure Appl. Geophys.* **2007**, *164*, 547–564.

3. Rikitake, B.T.; Aida, I. Tsunami hazard probability in Japan. *Bull. Seismol. Soc. Am.* **1988**, *78*, 1268–1278.

4. Tonini, R.; Pagnoni, G.; Tinti, S. Modeling the 2004 Sumatra tsunami at Seychelles Islands: site-effect analysis and comparison with observations. *Nat. Haz.* **2014**, *70*, 1507–1525.

5. Burbidge, D.; Cummins, P.R.; Mleczko, R.; Thio, H.K. A probabilistic tsunami hazard assessment for Western Australia. *Pure Appl. Geophys.* **2008**, *165*, 2059–2088.

6. Burroughs, S.M.; Tebbens, S.F. Power-law scaling and probabilistic forecasting of tsunami runup heights. *Pure Appl. Geophys.* **2005**, *162*, 331–342.

7. Tinti, S.; Armigliato, A.; Tonini, R.; Maramai, A.; Graziani, L. Assessing the hazard related to tsunamis of tectonic origin: A hybrid statistical-deterministic method applied to Southern Italy coasts. *ISET J. Earthq. Tech.* **2005**, *42*, 189–201.

8. Orfanogiannaki, K.; Papadopoulos, G. Conditional probability approach of the assessment of tsunami potential: Application in three tsunamigenic regions of the Pacific Ocean. *Pure Appl. Geophys.* **2007**, *164*, 593–603.

9. Geist, E.L.; Parsons, T. Probabilistic Analysis of Tsunami Hazards. *Nat. Haz.* **2006**, *37*, 277–314.

10. Gonzalez, F.L.; Geist, E.L.; Jaffe, B.; Kanoglu, U.; Mofjeld, H.; Synolakis, C.E.; Titov, E.E.; Arcas, D.; Bellomo, D.; Carlton, D.; *et al.* Probabilistic Tsunami Hazard Assessment at Seaside, Oregon, for near- and far-field seismic sources. *J. Geophys. Res.* **2009**, *114*, 1–19.

11. Sörensen, M.B.; Spada, M.; Babeyko, A.; Wiemer, S.; Grünthal, G. Probabilistic tsunami hazard in the Mediterranean Sea. *Geophys. Res. Lett.* **2012**, *117*, B01305.

12. Grezio, A.; Marzocchi, W.; Sandri, L.; Gasparini, P. A Bayesian procedure for Probabilistic Tsunami Hazard Assessment. *Nat. Haz.* **2010**, *53*, 159–174.

13. Grezio, A.; Sandri, L.; Marzocchi, W.; Argnani, A.; Gasparini, P. Probabilistic Tsunami Hazard Assessment for Messina Strait Area (Sicily-Italy). *Nat. Haz.* **2012**, *64*, 329–358.

14. Tatsumi, D.; Calder, C.A.; Tomita, T. Bayesian near-field tsunami forecasting with uncertainty estimates. *J. Geophys. Res. Oceans* **2014**, *119*, 2201–2211.

15. Annaka, T.; Satake, K.; Sakakiyama, T.; Yanagisawa, K.; Shuto, N. Logic-tree approach for Probabilistic Tsunami Hazard Analysis and its applications to the Japanese Coasts. *Pure Appl. Geophys.* **2007**, *592*, 164–577.

16. Heidarzaed, M.; Kijko, A. A Probabilistic Tsunami Hazard Assessment for the Makran subduction zone at the Northwestern Indian Ocean. *Nat. Haz.* **2011**, *56*, 577–593.

17. Grilli, S.T.; Taylor, S.O.; Baxter, C. D.P.; Maretzki, S. A probabilistic approach for determining submarine landslides tsunami hazard along the upper East Coast of the United States. *Mar. Geol.* **2009**, *264*, 74–97.

18. ten Brink, U.S.; Geist, E.L.; Andrews, B.D. Size distribution of submarine landslides and its implication to tsunami hazard in Puerto Rico. *Geophys. Res. Lett.* **2006**, *33*, L11307.

19. Gisler, G.; Weaver, R.; Mader, C.; Gittings, M. Two-dimensional simulations of explosive eruptions of Kick'emjenny and other submarine volcanoes. *Sci. Tsunami Haz.* **2006**, *25*, 34–41.

20. Heinrich, P.; Mangeney, A.; Guibourg, S.; Roche, R. Simulation of water waves generated by a potential debris avalanche in Montserrat, Lesser Antilles. *Geophys. Res. Lett.* **1998**, *25*, 3697–3700.

21. Lövholt, F.; Pedersen, G.; Gisler, G. Oceanic propagation of a potential tsunami from the La Palma Island. *J. Geophys. Res.* **2008**, *113*, 9–26.

22. Mader, C. Modeling the La Palma Landslide. *Sci. Tsunami Haz.* **2008**, *19*, 150–170.

23. Strunz, G.; Post, J.; Zosseder, K.; Wegscheider, S.; Muck, M.; Riedlinger, T.; Mehl, H.; Dech, S.; Birkmann, J.; Gebert, N.; *et al.* Tsunami risk assessment in Indonesia. *Nat. Haz. Earth Syst. Sci.* **2004**, *11*, 67–82.

24. Ward, S.N.; Day, S. Cumbre Vieja volcano–Potential collapse and tsunami at La Palma, Canary Islands. *Geophys. Res. Lett.* **2003a**, *28*, 3397–3400.

25. Poisson, B.; Pedreros, R. Numerical modelling of historical landslide-generated tsunamis in the French Lesser Antilles. *Nat. Haz. Earth Syst. Sci.* **2010**, *10*, 1281–1292.

26. Choi, B.H.; Pelinovsky, E.S.; Kim, K.O.; Lee, J.S. Simulation of the trans-oceanic tsunami propagation due to the 1883 Krakatau volcanic eruption. *Nat. Haz. Earth Syst. Sci.* **2003**, *3*, 321–332.

27. Ward, S.N.; Day, S. Ritter Island volcano: Lateral collapse and tsunami of 1888. *Geophys. J. Int.* **2003b**, *154*, 891–902.

28. Braathen, A.; Blikra, L.H.; Berg, S.; Karlsen, F. Rock-slope failures in Norway; type, geometry and hazard. *Nor. J. Geol.* **2004**, *84*, 67–88.

29. Fritz, H.M.; Hager, W.H.; Minor, H.E. Lituya bay case: Rockslide impact and wave run-up. *Sci. Tsunami Haz.* **2001**, *19*, 3–22.

30. Fritz, H.; Mohammed, F.; Yoo, J. Lituya Bay landslide impact generated mega-tsunami 50th anniversary. *Pure Appl. Geophys.* **2009**, *175*, 166–153.

31. Harbitz, C.; Pedersen, G.; Gjevik, B. Numerical simulations of large water waves due to landslides. *J. Hydr. Eng.* **1993**, *119*, 1325–1342.

32. Miller, D. Giant waves in Lituya Bay, Alaska. *U.S. Geol. Surv.* **1960**, *354-C*, 51–84.

33. Bondevik, S.; Lövholt, F.; Harbitz, C.; Mangerud, J.; Dawson, A.; Svendsen, J. The storegga slide tsunami comparing field observations with numerical simulations. *Ormen Lange Spec. Issue Marine Petrol. Geol.* **2005**, *208*, 22–195.

34. Fine, I.V.; Rabinovich, A.B.; Bornhold, B.D.; Thompson, R.E.; Kulikov, E.A. The Grand Banks landslide-generated tsunami of November 18, 1929: Preliminary analysis and numerical modelling. *Mar. Geol.* **2005**, *215*, 45–57.

35. Haflidason, H.; Sejrup, H.P.; Nygard, A.; Mienert, J.; Bryn, P.; Reidar, L.; Forsberg, C.F.; Berg, K.; Masson, D. The Storegga Slide: Architecture, geometry and slide development. *Mar. Geol.* **2004**, *213*, 201–234.

36. Tappin, D.; Watts, P.; Grilli, S. The Papua New Guinea tsunami of 17 July 1998: Anatomy of a catastrophic event. *Nat. Haz. Earth Syst. Sci.* **2008**, *8*, 243–266.

37. Leonard, L.J.; Rogers, G.C.; Mazzotti, S. Tsunami hazard assessment of Canada. *Nat. Haz.* **2014**, *70*, 237–274.

38. Chesley, S.; Ward, S. A quantitative assessment of the human and economic hazard from impact-generated tsunami. *Nat. Haz.* **2006**, *38*, 355–374.

39. Marzocchi, W.; Garcia-Aristizabal, A.; Gasparini, P.; Masellone, M.L.; di Ruocco, A.; Novelli, P. Basic principles of multi-risk assessment: A case study in Italy. *Nat. Haz.* **2012**, *62*, 551–573.

40. Selva, J. Long-term risk assessment: Statistical treatment of interaction among risks. *Nat. Haz.* **2013**, *67*, 701–722.

41. Grünthal, G.; Thieken, A.H.; Schwarz, J.; Radtke, K.S.; Smolka, A.; Merz, B. Comparison Risk Assessments for the city of Cologne - Storm, Floods, Earthquakes. *Nat. Haz.* **2006**, *30*, 21–44.

42. Gusiakov, V. Tsunami History: Recorded. In *The Sea*; Bernard, E.N., Robinson, A.R., Eds.; Harvard University Press: Boston, MA, USA, 2009; p. 15.

43. McGuire, R. Probabilistic seismic hazard analysis: Early history. *Earthq. Eng. Struct. Dyn.* **2008**, *37*, 329–338.

44. Marzocchi, W.; Sandri, L.; Selva, J. BET_VH: A probabilistic tool for long term volcanic hazard assessment. *Bull. Volcan.* **2010**, *72*, 705–716.

45. Cornell, C.A. Engineering seismic risk analysis. *Bull. Seismol. Soc. Am.* **1968**, *58*, 1583–1606.

46. Geist, E.L.; Parson, T. Distribution of tsunami inter-event times. *Geophys. Res. Lett.* **2008**), *35*, L02612.

47. Titov, V.; Moore, C.; Greenslade, D.; Pattiaratchi, C.; Badal, R.; Synolakis, C.; Kânoğlu, U. A new tool for inundation modeling: Community modeling interface for tsunamis. *Pure Appl. Geophys.* **2011**, *168*, 2121–2131.

48. Marzocchi, W.; Lombardi, A A double branching model for earthquake occurrence. *J. Geophys. Res.* **2008**, *113*, B08317.

49. Tinti, S.; Maramai, A.; Graziani, L. The new Catalogue of Italian Tsunamis *Nat. Haz.* **2004**, *33*, 439–465.

50. Helmstetter, A.; Kagan, Y.Y.; Jackson, D.D. Comparison of short-term and time-dependent earthquake forecast models for Southern California. *Bull. Seism. Soc. Am.* **2006**, *96*, 90–106.

51. Papadopoulos, G.A. Tsunami hazard in the eastern Mediterranean: strong earthquakes and tsunamis in the Corinth Gulf, Central Greece. *Nat. Haz.* **2003**, *29*, 437–464.

52. Basili, R.; Valensise, G.; Vannoli, P.; Burrato, P.; Fracassi, U.; Mariano, S.; Tiberti M.; Boschi, E. Database of individual seismogenic sources (DISS), version 3: Summarizing 20 years of research on Italy's earthquake geologys. *Tectonophysics* **2008**, 20–43.

53. Wells, D.L.; Coppersmith, K.J. New empirical relationships among magnitude, rupture length, rupture width, rupture area and surface displacement. *Bull. Seismol. Soc. Am.* **1994**, *84*, 974–1002.

54. Ascione A.; Caiazzo C.; Cinque A. Recent faulting in Southern Apennines (Italy): Geomorphic evidence, spatial distribution and implications for rates of activity. *Boll. Soc. Geol. Ital. (Ital. J. Geosci.)* **2007**, *126*, 293–305.

55. Galadini F.; Meletti C.; Vittori E Major active faults in Italy: Available surficial data. *Neth. J. Geosci.* **2001**, *80*, 273–296.

56. Okada, Y. Internal deformation due to shear and tensile faults in a half-space. *Bull. Seismol. Soc. Am.* **1982**, *82*, 1018–1040.

57. Kamigaichi, O. Tsunami forecasting and warning. In *Encyclopedia of Complexity and System Science*; Springer: New York, NY, USA, 2009; pp. 9592–9617.

58. Gutenberg, B.; Richter, C. *Seismicity of the Earth and Associated Phenomena*, 2nd ed.; Princeton University Press: Princeton, NJ, USA,1954.

59. Cinque, A.; Ascione, A.; Caiazzo, C. Distribuzione spazio-temporale e caratterizzazione della fagliazione quaternaria in Appennino meridionale. *Volume: Le ricerche del GNDT nel campo della pericolosita' sismica (1996–1999). A cura di Galadini* **2003**, *160*, 1–16.

60. DISS-Working-Group. Database of individual seismogenic sources (DISS), version 3.1.1: A compilation of potential sources for earthquakes larger than M 5.5 in Italy and surrounding areas. INGV 2010 - Istituto Nazionale di Geofisica e Vulcanologia - All rights reserved; Doi:10.6092/INGV.IT-DISS3.1.1. Available online: http://diss.rm.ingv.it/diss/ (accessed on 1 March 2012).

61. Masson, D.; Harbitz, C.; Wynn, R.; Pedersen, G.; Lovholt, F. Submarine landslides: processes, triggers and hazard prediction. *Phil. Trans. R. Soc. A* **2006**, *364*, 2009–2039.

62. Martel, S. Mechanism of landslide initiation as a shear fracture phenomenon. *Mar. Geol.* **2004**, *203*, 319–339.

63. Biscontin, G.; Pestana, J.; Nadim, F. Seismic triggering of submarine slides in soft cohesive soil deposits. *Mar. Geol.* **2004**, *354*, 203–341.

64. Booth, J.; Sangrey, D.; Fugate, J. A nomogram for interpreting slope stability of fine-grained deposits in modern and ancient marine environments. *J. Sediment Petrol.* **1985**, *36*, 29–36.

65. Kastens, K.; Mascle, J.; Auroux, C.; Bonatti, E.; Broglia, C.; Channel, J.; Curzi, P.; Emeis, K.; Glacon, G.; Hasegawa, S.; *et al.* ODP leg 107 in the Tyrrhenian Sea: Insights into passive margin and back-arc basin evolution. *Geol. Soc. Am. Bull.* **2012**, *7*, 1140–1156.

66. Biscontin, G.; Pestana, J.M. Factors affecting seismic response of submarine slopes *Nat. Haz. Earth Syst. Sci.* **2006**, *6*, 97–107.

67. Murty, T.S. Tsunami wave height dependence on landslide volume. *Pure Appl. Geophys.* **2003**, *160*, 2147–2153.

68. Calvari, S.; Spampinato, L.; Lodato, L.; Harris, A.L.; Patrick, M.; Dehn, J.B.; Andronico, D. Chronology and complex volcanic processes during the 2002–2003 flank eruption at Stromboli volcano (Italy) reconstructed from direct observations and surveys with a handheld thermal camera. *J. Geophys. Res.* **2005**, *110*, B02201.

69. Fabbri, A.; Gallignani, P.; Zitellini, N. *Geologic Evolution of the Peri-Tyrrhenian Sedimentary Basins*; Wezel, F.C., Ed.; Sedimentary basins of the Mediterranean margins, Technoprint: Bologna, Italy, 1981; pp. 101–126.

70. Grilli, S.T.; Watts, P. Tsunami generation by submarine mass failure. I: Modeling, experimental validation, and sensitivity analyses. *J. Waterway Port Coast. Ocean Eng.* **2005**, *131*, 283–297.

71. Watts, P.; Grilli M.; Tappin, D.R.; Fryer, G.J. Tsunami generation by submarine mass failure. II: Predictive equations and case studies. *J. Waterway Port Coast. Ocean Eng.* **2005**, *131*, 298–310.

72. Synolakis, C.E. The run-up of solitary waves. *J. Fluid Mech.* **1987**, *185*, 523–545.

73. Selva, J.; Orsi, G.; di Vito, M.; Marzocchi, W.; Sandri, L. Probability hazard map for future vent opening at the Campi Flegrei caldera, Italy. *Bull. Volcanol.* **2012**, *74*, 497–510.

74. Milia, A.; Torrente, M.; Giordano, F. Gravitational instability of submarine volcanoes offshore Campi Flegrei (Naples Bay, Italy). In *Volcanism in the Campania Plain: Vesuvius, Campi Flegrei and Ignimbrites*; De Vivo, B., Ed.; Elsevier, Developments in Volcanology, Universita' di Napoli Federico II: Naples, Italy, 2006; pp. 69–83.

75. de Alteriis, G.; Insinga, D.; Morabito, S.; Morra, V.; Chiocci, F.; Terrasi, F.; Lubritto, C.; di Benedetto, C.; Pazzanese, M. Age of submarine debris avalanches and tephrostratigraphy offshore Ischia Island, Tyrrhenian Sea, Italy. *Marine Geol.* **2010**, *278*, 1–18.

76. Della Seta, M.; Marotta, E.; Orsi, G.; de Vita, S.; Sansivero, F.; Fredi, P. Slope instability induced by volcano-tectonics as an additional source of hazard in active volcanic areas: The case of Ischia Island (Italy). *Bull. Volcanol.* **2012**, *106*, 74–79.

77. Kagan, Y. Likelihood analysis of earthquake catalogues. *Geophys. J. Int.* **1991**, *106*, 135–148.

78. Kagan, Y.; Knopoff, L. Stocastic synthesis of earthquake catalogues. *Geophys. J. Int.* **1981**, *86*, 2853–2862.

79. Ogata, Y. Statistical models for earthquake occurrences and residual analysis for point processes. *J. Am. Statist. Assoc.* **1988**, *83*, 9–27.

80. Ogata, Y. Space-time-point-process for earthquake occurence. *Ann. Inst. Statist. Math.* **1998**, *50*, 379–402.

81. Console, R.; Murru M.; Lombardi, A. Refining earthquake clustering models. *J. Geophys. Res.* **2003**, *108*, 2468–2473.

82. Lombardi, A.M.; Marzocchi, W. Evidence of clustering and *non*-stationarity in the time distribution of large worldwide earthquakes. *J. Geophys. Res.* **2007**, *112*, B02303.

83. Omori, F. On the aftershocks of earthquakes. *J. Coll. Sci. Imp. Univ. Tokio* **1894**, *7*, 111–120.

84. Utsu, T. A statistical study on the occurrence of aftershocks. *Geophys. Mag.* **1961**, *30*, 521–605.

85. Lombardi, A.M.; Marzocchi, W. The ETAS model for daily forecasting of Italian seismicity in the CSEP experiment. *Ann. Geophys.* **2010**, *53*, 3.

86. Chiocci, F.L.; de Alteriis, G. The Ischia debris avalanche: First clear submarine evidence in the Mediterranean of a volcanic island prehistorical collapse. *Terra Nova* **2006**, *18*, 202–209, doi:10.1111/j.1365-3121.2006.00680.x.

87. Tinti S.; Chiocci F.L.; Zaniboni F.; Pagnoni G.; de Alteriis G. Numerical simulation of the tsunami generated by a past catastrophic landslide on the volcanic Island of Ischia, Italy. *Mar. Geophys. Res.* **2011**, *32*, 287–297, doi:10.1007/s11001-010-9109-6.

88. Okal, E.A.; Synolakis, C.E. Source discriminants for near-field tsunamis. *Geophys. J. Int.* **2004**, *158*, 899–912.

The Tsunami Vulnerability Assessment of Urban Environments through Freely Available Datasets: The Case Study of Napoli City (Southern Italy)

Ines Alberico, Vincenzo Di Fiore, Roberta Iavarone, Paola Petrosino, Luigi Piemontese, Daniela Tarallo, Michele Punzo and Ennio Marsella

Abstract: The analysis of tsunami catalogues and of data published on the NOAA web site pointed out that in the Mediterranean basin, from 2000 B.C. to present, about 480 tsunamis occurred, of which at least a third involved the Italian peninsula. Within this framework, a GIS-aided procedure that takes advantage of spatial analysis to apply the Papathoma Tsunami Vulnerability Assessment model of urban environments is presented, with the main purpose of assessing the vulnerability of wide areas at spatial resolution of the census district. The method was applied to the sector of Napoli city enclosed between Posillipo Hill and the Somma-Vesuvio volcano because of the high population rates (apex value of 5000 inh/km^2) and potential occurrence of hazardous events such as earthquakes, volcanic eruptions and mass failures that can trigger tsunamis. The vulnerability status of the urban environment was depicted on a map. About 21% of the possibly inundated area, corresponding with the lowlands along the shoreline, shows a very high tsunami vulnerability. High vulnerability characterizes 26% of inundable zones while medium-low vulnerability typifies a wide area of the Sebeto-Volla plain, ca 800 m away from the shoreline. This map represents a good tool to plan the actions aimed at reducing risk and promoting resilience of the territory.

Reprinted from *J. Mar. Sci. Eng.* Cite as: Alberico, I.; Di Fiore, V.; Iavarone, R.; Petrosino, P.; Piemontese, L.; Tarallo, D.; Punzo, M.; Marsella, E. The Tsunami Vulnerability Assessment of Urban Environments through Freely Available Datasets: The Case Study of Napoli City (Southern Italy). *J. Mar. Sci. Eng.* **2015**, 3, 981–1005.

1. Introduction

The tsunami was considered as a secondary hazard included with earthquake [1,2] or volcanic eruption archives [3,4]. The first tsunami catalogue of the Eastern Mediterranean Sea was published by Antonopoulos *et al.* [5] and only at the end of the 1990s, thanks to the results of the European projects GITEC (1992–1995) and GITEC TWO (1998–2001), a database of tsunamis that occurred in the Mediterranean basin was published.

The analysis of available data [6–8] revealed that from 2000 B.C. to present, about 476 tsunamis occurred in the Mediterranean Sea; as far as their causes are concerned, 380 were triggered by earthquakes and only 20 by volcanic eruptions, the remaining were ascribed to mass failure, earthquakes, landslides, volcanic eruptions or unknown causes.

These natural events become dangerous when the tsunamis hit a densely inhabited area, where they can pose hazards and turn into disasters. Risk is defined by UNISDR [9] as the combination of the probability of occurrence of an event and its negative consequences. Mostly for geophysical risks [10–17], the notation proposed by UNESCO [18] and Fournier d'Albe [19], Risk = Hazard × Vulnerability × Exposure, is adopted. Hazard is the likely frequency of occurrence of a dangerous event in a fixed future time, Exposure measures people, property, systems, or other elements present in hazard zones that are thereby subject to potential losses, Vulnerability is the proportion of lives or goods likely to be lost, and accounts for the features of a system or asset that make it susceptible to the damaging effects of a hazard. This definition identifies vulnerability as a characteristic of the element of interest (community, system or asset) which does not depend on its exposure [9]. Ultimately, the notation points out that the risk can be reduced both by lowering exposure and acting on vulnerability.

Traditionally, the vulnerability of a territorial system affected by tsunami events was assumed to be invariable within the flood zone. e.g., [20,21]. Recent papers, thanks to the analysis of damages caused by tsunamis in many areas around the world, evidenced that tsunami vulnerability changes within the flood zone in response to several parameters mainly reflecting the features of urban environment [22–27].

Starting from this idea, to evaluate the territorial vulnerability, Papathoama et al. [22] developed the Papathoma Tsunami Vulnerability Assessment Model (PTVA). They built a database recording information on building features, sociological, economic, environmental and physical data and assigned to each a weight to quantify its contribution to the vulnerability assessment. Two indices, the building vulnerability and human vulnerability, defined as weighted average of data recorded into the database were assessed to show into maps the spatial and temporal variability of tsunami vulnerability over the inundated area [22,24,28].

Dall'Osso et al. [29] improved the PTVA Model, renamed PTVA-3, introducing a new set of attributes, related to the water intrusion effects, and the use of the Analytic Hierarchy Process (AHP) to evaluate the building vulnerability. This multi-criteria approach limits the subjective ranking and displays the contribution of the single attribute to the overall buildings vulnerability.

In this paper, the PTVA-3 model was partly modified to implement it in a Geographic Information System (GIS), as a procedure to assess the Relative

Vulnerability Index (RVI) of an urban environment (*i.e.*, of wide zones with high building density) without losing the information associated with individual buildings, to which the procedure had originally been applied. The implemented workflow takes advantage of the experiences gained in the areas recently affected by tsunamis and of freely available datasets as the ISTAT (Italian National Institute for Statistics) [30] data and the Google Earth satellite images.

The main outcome of the present research is the urban vulnerability map; it represents a good tool to limit the physical damage of tsunami propagation and to improve social preparedness, essential to favor the development of a resilient community. The method was tested on Napoli megacity because it is exposed to many natural events (seismic, volcanic, landslide) that could trigger tsunamis.

2. Method for Tsunami Vulnerability Assessment at Urban Scale

In the present paper, the PTVA-3 model previously applied to the single buildings of several areas [29,31–34] was partially modified to work at the regional scale, assessing the urban vulnerability of wide areas by using remote sensing and National Institute for Statistics data. The procedure was implemented in a GIS environment to take full advantage of the spatial analysis algorithms for analyzing and integrating data with different spatial resolutions.

Figure 1 shows the input data and the sequence of steps required for calculation of Relative Vulnerability Index (RVI); the ISTAT [30] geodataset and the satellite images represent the main source of data needed to implement the proposed procedure.

The censual district (ISTAT 2011, [30]) was assumed as the smallest geographical feature (spatial resolution) while the associated dataset (number of buildings, typologies of buildings, number of floors, age of building) provided part of the parameters required for the vulnerability assessment. The distance from the shoreline, shape and orientation of buildings, building rows and the presence of seawalls were identified thanks to the analysis of Google Earth satellite images of 2015 (Figure 1). The method adopted to calculate the inundation height is described in the hazard assessment section.

The RVI was defined as the sum of two factors:

$$RVI = 1/3 * WV + 2/3 * SV \qquad (1)$$

The *WV* (Water Vulnerability) is the vulnerability of buildings linked to the impact of water and *SV* (Structural Vulnerability) is the carrying capacity of buildings associated with horizontal hydrodynamic force of water flow. The *WV* factor was calculated by comparing the mean building elevation with the mean inundation height.

160

The mean building elevation at census district scale (B_h) was calculated for the four building categories defined by the ISTAT [30] (buildings with one floor, two floors, three floors, four and more than four floors) and assuming a mean floor height (h_i) of 3 m, as follows:

$$B_h = \frac{\sum_{i=1}^{n} B\, f_n\, h_i}{N} \tag{2}$$

where B represents the number of buildings pertaining to the four ISTAT [30] categories, f_n is the number of floors and N is the total number of buildings located in the census district.

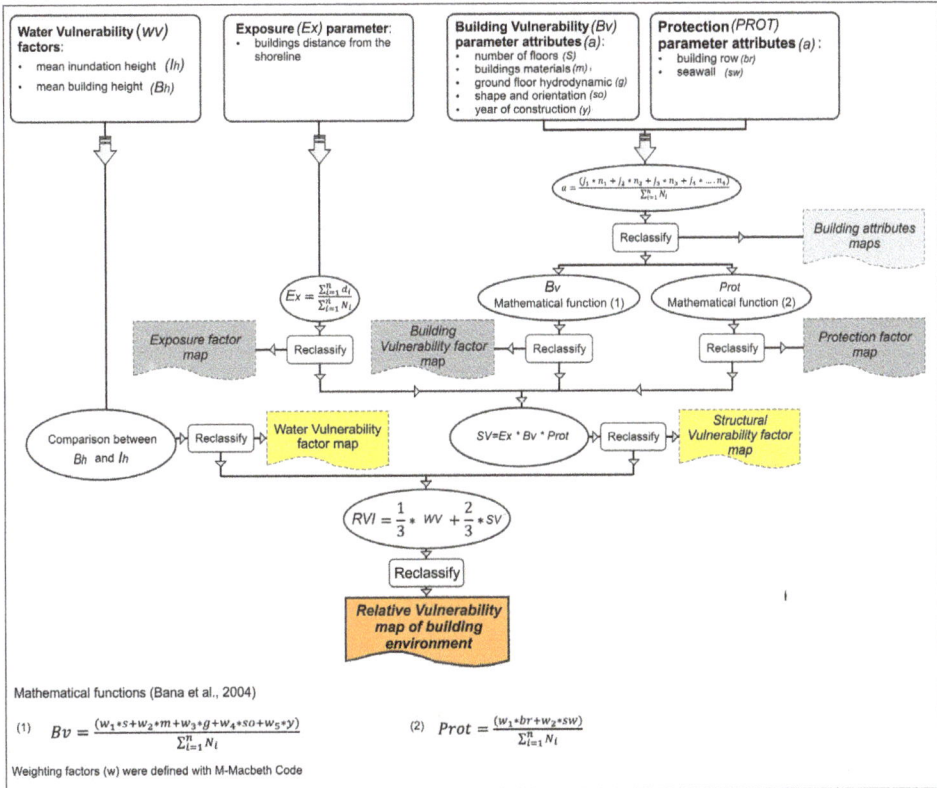

Figure 1. Flow chart. Diagram reporting the procedure for census districts vulnerability assessment: white rounded rectangles enclose the factors and the attributes; ellipses indicate the algorithm adopted; the colored boxes indicate the maps drawn in the different steps of the procedure.

Similarly to the last parameter, the inundation height was calculated, at census district scale, as mean of inundation values assigned to each building from the inundation dataset.

The water vulnerability was evaluated through the comparison of the mean inundation height and the mean building elevation; the census districts with inundation height higher than building elevation were identified as the most vulnerable. Following this premise we identified four categories of vulnerability (1 floor flooded /4 floors, 2 floors flooded /4 floors, 3 floors flooded/4 floors, buildings totally flooded) with a weight varying linearly from 0.25 to 1.

The SV factor of (1), was defined as the product of Ex, Bv and $Prot$ parameters.

The Ex (Exposure) parameter depends on the water depth and on the location of the building.

The Bv (Building Vulnerability) parameter depends on the building features and reflects the resistance to flooding.

The $Prot$ (Protection) parameter expresses the capability of manmade structures to protect the coastal zone from the tsunami during its inland propagation.

For the Ex definition, in accordance with Fritz et $al.$ [35], the assumption that exposure is directly related to the depth of the flow and decreases away from the coast was adopted. This parameter was calculated as average of building distance from the shoreline assigned to each census district.

For the Bv and $Prot$ assessment were used the following relations:

$$Bv = \frac{(w_1 * s + w_2 * m + w_3 * g + w_4 * so + w_5 * y)}{\sum_{i=1}^{n} N_i} \quad (3)$$

$$Prot = \frac{(w_1 * br + w_2 * sw)}{\sum_{i=1}^{n} N_i} \quad (4)$$

where s is the number of floors, m is the building materials, g is the ground floor hydrodynamics, so is the shape and orientation, y is the age of building, br is the building row, sw is the seawall, $w_{1....n}$ are the weighting coefficients of each attribute and N is the total number of buildings pertaining to each census district.

The weighting coefficients (w) reported in Table 1, the only available for the building vulnerability assessment, are those defined by Dall'Osso et $al.$ [29] through the Analytic Hierarchy Process (AHP) accessible in the M-Macbeth environment [36,37]. The attributes s, m, y, were derived from the ISTAT geodataset (2011, [30]) and are linked to geographical element of census districts, whereas g, so, br and sw attributes were acquired through the visual analysis of Google Earth satellite images and are related to the single buildings (Table 2). Since the census district was assumed as the smallest geographical element available for analysis at urban scale, a procedure to summarize the different features of single attributes ($i.e.$, for so there are four classes: regular floor plan inscribed in a square, regular floor plan with the short side oriented perpendicularly to the flow direction, $etc.$) and transfer the information from the buildings to the censual district was defined.

Table 1. Weights of attributes. List of attributes weighting factors used for the *Bv* and *Prot* assessment (from Dall'Osso *et al.* [29]).

Building Vulnerability (*Bv*)		Protection (*Prot*)	
Attributes	**Weighting Factors (%)**	**Attributes**	**Weighting Factors (%)**
s (number of floors)	$w_1 = 100$	*br* (building row)	$w_1 = 100$
m (building materials)	$w_2 = 80$	*sw* (seawall)	$w_2 = 73$
g (ground floor hydrodynamics)	$w_3 = 60$		
so (shape and orientation)	$w_4 = 46$		
y (year of construction)	$w_5 = 23$		

Table 2. Indices of building vulnerability attributes. List of indices assigned to the single attribute influencing the structural vulnerability of a building (*Bv*). The values ranging from 0.25 to 1 indicate an increase of the average building vulnerability ([29]; modified).

Attributes	Indices			
	0.25	**0.50**	**0.75**	**1.00**
number of floors (*s*)	≥4	3	2	1
building materials (*m*)	reinforced concrete	double brice	single brice (tuff)	timber
ground floor hydrodynamics (*g*)	open plan	50% open plan	not open plan but many windows	not open plan
shape and orientation (*so*)	regular floor plan inscribed in a square or with the short side oriented perpendicularly to the direction of flow	regular floor plan, which does not fit into a square, with the long side oriented perpendicularly or diagonally to the direction of flow	irregular complex floor plan, with the short side oriented perpendicularly to the direction of flow	irregular complex floor plan, with the long side oriented perpendicularly or diagonally to the direction of flow
year of construction (*y*)	before 1982	between 1962 and 1981	between 1919 and 1961	after 1919

Three main steps were followed: (a) the *so, br, g* and *sw* attributes were identified for the single building; (b) these values were successively ranked and to each class an index j_i ranging between 0.25 and 1 was assigned (Table 2), where the higher index accounts for the higher vulnerability; (c) these indices were introduced in the following formula to calculate the value of each attribute as arithmetic mean:

$$a = \frac{(j_1 * n_1 + j_2 * n_2 + j_3 * n_3 + j_4 * \ldots . n_4)}{\sum_{i=1}^{n} N_i} \tag{5}$$

where (*a*) is one of the seven attributes listed above (for example: shape and orientation), (*n*) symbolizes the number of buildings with specific features (for example: regular floor plan inscribed in a square, regular floor plan with the short side oriented perpendicularly to the flow direction, *etc.*) and N is the number of buildings pertaining to each census district. The procedure is used to define the average of ISTAT [30] attributes as well, because the single census district has different classes for each attribute type (*i.e.*, for the number of floors).

The criteria used to define the j_i indices are:

- For the "*s*" attribute - starting from the idea that multi-floors buildings have a good possibility of not being completely submerged by water and could promote the vertical evacuation, the lowest vulnerability was assigned to buildings with four and more than four floors and the highest to buildings with one floor (Table 2).
- For the "*m*" attribute - the reinforced concrete buildings resist the tsunami effects better than the other building typologies (Table 2).
- For the "*g*" attribute - the survey of damages to buildings as a consequence of several tsunamis, points out that buildings with openings on the ground floor suffered a lower damage than the other buildings typologies [38] (Table 2).
- For the "*so*" attribute - the buildings with long rectangular shape or "L" shaped with the main wall oriented perpendicular to the flow direction showed high vulnerability [24,39]. The buildings with the lowest damages are those having a shape fitting into a square, because they react equally to the flows notwithstanding the direction of impact [40]. Furthermore, the degree of vulnerability decreases progressively from buildings diagonal to perpendicular to parallel to the coastline [24].
- For the "*y*" attribute - since the older a building the higher is its vulnerability, the elapsed time from its construction was used as indicator of good/worst condition (Table 2).

- For the "*br*" attribute - the number of structures located between each building and the coastline is one of the most important protection factors against the tsunami impact. The lower the value of this attribute, the higher is vulnerability (Table 3).
- For the "*sw*" attribute - Darlymple and Kriebel [38] stated that buildings protected by enclosure walls were significantly less damaged (Table 3).

Moreover, for the formulae (3) and (4) some simplifications were needed at the scale adopted in the present work. In detail, the sub-parameter age of building, displaying the state of preservation of buildings, substitutes the foundation type and the preservation condition considered by Dall'Osso *et al.* [29] because this data is not available at census district scale. The attribute "seawall" was not considered because this information was too detailed for an analysis at urban scale.

3. The Case Study of the City of Napoli

Napoli megacity, with about one million people, has one of the highest population rates of all of Europe (population rate reaching apex values of 5000 inh/km^2). The long history of this city can be read by both punctual architectures and large zones displaying the historical transition from ancient to modern time. We can read the signs of the first settlements at Echia Mount (Pizzofalcone), the site of the first Greek inhabitants of Parthenope and Neapolis; the Roman town; the modern city, which grew in size and population density; Napoli as capital in Byzantine age and the urban zones built after the Second World War [41]. Today, these territorial transformations result in a complex urban texture and space configuration that schematically divide the city into five macro-zones. The old town includes all districts sourced until the end of last century. The consolidated urban zones coincide with the hills bordering the core of the city where the population has moved from the overcrowded central urban area after the First World War. The western area includes the districts falling inside the Campi Flegrei volcanic area, whose original agricultural vocation was gradually transformed into industrial and then tertiary activities in the last decade. The suburbs include the northern and eastern zones of the city, the former as a symbol of the urban distress characterizing the megacity while the latter as a symbol of degradation mainly linked to the progressive abandon of the factories once well developed in these zones [41].

This urban landscape, representing about 78% of Napoli city, was conditioned by the morphology (Figure 2). It is quite various, since several monogenetic volcanic edifices like Nisida Island, La Pietra, Monte Spina and Fossa Lupara characterize the area. The highest peak of the city is the Camaldoli hill (454 m), bordered to the west by Pianura (175 m a.s.l.) and Soccavo (100 m a.s.l.) inland plains, and gradually dipping towards the Agnano-Bagnoli coastal plain. The Camaldoli hill passes to the East to the Vomero area, where the narrow tuffaceous belt of the Posillipo hill

(150 m a.s.l) starts, and gently dips towards E-NE in the Sebeto-Volla plain [42]. This morphology is linked to the growth of Napoli city in a complex multi-source active volcanic area sandwiched between the Campi Flegrei volcanic field and the Somma-Vesuvio district. It is situated in a large graben, broken into smaller horst and graben structures, connected to the Pliocene-Quaternary Tyrrhenian margin evolution [42–47]. In particular, the western sector of the city is set on the horst-type structure of Campi Flegrei caldera, the eastern one is located in a smaller graben, bounded to the east by another horst which hosts the Somma-Vesuvio volcano. The substratum of the city is made up of volcanic rocks, marine and alluvial deposits in response to regional, volcano tectonics and sea level changes occurred during the Holocene [48].

Neapolitan Yellow Tuff (NYT, 15 ky, [49]) forms the framework of most of the hills in Napoli and is overlain by the fall products of the last 10 ka explosive activity of Campi Flegrei. The same products are found in sparse boreholes of the Agnano, Bagnoli-Fuorigrotta, Pianura and Soccavo plains, with a mean thickness of 30 m [50,51]). The Agnano plain, generated by a volcano-tectonic collapse following the Agnano-Monte Spina eruption (4.1 ka [52]), has a poly-crateric morphology linked to several eruptive events [53]. In the subsurface also marine and swampy deposits are found, related to the volcano-tectonic evolution which occurred in the middle of the Holocene, when the northern half of the caldera was uplifted, while the southern half remained under the sea level to form the present bay of Pozzuoli [54]. The Bagnoli-Fuorigrotta volcano-tectonic plain is filled with the products of the post-NYT explosive activity. In the southern part of the Bagnoli-Furigrotta plain, as in the Agnano plain, the volcanic products are interbedded to marine deposits [48,50,51].

Several drill holes performed in the eastern part of the city, corresponding to the Sebeto-Volla Holocene coastal plain, encountered the effusive and explosive products of Somma-Vesuvio and the explosive products of Campi Flegrei piled up during the last 15 ka. The stratigraphic successions of over one hundred drill holes were investigated [51,55] and made it possible to deduce the presence of marine and alluvial deposits embedded to volcanic products. In the western sector of Sebeto-Volla plain prevails the NYT that thins toward the eastern sector, where lava flows and pyroclastic flow deposits of the Somma-Vesuvio eruptions are more common.

This complex geological and morphological on-set coupled with several sources of hazard (earthquakes, landslides, pyroclastic flows) that could trigger the tsunamis as cascade events and the intense urbanization make the Napoli city a good test area for the RVI assessment. For the case study area, firstly the tsunami hazard zone was defined by applying a deterministic approach based on a worst case scenario [56] and then the urban vulnerability was assessed.

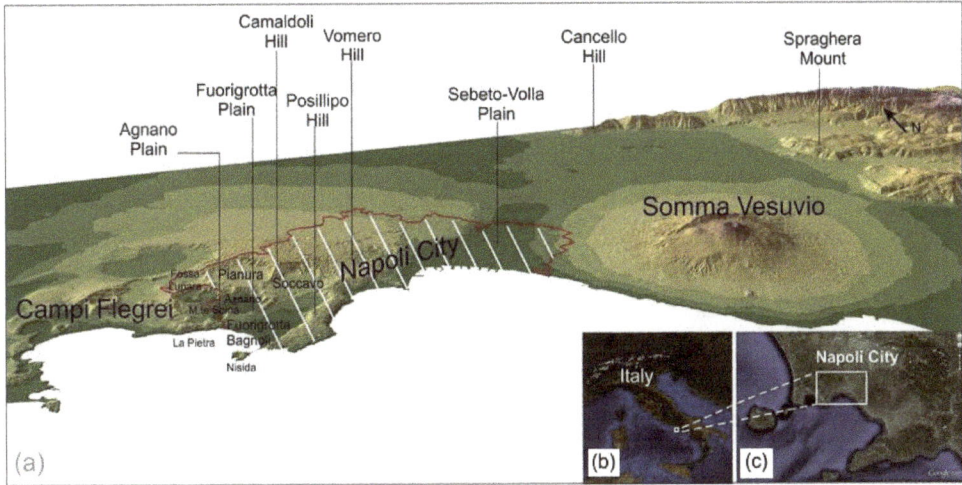

Figure 2. Location of study area. In the 3D view the morphology of area close to the Napoli municipality is reported (red line) (**a**); The Napoli city is drawn at national (**b**) and regional (**c**) scale.

3.1. Tsunami Hazard in Napoli Gulf

The occurrence of past tsunami events in the Napoli Bay is testified to in historical sources. In his *Geographica* (source in Buchner, [57]), the Greek writer Strabo (64 B.C.–21 A.D.) gives the description of a sudden collapse of a sector of the Ischia island with an associated tsunami wave aged before the Greek period and recently dated between ~3 ka B.P. and 2.4 ka B.P. [58]. The possible consequences of this event were assessed by Tinti *et al.* [59], suggesting that this could have produced a tsunami that propagated in the Gulf of Napoli in 8 to 15 minutes time and reached the present city site with a water elevation of 20 m.

The mass failures may be responsible for the generation of tsunami waves [60]; the largest events may have an effect at the scale of the whole Mediterranean basin and bear a high tsunamigenic potential regardless of the water depth at which they occur. Landslides exceeding 10 km^3 in volume can still bear significant tsunamigenic potential at regional scale, and therefore their probability of occurrence should be carefully considered.

Tsunamis linked to the volcanic activity of Somma-Vesuvio are rare, nevertheless, the historical documents report the occurrence of anomalous waves in the Gulf of Napoli not only in correspondence of the largest explosive eruptions (79 A.D., 1631) but also of the low-size eruptive events of 14 May 1698, 17 May 1813 and 4 April 1906 [61,62]. Tinti *et al.* [62] analyzed through numerical simulations the sea perturbation provoked by the lightest component of the pyroclastic flows travelling down the flank of the Vesuvio volcano. They pointed out that the size of tsunamis that could reach Napoli is moderate and that they would produce effects similar to those triggered by the pyroclastic flows of the 1631 eruption [63–66].

Tsunami hydrodynamic modeling includes three main steps: generation, propagation and inundation [67]. The thorniest issue in tsunami calculation is not deep water propagation, but rather the point where the waves are generated and the point where the waves run into shallow water and then onto land. The first part is critical to assess the intrinsic level of hazard, the second is important to map site-specific hazard [68].

The present work mainly deals with the inland propagation of water. The results of numerical model applied by Tinti *et al.* [59] that explores the consequences of the Ischia Debris Avalanche (IDA) which occurred in historical times, was used as input data to assess the maximum inland inundation. A giant sea wave triggered by the flank collapse of Ischia Island may have affected all the coasts of Ischia and of the Gulf of Napoli. It can be considered an extreme event for the Napoli Bay, in fact those occurred in historical times were characterized by a lower wave height [59,61–65,69]. The mechanism that has led to overload-oversteepening, and thus to failure, is the volcano-tectonic uplift of the Mt Epomeo and the recurrent seismicity of the area may ultimately have triggered the mass failure [58].

Figure 3. Topographic barriers and areas prone to inundation. The zones with slope exceeding 5° (brown color) and those prone to inundation (white color, slope lower than 5°) are shown. The dashed lines indicate the areas more than 1200 m away from the shoreline.

168

This event, which occurred about 36 km away from Napoli city, may have determined a maximum wave height of about 20 m along the coast of the study area. The identification of areas likely to be flooded was carried out assuming that the first wave reaches the maximum height and keeps the free horizontal surface condition. This hypothesis, as a first approach to the problem, acts for safety advantage and completely neglects the loss of energy to which the mass of water undergoes during propagation on the mainland. Accepting this hypothesis, the inland tsunami propagation was defined through a run up formula $\left(I = \frac{R_u}{\tan\alpha}\right)$ [70], where "I" is the maximum inland inundation, R_u is the maximum water height calculated by Tinti *et al.* [59] and α is the slope characterizing the coastal zones of the study area. In order to consider the influence of inland morphology on the maximum inland propagation [71–74], a slope zoning of coastal zone was performed from a detailed Digital Elevation Model (DEM) [75]. Firstly a Boolean map was drawn to point out the area with high spatial probability to be inundated (white zones in Figure 3), and those with slope exceeding 5° that represent topographic barriers against the tsunami inland propagation (dark brown in Figure 3); in fact, since a value of $\alpha = 5°$ and a water elevation of 20 m, deduced for IDA [29], were taken into account the maximum depth is about 200 m. For the Sebeto-Volla and Chiaia zones a mean slope of 1° and 2° was respectively calculated and used in the run-up formula to delineate the maximum inundation depth (Figure 4). Moreover, the inundation height (I_h) was assessed by applying the relation $I_h = W_h - T_h$, where W_h is the water height and T_h is the topographic height [70,76], the inundability becomes zero when topographic height is larger than the water height.

3.2. Urban Vulnerability of Napoli City

Starting from the method described in section 2, in the following the criteria adopted to define the role of attribute, parameters and factors necessary to calculate the RVI for urban areas of Napoli city are explained.

For the number of floors (s), the lowest vulnerability was assigned to the buildings with four and more than four floors and the highest to buildings with one floor considering that several floors of a multi-floor building could be safe from inundation and consequently host the people climbing upstairs from the lower floors (Table 2, Figure 5A1).

As far as the building materials (m) are concerned, Napoli is characterized by reinforced concrete and masonry buildings; a level of vulnerability higher than reinforced concrete was assigned to masonry because of its excellent resistance to vertical stresses, but poor ability to react to horizontal forces (Table 2, Figure 5A2).

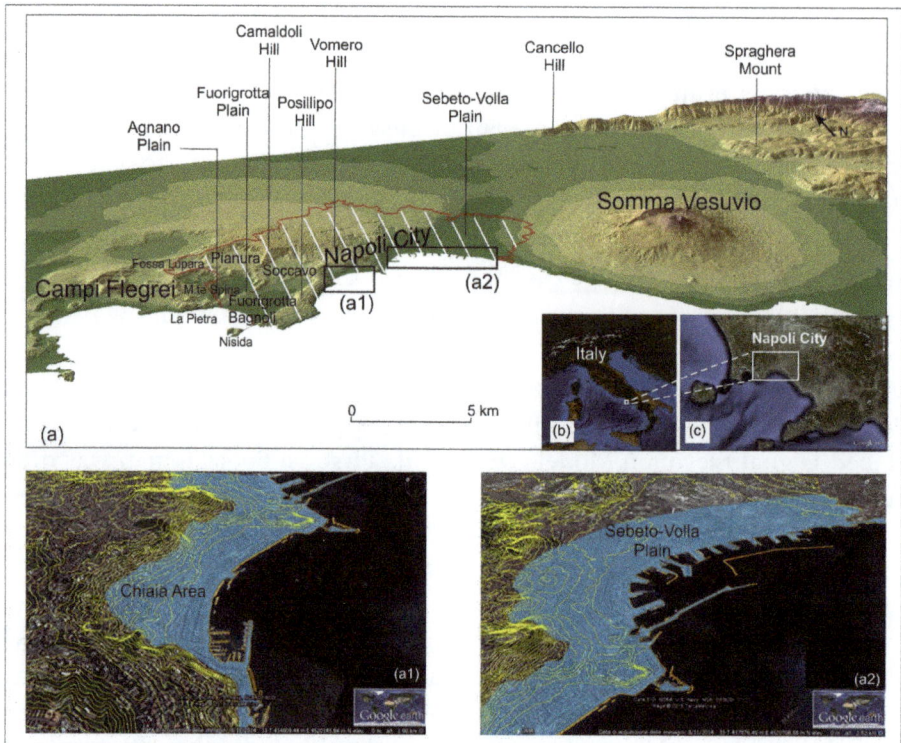

Figure 4. Location of zones that could be reached by a tsunami. The black boxes in (**a**) indicate the two inland zones prone to tsunami inundation. In (**a₁**) and (**a₂**) the light blue zones point out the maximum tsunami inland propagation and the green lines represent the topographic contours with 10 m interval. The Napoli city is drawn at national (**b**) and regional (**c**) scale.

As to the ground floor hydrodynamics (g), buildings with openings on the ground floor showed lower damage than the other building typologies. In the study area, there are few buildings displaying these features, therefore a unique index was assigned to all buildings (Table 2, Figure 5A3).

For the shape and orientation of buildings (so), an index value that increases from buildings inscribed into a square to buildings with irregular floor plan and with the long side normal to the direction of flow was defined (Table 2, Figure 5A4).

For the year of construction (y), an index taking into account the age of the building (Table 2, Figure 5A5) was calculated.

170

Table 3. Indices of protection attributes. List of indices assigned to the single attribute influencing the level of protection of a building (*Prot*). The values from 0.25 to 1 indicate an increase of the average building vulnerability ([29], modified).

Attributes	Indices			
	0.25	0.50	0.75	1.00
building row (*br*)	>VII	IV–VI	II–III	I
seawall (*sw*)	height exceeding 3 m	height between 1.5 and 3 m	height between 0 and 1.5 m	Absent

For building row (*br*), the number of structures located between each building and the coastline is one of the most important protection factors against the impact of tsunami. For the "Porto di Napoli" and the area close to "Via Marina", the buildings are spread over the territory and are directly exposed to the tsunami propagation; the maximum vulnerability was assigned to them. To each building located at the back of Via Marina a decreasing vulnerability was assigned in response to the idea that moving away from the coast the buildings are protected by an increasing numbers of building rows (Table 3, Figure 5B1).

For seawall (*sw*), in the Gulf of Napoli there are just two breakwaters with rectangular shape with the main wall perpendicular to the flood. The height of these protections was used to calculate the related index value, the higher the protection the lower is the vulnerability. The *sw* was the only attribute defined at a scale larger than the census district. The area exposed to tsunamis was divided into three sectors: (a) the zone without seawall protection; (b) the zone protected by a seawall with a mean height of 1.20 m a.s.l.; (c) the zone protected by a seawall with mean height of 0.9 m a.s.l. (Table 3, Figure 5B2).

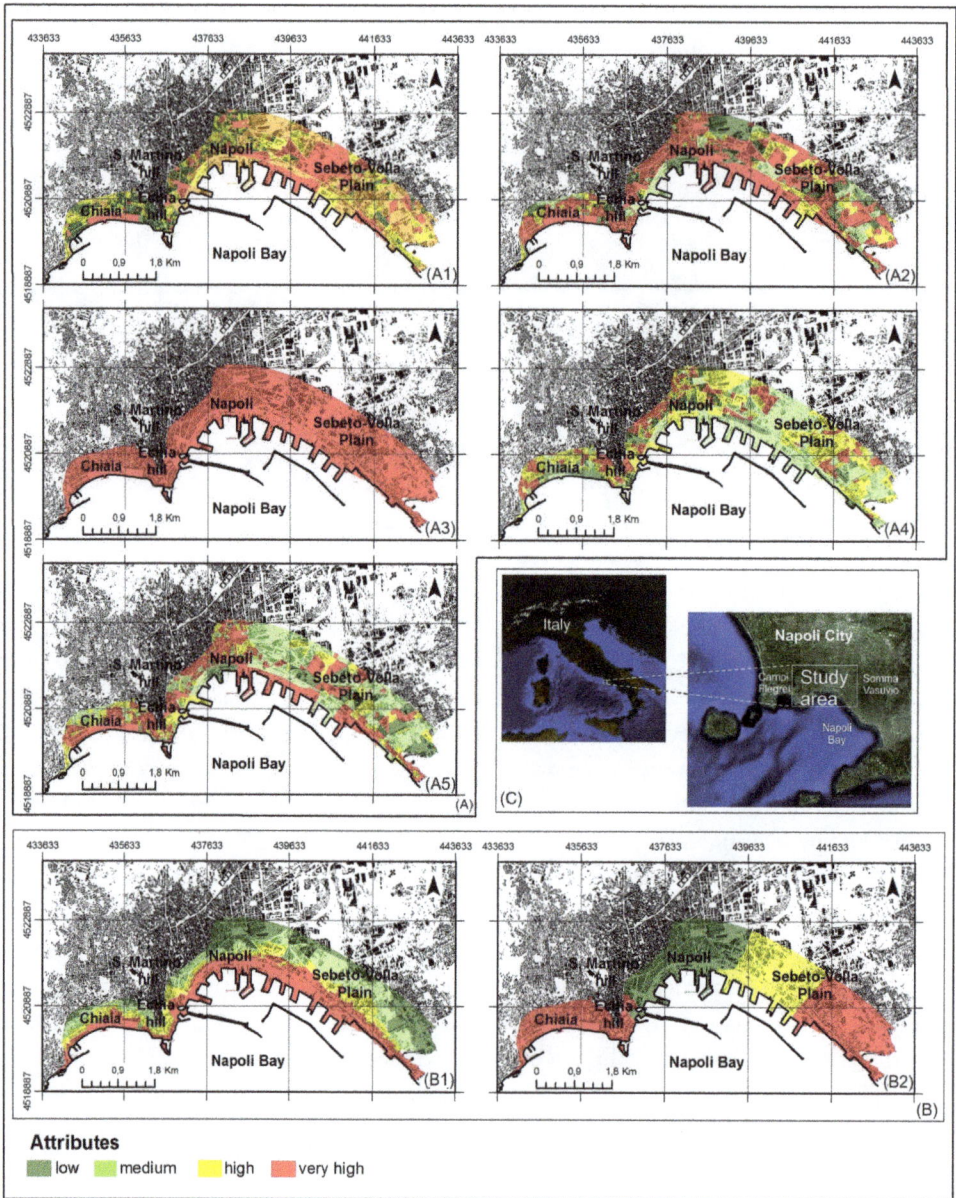

Figure 5. Attributes maps. Frameworks displaying the spatial distribution of building vulnerability (A1–A5) and protection factors (B1-B2): (**A1**) number of floors; (**A2**) buildings material; (**A3**) capability of the ground floor to resist to water flow pressure; (**A4**) shape and the orientation of building; (**A5**) year of building; (**B1**) buildings row; (**B2**) seawall protection; In the picture (**C**) the framework of the study area is shown.

The attributes were ranked into several classes and labeled with specific indices to figure out their contribution to the vulnerability assessment at census scale (Table 4). The ranking of values was performed according to the natural break method; it identifies groups and patterns of data in a set by defining the separation points among the frequency values and minimizing the variance in each class [77]. These values and the weighting factors (w_i) were entered in equations (3) and (4) to define the Bv and $Prot$ parameters and consequently the Sv factor of equation (1).

Moreover, to easily and entirely visualize the spatial distribution of the attributes, the sub-parameters and factors used in the RVI assessment were depicted into easily readable maps.

Table 4. List of attributes and parameters structural vulnerability assessment. The attributes and parameters were classified into four vulnerability classes.

Parameters and Attributes	Range			
	Low (1)	Medium (2)	High (3)	Very high (4)
Building vulnerability (Bv)	1.00–2.25	2.25–2.83	2.83–3.45	3.45–4.00
Number of stories (s)	0.00–0.35	0.35–0.45	0.45–0.79	0.79–1.00
Building materials (m)	0.00–0.32	0.32–0.54	0.54–0.68	0.68–0.75
Ground floor hydrodynamics (g)				1.00
Shape and orientation (so)	0.00–0.54	0.54–0.72	0.72–0.89	0.89–1.00
Year of construction (y)	0.00–0.58	0.58–0.81	0.81–0.93	0.93–1.00
Protection (Prot)	0.00–1.43	1.43–2.30	2.30–3.14	3.14–4.00
Building row (br)	0.00–0.35	0.35–0.59	0.59–0.82	0.82–1.00
Seawall (sw)		0.5	0.75	1.00
Exposure (Ex)	1200–885	885–590	590–295	⩽ 295

4. Results and Discussion

The main goal of this study was to draw a map showing the vulnerability of a wide urban environment to the occurrence of tsunamis thought as a tool for the territorial planning. In the following the spatial distribution of all attributes and parameters used to assess the tsunamis vulnerability of Napoli city is recalled.

4.1. Attributes Maps

In Figure 5A,B the spatial distribution of attributes required to define the building vulnerability (frames A1 to A5) and protection parameters (frames B1 and B2) are respectively shown.

The frames from A1 to A5 point out some differences between the western part of the city, which represents the oldest suburb, and the eastern one, mainly characterized by industrial activities.

The frame A1 refers to the number of floors attribute. It shows that the western area has a medium-low vulnerability (yellow to green color) due to the presence of many buildings with four or more floors; the eastern area, on the counterpart, is characterized by buildings with few floors (orange to red color = medium-high vulnerability).

The frame A2 shows the distribution of type of material attribute, and a high vulnerability results for the western part of the city as a consequence of the wide presence of masonry buildings. The eastern zone is characterized by the presence of both reinforced concrete and masonry buildings; this condition causes a level of vulnerability variable and proportional to the type of buildings characterizing the single census district.

The frame A3 depicts the distribution of hydrodynamics of the ground floor attributes; a unique value of high vulnerability was assigned to the whole study area because most buildings have the ground floors closed by thick walls.

The frame A4 provides information on the distribution of shape and orientation of buildings attribute. The western part of the study area is more vulnerable than the eastern one, because in the former the buildings with the long side normal to water flow prevail.

The frame A5 shows the spatial distribution of the year of construction attribute; the census districts with low vulnerability (shades of green) are mostly made up of edifices built after 1982 and are concentrated in the eastern area.

The frame B1 outlines the areas with a different number of building rows attributes. Four zones with a level of vulnerability decreasing proportionally to the increasing distance from the shoreline are identified. The highest vulnerability pertains to the buildings located in the first row facing the shoreline, conversely the buildings far away from the shoreline (green color) are protected from the tsunami effect by a high number of building rows.

The frame B2 points out the protection role of seawalls. The western and the eastern zones have the highest vulnerability because of the absence of a seawall, whereas the central area shows a good level of protection due to the presence of two man-made protection structures.

4.2. The Bv, Prot and Ex Parameters Maps

The maps of the single attributes were spatially combined, following the relation (3) and (4), to outline the distribution of *Bv* and *Prot* (Figure 6 A,B) parameters into two new maps. Figure 6A shows the vulnerability of urban environment in response to buildings features. Most of the investigated area shows a medium to high level of vulnerability, mainly due to the diffused presence of large masonry buildings. In detail, the low vulnerability of the western zone is linked to the number of floors and the medium-high vulnerability to shape and orientation and year of construction, whereas the eastern sector shows opposite conditions.

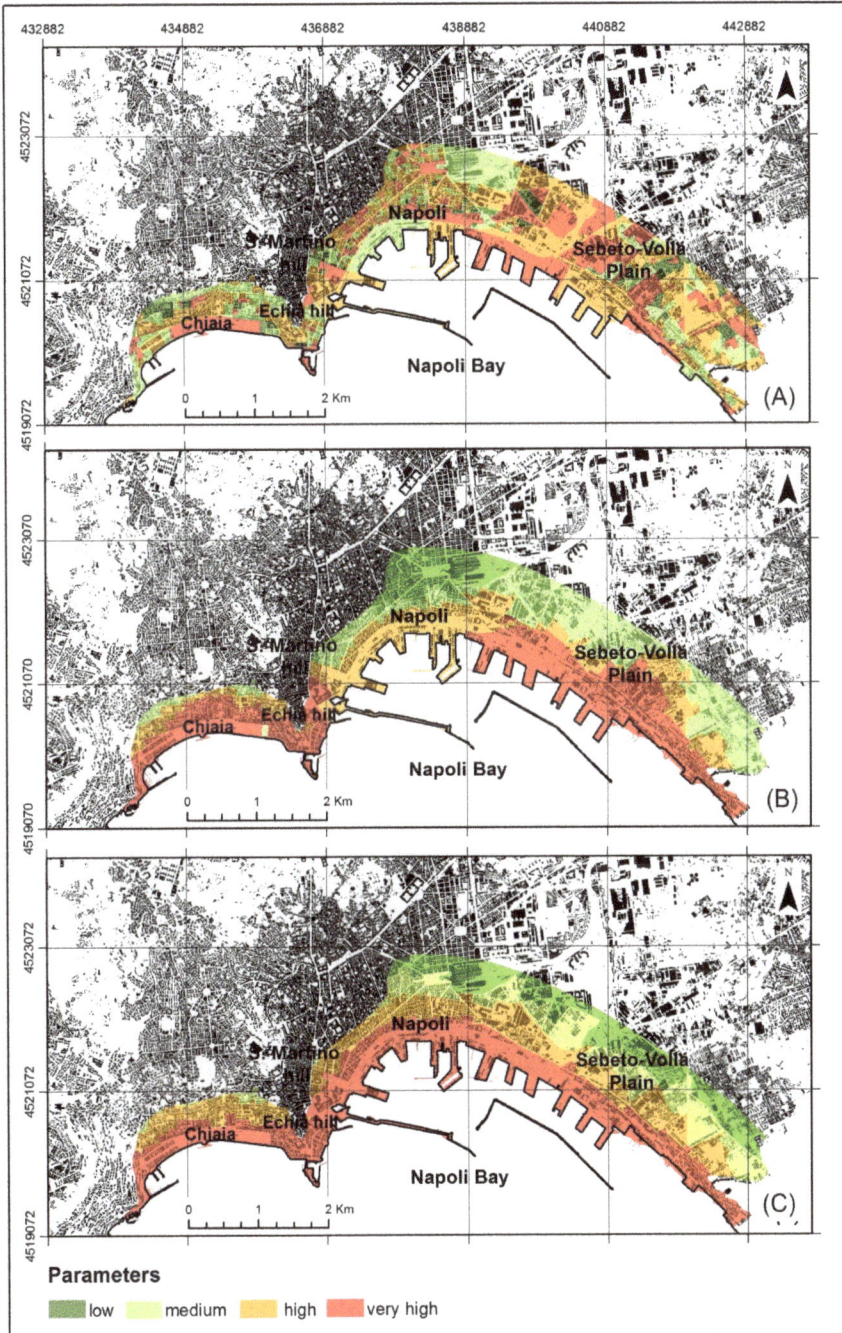

Figure 6. Parameters maps. The maps show the distribution of classes for building vulnerability (**A**), protection (**B**) and exposure (**C**) parameters.

Figure 6B summarizes the effects of number of building rows and seawall attributes on the protection parameter assessment. The map points out that the more a vulnerable area is close to the shoreline and that at the same distance from it the central area of the city is less vulnerable (orange color) than the two side zones (red color) due to the presence of protection barriers. Yellow and green colors characterize the zones far from the shoreline.

Figure 6C depicts the contribution of water flow (*Ex* parameter) at the site where the building is located; since the pressure applied to the building decreases with distance from the coastline, the census district vulnerability decreases with increasing distance from it.

4.3. The Sv and Wv Factors Maps

The structural vulnerability map (Figure 7A) zones the territory through the combined effect of buildings vulnerability, protection and exposure parameters. The highest *Sv* values characterize the Chiaia area close to the shoreline and the eastern sector of Sebeto-Volla plain. A sea protection favored the occurrence of a medium vulnerability class in the central sector of the city. Moving away from the shoreline medium and low structural vulnerability characterizes all the territory. In detail, the percentage of area pertaining to the four *Sv* classes evidenced that 36% of the inundated area is characterized by low vulnerability, 48% by high and very high vulnerability and only 16% is characterized by medium values.

The territorial zoning of water vulnerability factor (*Wv*) evidences that the buildings close to the shoreline, where the inundation height is the highest, are the most vulnerable (red color in Figure 7B). This factor is linked to the magnitude of the tsunami event simulated to define a vulnerability scenario. Considering a water elevation of 20 m [59], all the buildings along the shoreline are submerged by the tsunami wave propagation. The Figure 7B points out the different extent of water vulnerability classes. In the Sebeto-Volla Plain the area displaying the highest vulnerability (red zone) extends inland for about 1 km while in the western sector this class has an extent of about 400 m. The distributions of medium and low water vulnerability classes are similar, occupying 14 and 8% of inundated zone, respectively.

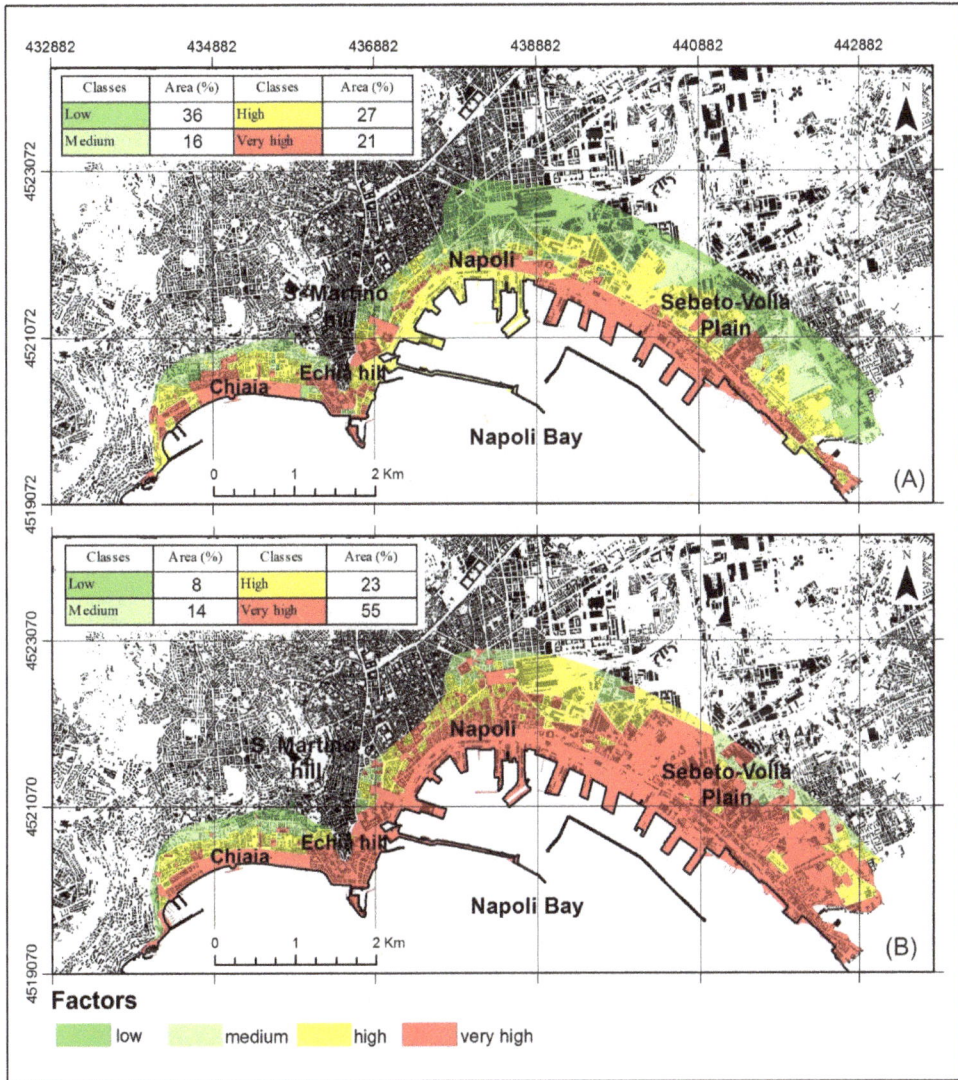

Figure 7. Maps of vulnerability factors. In **A** the vulnerability of the carrying capacity of the buildings structures (Structural Vulnerability) is depicted while in **B** the vulnerability of buildings as a consequence of their contact with tsunami waves (Water Vulnerability) is shown.

4.4. The RVI Map

The Relative Vulnerability Index map summarizes the contribution of structural and water vulnerability factors. In fact, in the territorial zoning shown in Figure 8 several sectors of coastal zones display medium structural vulnerability (Figure 8) are classified as very highly vulnerable in the *RVI* map thanks to the contribution of the *Wv* factor. The coastal zones show the highest vulnerability of the whole study area, with a depth that increases moving from Chiaia to Sebeto-Volla plain. A similar spatial distribution characterizes the other vulnerability classes evidencing the key role of morphology. The gradient of area close to the narrow plain zone of Chiaia and the slopes of San Martino hill prevent the propagation of the tsunami that remains confined within the flat zone. In contrast, the gentle slope of Sebeto-Volla plain favors the propagation of tsunamis for a depth exceeding 1 km.

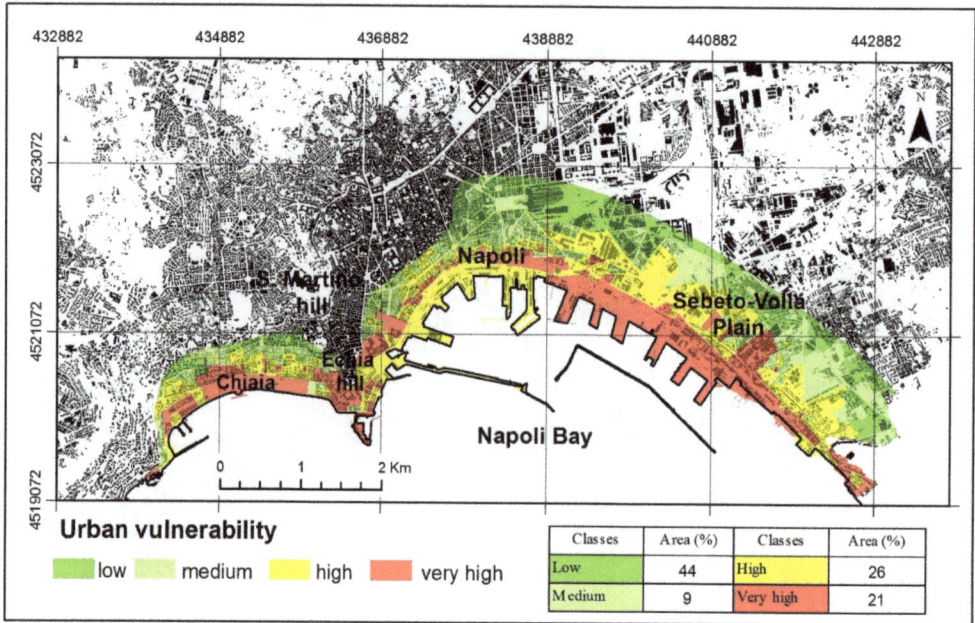

Figure 8. The Relative Vulnerability Index Map of the study area. The colors from red to green indicate a decreasing level of vulnerability.

5. Conclusions

This paper proposes a workflow able to define the tsunami vulnerability of wide urban zones taking advantage of freely available datasets as the ISTAT [30] urban dataset and the regularly up-dated Google Earth satellite images. The census district was assumed as the smallest geographical element for RVI (Relative Vulnerability index) assessment, which summarizes the contribution of two factors, the Structural Vulnerability (*Sv*) and the Water Vulnerability (*Wv*). The application of the procedure to the censual districts ensures a speedy data processing and a fast map visualization and can hence be successfully used both on the long time for territorial planning and on the short time for possible crisis management.

The method was applied to the case study of Napoli megacity because it is exposed to both volcanotectonic earthquakes, sourcing at Campania active volcanoes, and to tectonic high magnitude earthquakes, sourcing in the Apennine chain. Mostly the landslide cascade events of those earthquakes, implying the remobilization of huge quantities of rocks and their interaction with sea-water, could cause tsunami events. Apart from the high hazard related to tsunamis, the peculiar topography of the city, forcing urban zones to grow in the flat areas along the coastline, makes it even more vulnerable. Furthermore, the intense urbanization and the high population rates contribute to increase the risk of the city and make it worth a specific investigation of urban vulnerability in case of a future tsunami event.

The analysis make it possible to infer that all the buildings located within the narrow land belt along the coastline display a high vulnerability mainly because of their proximity to the sea and of the main occurrence of masonry building. In particular:

- A high value of Sv characterizes the Chiaia area close to the shoreline and the eastern sector of the Sebeto-Volla plain; sea protections favored the presence of a medium vulnerability class in the central sector of the city and moving away from the shoreline the medium and the low structural vulnerability prevails.

- A high value of Wv typifies a wide area of the Sebeto-Volla Plain, at depth more than 1 km, whereas in the western part Wv is controlled by the presence of a 70 m high hill, located only at 400 m from the coastline, that prevents the inland propagation of the tsunami wave.

- The same spatial distribution of high vulnerability class is shown by the medium and low vulnerability classes that are progressively located farther from the coastline.

- RVI map, as the sum of Sv and Wv, shows that very high and high vulnerability (47%) typifies the areas along the shoreline for a depth of about 300 m in the western sector and of about 700 m in the eastern one. This final RVI map reflects the magnitude of the type-event on which the procedure is based, that for the Napoli city case study is a high magnitude event (IDA, [59]), but one of the advantages of the workflow implemented in a GIS environment is to make it possible to routinely change the tsunami scenario, by changing the Wv factor, and rapidly redraw a new urban vulnerability map.

The RVI map here proposed is conceived as a basic tool for territorial planning in areas exposed to tsunami hazard, where laws and regulations are needed to promote the use of a building code with the aim of increasing durability of residential buildings and reducing the economic losses in case of a destructive event. In the present situation, for Napoli city, there is no tsunami risk program, either in the short term, finalized to a possible crisis management, or in the long term, for increasing resilience in population.

The possibility to identify the tsunami building vulnerability at the censual district scale, supplied by the noticeable flexibility of the GIS-aided procedure here coded, offers vital tools for spatial planning purposing to:

- identify the buildings with features that make them suitable for the vertical evacuation.

- hypothesize for the single areas a building code to reduce the future damage linked to a tsunami phenomenon (e.g., habitable space in building structures must be elevated above the regulatory flood elevation, walls and partitions are required to break away so as not to induce excessive loads on the structural frame, a stable refuge for the vertical evacuation should be provided, *etc.*).

- address resource allocation towards specific interventions aimed at reducing the urban vulnerability.

Acknowledgments: The authors wish to thank three anonymous reviewers whose suggestions greatly improved the manuscript. A grateful acknowledgement goes to the editor of the journal for his valuable suggestions and kind availability.

Author Contributions: Ines Alberico and Paola Petrosino outlined the working methodology and assessed the areas exposed to tsunami hazard. Roberta Iavarone and Luigi Piemontese carried out the detailed characterization of urban environment in the light of exposure to tsunamis hazard. Vincenzo Di Fiore, Daniela Tarallo and Michele Punzo dealt with historical accounts and reviewed scientific literature to assess tsunami hazard in the Napoli area. Ines Alberico and Ennio Marsella coordinated the research group.

Conflicts of Interest: The authors declare no conflict of interest.

References

1. Mallet, R. Catalogue of recorded earthquakes from 1606 B.C. to 1850 A.D. Available online: http://www.biodiversitylibrary.org/item/46640#page/5/mode/1up (accessed on 26 August 2015).
2. Baratta, M. *I terremoti d'Italia*; Fratelli Bocca: Torino, Italy, 1901.
3. Mercalli, G. *Vulcani e fenomeni vulcanici in Italia*; Francesco Vallardi: Milano, Italy, 1883.
4. Baratta, M. *Il Vesuvio e le sue eruzioni. Dall'anno 79 d.C. al 1896*; Società Editrice Dante Alighieri: Roma, Italy, 1897.
5. Antonopoulus, J. Data for investigating tsunami activity in the Mediterranean Sea. *Int. J. Tsu. Soc.* **1990**, *8*, 38–53.
6. Tinti, S.; Maramai, A.; Graziani, L. The New Catalogue of Italian Tsunamis. *Nat. Hazards* **2004**, *33*, 439–465.
7. Pasaric, M.; Brizuela, B.; Graziani, L.; Maramai, A.; Orlic, M. Historical tsunamis in the Adriatic Sea. *Nat. Hazards* **2012**, *61*, 281–316.
8. NOAA, (National Geographic Data Center). Available online: http://www.ngdc.noaa.gov/hazard/tsu_db.shtml (accessed on 16 December 2012).
9. *Terminology on Disaster Risk Reduction*; UNISDR United Nations International Strategy for Disaster Reduction: Genevra, Swetzerland, 2009.

10. Sleiko, D. Considerations on seismic risk. *Ann. Geofis.* **1993**, *36*, 169–175.

11. Glade, T. Landslide occurrence as a response to land use change: A review of evidence from New Zealand. *Catena* **2003**, *51*, 297–314.

12. Petrosino, P.; Alberico, I.; Caiazzo, S.; Dal Piaz, A.; Lirer, L.; Scandone, R. Volcanic risk and evolution of the territorial system in the volcanic areas of Campania. *Acta Vulcanol.* **2004**, *16*, 163–178.

13. Marzocchi, W.; Selva, J.; Sandri, L. Probabilistic Volcanic Hazard Assessment and Eruption Forecasting: The Bayesian Event Tree approach. In *Conception, Verification and Application of Innovative Techniques to Study Active Volcanoes*; INGV: Bologna, Italy, 2008; pp. 77–91.

14. Pesaresi, M.; Gerhardinger, A.; Haag, F. Rapid damage assessment of build-up structures using VHR Satellite Data in Tsunami Affected Area. *Int. J. Remote Sensing* **2007**, *28*, 3013–3036.

15. Lirer, L.; Petrosino, P.; Alberico, I. Volcanic hazard assessment at volcanic fields: The Campi Flegrei case history. *J. Volcanol. Geotherm. Res.* **2001**, *101*, 55–75.

16. Lirer, L.; Petrosino, P.; Alberico, I. Hazard and risk assessment in a complex multi-source volcanic area: the example of the Campania region, Italy. *Bull. Volcanol.* **2010**, *72*, 411–429.

17. Grezio, A.; Gasparini, P.; Marzocchi, W.; Patera, A.; Tinti, S. Tsunami risk assessments in Messina, Sicily, Italy. *Nat. Hazards Earth Syst. Sc.* **2012**, *12*, 151–163.

18. *Consultative Meeting of Experts on the Statistical Study of Natural Hazards and Their Consequences*; UNESCO: Paris, France, 1972; pp. 1–11.

19. Fournier d'Albe, E.M. Objective of volcanic monitoring and prediction. *J. Geol. Soc.* **1979**, *136*, 321–326.

20. Papadopoulos, G.A.; Dermentzopoulos, T. A tsunami risk management pilot study in Heraklion, Crete. *Nat. Hazards* **1998**, *18*, 91–118.

21. Ganas, A.; Nikolaou, E.; Dermentzopoulos, Th.; Papadopoulos, G.A. A GIS for tsunami risk mapping and management. In Presented at NATO Advanced Research Workshop "Underwater Ground Failures on Tsunami Generation Modeling, Risk and Mitigation", Istanbul, Turkey, 23–26 May 2001; pp. 249–252.

22. Papathoma, M.; Dominey-Howes, D. Tsunami vulnerability assessment and its implications for coastal hazard analysis and disaster management planning, Gulf of Corinth, Greece. *Nat. Hazards Earth Syst. Sci.* **2003**, *3*, 733–747.

23. Leone, F.; Denain, J.C.; Vinet, F.; Bachri, S. *Analyse spatiale des dommages au bâti de Banda Aceh (Sumatra, Indonésie), Contribution à la connaissance du phénomène et à l'élaboration de scénarios de risque tsunami, in LAVIGNE F. & PARIS R. (dir.)*; Rapport scientifique du programme Tsunarisque (2005–2006) sur le tsunami du 26 décembre 2004 en Indonésie—Délégation Interministérielle pour l'aide Post-Tsunami (DIPT) Ambassade de France en Indonésie, Paris; CNRS: Paris, France, 2006; pp. 69–94.

24. Dominey-Howes, D.; Papathoma, M. Validating a Tsunami Vulnerability Assessment Model (the PTVA Model) Using Field Data from the 2004 Indian Ocean Tsunami. *Nat. Hazards* **2007**, *40*, 113–136.

25. Reese, S.; Cousins, W.J.; Power, W.L.; Palmer, N.G.; Tejakusuma, I.G.; Nugrahadi, S. Tsunami vulnerability of buildings and people in South Java—Field observations after the July 2006 Java tsunami. *Nat. Hazards Earth Syst. Sci.* **2007**, *7*, 573–589.

26. Garcin, M.; Desprats, J.F.; Fontaine, M.; Pedreros, R.; Attanayake, N.; Fernando, S.; Siriwardana, C.H.E.R.; de Silva, U.; Poisson, B. Integrated approach for coastal hazards and risks in Sri Lanka. *Nat. Hazards Earth Syst. Sci.* **2008**, *8*, 577–586.

27. Omira, R.; Baptista, M.A.; Miranda, J.M.; Toto, E.; Catita, C.; Catalao, J. Tsunami vulnerability assessment of Casablanca—Morocco using numerical modeling and GIS tools. *Nat. Hazards* **2010**, *54*, 75–95.

28. Dominey-Howes, D.; Dunbar, P.; Verner, J.; Papathoma, M. Estimating probable maximum loss from a Cascadia tsunami. *Nat. Hazards* **2010**, *53*, 43–61.

29. Dall'Osso, F.; Gonella, M.; Gabbianelli, G.; Withycombe, G.; Dominey-Howes, D. A revised (PTVA) model for assessing the vulnerability of buildings to tsunami damage. *Nat. Hazards Earth Syst. Sci.* **2009**, *9*, 1557–1565.

30. ISTAT. Available online: http://www.istat.it/it/ (accessed on 26 August 2015).

31. Dall'Osso, F.; Gonella, M.; Gabbianelli, G.; Withycombe, G.; Dominey-Howes, D. Assessing the vulnerability of buildings to tsunami in Sydney. *Nat. Hazards Earth Syst. Sci.* **2009**, *9*, 2015–2026.

32. Dall'Osso, F.; Dominey-Howes, D. Public assessment of the usefulness of "draft" tsunami evacuation maps from Sydney, Australia—Implications for the establishment of formal evacuation plans. *Nat. Hazards Earth Syst. Sci.* **2010**, *10*, 1739–1750.

33. Dall'Osso, F.; Maramai, A.; Graziani, L.; Brizuela, B.; Cavalletti, A.; Gonella, M.; Tinti, S. Applying and validating the PTVA-3 Model at the Aeolian Islands, Italy: Assessment of the vulnerability of buildings to tsunamis. *Nat. Hazards Earth Syst. Sci.* **2010**, *10*, 1547–1562.

34. Tarbotton, C.; Dominey-Howes, D.; Goff, J.R.; Papathoma-Kohle, M.; Dall'Osso, F.; Turner, I.L. *GIS-Based Techniques for Assessing the Vulnerability of Buildings to Tsunami: Current Approaches and Future Steps*; The Geological Society: London, UK, 2012.

35. Fritz, M.; Borrero, J.C.; Synolakis, C.E.; Yoo, J. 2004 Indian Ocean tsunami flow velocity measurements from survivors videos. *Geophys. Res. Lett.* **2006**, *33*.

36. Bana e Costa, C.A.; Chargas, M.P. An example of how to use MACBETH to build a quantitative value model based on qualitative value judgements. *Eur. J. Oper. Res.* **2004**, *153*, 323–331.

37. Bana e Costa, C.A.; Da Silva, P.A.; Correia, F.N. Multicriteria Evaluation of Flood Control Measures: The case of Ribeira do Livramento. *Water Resour. Manag.* **2004**, *18*, 263–283.

38. Dalrymple, R.A.; Kriebe, D.L. Lessons in Engineering from the Tsunami in Thailand. *Bridge* **2005**, *35*, 4–13.

39. Warnitchai, P. Lessons Learned from the 26 December 2004 Tsunami Disaster in Thailand. In Proceedings of the 4th International Symposium on New Technologies for Urban Safety of Mega Cities in Asia, Singapore, Asia, 18–19 October 2005.

40. Van de Lindt, J.W.; Gupta, R.; Garcia, R.A.; Wilson, J. Tsunami bore forces on a compliant residential building model. *Eng. Struct.* **2009**, *31*, 2534–2539.

41. PUC of Napoli city. 2004. Available online: http://www.comune.napoli.it/flex/cm/pages/ServeBLOB.php/L/IT/IDPagina/1022 (accessed on 26 August 2015).

42. Carrara, E.; Iacobucci, F.; Pinna, E.; Rapolla, A. Gravity and magnetic survey of the Campanian Volcanic Area, Southern Italy. *Boll. Geofis. Teor. Appl.* **1973**, *15*, 39–51.

43. Carrara, E.; Iacobucci, F.; Pinna, E.; Rapolla, A. Interpretation of gravity and magnetic anomalies near Naples, Italy using computer techniques. *Bull. Volcanol.* **1974**, *38*, 458–467.

44. Ippolito, F.; Ortolani, F.; Russo, M. Struttura marginale tirrenica dell'Appennino Campano: Reinterpretazione di dati di antiche ricerche di idrocarburi. *Mem. Soc. Geol. Ital.* **1973**, *7*, 227–250.

45. D'Argenio, B.; Pescatore, T.S.; Scandone, P. Schema geologico dell'Appennino Meridionale. *Accad. Naz. Lincei Quad.* **1973**, *183*, 49–72.

46. Finetti, I.; Morelli, C. Esplorazione sismica a riflessione dei golfi di Napoli e Pozzuoli. *Boll. Geofis. Teor. Appl.* **1974**, *16*, 175–222.

47. *Geologia e geofisica del Sistema geometrico dei Campi Flegre*; AGIP: Rome, Italy, 1987.

48. Cinque, A.; Rolandi, G.; and Zamparelli, V. L'estensione dei depositi marini olocenici nei Campi Flegrei in relazione alla vulcano-tettonica. *Boll. Soc. Geol. Ital.* **1985**, *104*, 327–348.

49. Deino, A.L.; Orsi, G.; de Vita, S.; Piochi, M. The age of the Neapolitan Yellow Tuff caldera-forming eruption (Campi Flegrei caldera—Italy) assessed by 40Ar/39Ar dating method. *J. Volcanol. Geotherm. Res.* **2004**, *133*, 157–170.

50. Di Vito, M.A.; Isaia, R.; Orsi, G.; Southon, J.; de Vita, S.; D'Antonio, M.; Pappalardo, L.; Piochi, M. Volcanism and deformation since 12,000 years at the Campi Flegrei caldera (Italy). *J. Volcanol. Geotherm. Res.* **1999**, *91*, 221–246.

51. Alberico, I.; Petrosino, P.G.; Zeni, F.; Lirer, L. Geocity: A drill hole geodatabase as a tool to investigate geological hazard in Napoli urban area. In Proceedings of the EUG Joint Assembly, Nice, France, 6–11 April 2003.

52. de Vita, S.; Orsi, G.; Civetta, L.; Carandente, A.; D'Antonio, M.; Deino, A.; di Cesare, T.; Di Vito, M.A.; Fisher, R.V.; Isaia, R.; *et al.* The Agnano-Monte Spina eruption (4100 years BP) in the restless Campi Flegrei caldera (Italy). *J. Volcanol. Geotherm. Res.* **1999**, *91*, 269–301.

53. Di Girolamo, P.; Ghiara, M.R.; Lirer, L.; Munno, R.; Rolandi, G.; Stanzione, D. Vulcanologia e Petrologia dei Campi Flegrei. *Boll. Soc. Geol. Ital.* **1984**, *103*, 349–413.

54. Cinque, A.; Aucelli, P.P.C.; Brancaccio, L.; Mele, R.; Milia, A.; Robustelli, G.; Romano, P.; Russo, F.; Russo, M.; Santangelo, N.; *et al.* Volcanism, tectonics and recent geomorphological change in the bay of Napoli. *Suppl. Geogr. Fis. Dinam. Quat.* **1997**, *3*, 123–141.

55. Bellucci, F. Nuove conoscenze stratigrafiche sui depositi vulcanici del sottosuolo del settore meridionale della Piana Campana. *Boll. Soc. Geol. Ital.* **1994**, *113*, 395–420.

56. Eckert, S.; Jelinek, R.; Zeurg, G.; Krausmann, E. Remote sensing-based assessment of tsunami vulnerability and risk in Alexandria, Egypt. *Appl. Geog.* **2012**, *32*, 714–723.

57. Buchner, G. Eruzioni vulcaniche e fenomeni vulcano-tettonici di età preistorica e storica nell' isola di Ischia. In *Tremblements de terre, éruptions volcaniques et vie des hommes dans la Campanie antique*; Albore-Livadie C.: Naples, Italy, 1986.

58. De Alteriis, G.; Insinga, D.; Morabito, S.; Morra, V.; Chiocci, F.L.; Terrasi, F.; Lubritto, C.; Di Benedetto, C.; Pazzanese, M. Age of submarine debris avalanches and tephrostratigraphy offshore Ischia island, Tyrrhenian sea, Italy. *Mar. Geol.* **2010**, *278*, 1–18.

59. Tinti, S.; Latino Chiocci, F.; Zaniboni, F.; Pagnoni, G.; de Alteriis, G. Numerical simulation of the tsunami generated by a past catastrophic landslide on the volcanic island of Ischia, Italy. *Mar. Geophys. Res.* **2011**, *32*, 287–297.

60. Roger Urgeles, R.; Camerlenghi, A. Submarine landslide softhe Mediterranean Sea: Trigger mechanisms, dynamics, and frequency-magnitude distribution. *J. Geophys. Res.* **2013**, *118*, 2600–2618.

61. Tinti, S.; Saraceno, A. Tsunamis related to volcanic activity in Italy. In *Tsunamis in the World*; Springer: Houten, The Netherlands, 1993; Volume 1, pp. 43–63.

62. Tinti, S.; Maramai, A. Catalog of tsunami generated in Italy and in Cote d'Aruz, France a step towards a unified calalogue of tsunamis in Europe. *Ann. Geophys.* **1996**, *39*, 6.

63. Relazione dell'incendio del Vesuvio del 1631. In *Documenti inediti*; Riccio, L., Giannini, e Figli, Eds.; ETH-Bibliothek Zürich: Napoli, Italy, 1631; pp. 513–521.

64. Braccini, G.C. *Dell'incendio fattosi sul Vesuvio a XVI Dicembre MDCXXXI e delle sue cause ed effetti, con la narrazione di quanto 'e seguito in esso per tutto marzo 1632 e con la storia di tutti gli altri incendi, nel medesimo monte avvenuti, Secondino Roncagliolo*; ETH-Bibliothek Zürich: Napoli, Italy, 1631.

65. Giuliani, G. *Trattato del Monte Vesuvio e de' suoi incendi*; ETH-Bibliothek Zürich: Napoli, Italy, 1632.

66. Mormile, J. *L'incendio del Monte Vesuvio, e delle stragi, e rovine, che ha fatto ne' tempi antichi e moderni, infine a 3 di marzo 1632, con nota di tutte le relazioni stampate fino ad oggi del Vesuvio, raccolte da Vincenzo Bove*; ETH-Bibliothek Zürich: Napoli, Italy, 1632.

67. Liu, P.L.F.; Cho, Y.S.; Woo, S.B.; Seo, S.N. Numerical simulations of the 1960 Chilean tsunami propagation and inundation at Hilo, Hawaii. In *Tsunami: Progress in Prediction, Disaster Prevention and Warning*; Springer: Houten, The Netherlands, 1994; pp. 99–115.

68. Ward, S.N.; Day, S. Tsunami Balls: A Granular Approach to Tsunami Runup and Inundation. *Commun. Comput. Phys.* **2008**, *3*, 222–249.

69. Tinti, S.; Pagnoni, G.; Piatanesi, A. Simulation of tsunamis induced by volcanic activity in the Gulf of Naples (Italy). *Nat. Hazards Earth Syst. Sci.* **2003**, *3*, 311–320.

70. Khomarudin, R.M.; Strunz, G.; Ludwig, R.; Zobeder, K.; Post, J.; Kongko, W.; Pranowo, W.S. Hazard analysis and estimation of people exposure as contribution tsunami risk assessment in the West Coast of Eumatra, the South Coast of Java and Bali. *Z. fur Geomorphol.* **2010**, *5*, 337–356.

71. Szczuciński, W.; Chaimanee, N.; Niedzielski, P.; Rachlewicz, G.; Saisuttichai, D.; Tepsuwan, T.; Lorenc, S.; Siepak, J. Environmental and Geological Impacts of the 26 December 2004 Tsunami in Coastal Zone of Thailand—Overview of Short and Long-Term Effects. *Pol. J. Environ. Studies* **2006**, *15*, 793–810.

72. Nakamura, Y.; Nishimura, Y.; Putra, P.S. Local variation of inundation, sedimentary characteristics, and mineral assemblages of the 2011 Tohoku-oki tsunami on the Misawa cost, Aomori, Japan. *Sediment. Geol.* **2012**, *282*, 216–277.

73. Kazuhisa, G.; Koji, F.; Daisuke, S.; Shigehiro, F.; Kentaro, I.; Ryouta, T.; Tomoya, A.; Tsuyoshi, H. Field measurements and numerical modeling for the run-up heights and inundation distances of the 2011 Tohoku-oki tsunami at Sendai Plain, Japan. *Earth Planets Space* **2012**, *64*, 1247–1257.

74. Gopinath, G.; Løvholt, F.; Kaiser, G.; Harbitz, C.B.; Srinivasa Raju, K.; Ramalingam, M.; Singh, B. Impact of the 2004 Indian Ocean tsunami along the Tamil Nadu coastline: Field survey review and numerical simulations. *Nat. Hazards* **2014**, *72*, 743–769.

75. Alberico, I.; Lirer, L.; Petrosino, P.; Zeni, G. Geologia e geomorfologia dell'area urbana di Napoli. In Proceedings of the Geoitalia 2001—Terzo Forum Italiano di Scienze Della Terra, Chieti, Italy, 5–8 settembre 2001; pp. 359–360.

76. Federici, B.; Bacino, F.; Cosso, T.; Poggi, P.; Rebaudengo Landó, L.; Sguerso, D. Analisi del rischio tsunami applicata ad un tratto della costa Ligure. *Geomat. Workb.* **2006**, *6*, 53–57.

77. Jenks, G.F.; Caspall, F.C. Error on choroplethic maps: Definition, measurement, reduction. *Ann. Assoc. Am. Geogr.* **1971**, *61*, 217–244.

Geological and Sedimentological Evidence of a Large Tsunami Occurring ~1100 Year BP from a Small Coastal Lake along the Bay of La Paz in Baja California Sur, Mexico

Terrence A. McCloskey, Thomas A. Bianchette and Kam-biu Liu

Abstract: The importance of small-scale seismic events in enclosed water bodies, which can result in large tsunami waves capable of affecting comprehensive damage over small, geographically-confined areas are generally overlooked, although recognizing the occurrence of such events is a necessary element in adequately assessing the risk of natural hazards at specific locations. Here we present evidence for a probable large localized tsunami that occurred within the Bay of La Paz, Baja California Sur, ~1100 year before present (BP), which resulted in the creation of a shelly ridge at an elevation of ~2 m above mean high water (MHW). This ridge consists of a continuous wedge of poorly mixed marine sands and shells ~50 cm in depth deposited along the entire seaward edge of the lake. The marine shells collected from terrestrial environments around the lake include species from a variety of environments, including offshore species with minimum preferred depths of >13 m. The evidence suggests that this material was likely deposited by a tsunami with a runup of 2–3.6 m above MHW, probably associated with the slumping of an island along the tectonically active eastern edge of the bay.

Reprinted from *J. Mar. Sci. Eng.* Cite as: McCloskey, T.A.; Bianchette, T.A.; Liu, K. Geological and Sedimentological Evidence of a Large Tsunami Occurring ~1100 Year BP from a Small Coastal Lake along the Bay of La Paz in Baja California Sur, Mexico. *J. Mar. Sci. Eng.* **2015**, *3*, 1544–1567.

1. Introduction

Two recent events have focused attention on the hazards associated with supra-regional tsunamis generated by mega earthquakes occurring along major fault lines in the open ocean. On 26 December 2004 a rupture along the fault boundary between the Indo-Australian and Eurasian plates resulted in the 9.1 magnitude Sumatra-Andaman earthquake. The ensuing tsunami, that generated waves up to 30 m high, caused over 283,000 deaths in 14 countries [1,2]. On 11 March 2011, a 9.0 magnitude earthquake associated with a 450 km rupture of the Japan Trench generated a tsunami [3,4] that devastated Japan and achieved wave heights of 3.94 m as distant as the southern coast of Chile [5]. However, the risks associated with locally-generated tsunamis in confined water bodies have generally been ignored [6].

186

One example is the 1958 Lituya Bay, Alaska tsunami. Although this event was associated with a large earthquake on the Fairweather Fault, the proximal cause was a rockslide at the head of the bay that generated a tsunami that removed soil and vegetation up to a height of 524 m [7]. Miller [7] identified five occurrences of giant waves over the last 200 years in Lituya Bay indicating that in tectonically-active areas devastating localized tsunamis can occur over short return intervals, thereby posing large, but underappreciated, hazards for the immediate vicinity. Other large localized tsunamis associated with rockslides into confined bodies of water include events at Spirit Lake, Washington in 1980 (260 m runup) [8] and at the Vajont Reservoir, Italy in 1963 (260 m runup) [9,10]. The lack of focus on tsunami risks is problematic all along Mexico's west coast, but especially throughout the Baja Peninsula where the economies of coastal communities are largely tied to water-based tourism. The risk of localized events is of particular import within the semi-enclosed Bay of La Paz, given the presence of the city of La Paz in the shallow SW corner. Here we present geologic evidence from a small coastal lake on the margins of that bay, suggesting the possible occurrence of a tsunami at ~1100 year before present (BP).

2. Study Site

2.1. Geologic Setting

The Baja California (BC) Peninsula is rocky and dry, separated from the rest of Mexico by the Gulf of California. During the Mesozoic, when it was the western edge of both the continent and the North American plate, the BC Peninsula received volcanic detritus from calc-alkaline volcanoes farther to the east [11]. During the Oligocene and Miocene the volcanoes migrated west onto what is currently the east coast of the Baja Peninsula, depositing the Comondú Formation [11], at first in a shallow marine basin, then onto non-marine fans. Some subduction on the west coast of BC occurred at this point, terminating by ~10 million years ago (mya) [12]. Between 12 and 4 mya, as the North American Plate overrode the East Pacific Rise, Baja California slowly transferred unto the Pacific Plate as the plate boundary moved into the Gulf of California [13–15]. Although this is a right-lateral strike-slip boundary dominated by long transform faults and narrow spreading [16,17], the western edge of BC is part of the surrounding extensional province, characterized by normal faulting. The BC Peninsula as a whole has been being uplifted at ~0.1 m/kyr since 1 mya [18,19].

Our study site is located in the Bay of La Paz, located on the southern third of the peninsula; ~1000 km south of the Mexican/United States border (Figure 1). Hausback [11] identified the La Paz Fault as a major NE-SW fault in the southern section of the peninsula, cutting across the entire peninsula just south of our study site. More recent studies have questioned the existence of this fault, and have instead identified the active normal San José del Cabo Fault as the dominant fault system [16]. North of this fault Coyan and others [17] have identified an array of active faults, the one closet to our site being the normal, northeast-dipping Saltito Fault, with scarp heights ranging from 7 to 14 m, which continues offshore. Saltito's 14 km onshore rupture length corresponds to a magnitude (M) 6.5 earthquake [20]. Researchers working in the Bay of La Paz commonly identify a major buried fault along the western edge of the islands lying along the eastern edge of the bay [21,22]. The La Paz Basin is a half graben, controlled by the east-dipping Centenario Fault along its western edge [17]. Although the majority of regional earthquakes occur along the spreading axes in the center of the Gulf of California, a number of earthquakes >5.0 M have occurred in the Bay of La Paz since 1900 [16,21,22], with magnitude 7.0 events having an estimated occurrence interval of ~1000 years [22].

Figure 1. Location of study site. Major faults on the southern section of the Baja California peninsula, as described by Coyan and others [17], are displayed on Google Earth imagery, with dots marking the downthrown blocks. The dashed line represents an unnamed buried fault inferred by previous studies [21,22]. The red dot marks our study site. Inset displays the location of the Baja California Peninsula in relation to the Pacific coast of Mexico.

Detailed data for post-Wisconsin sea level changes are almost entirely lacking for the Pacific coast of Mexico. The single relevant curve, published in 1969 [23], shows a smooth rise of ~40 m to ~15 m below present sea level from 10,000 to 8000 BP, after which the rate of sea level rise steadily decreases, reaching ~1 m below present sea level by 3000 BP. There are no published records of relative sea level for the La Paz area.

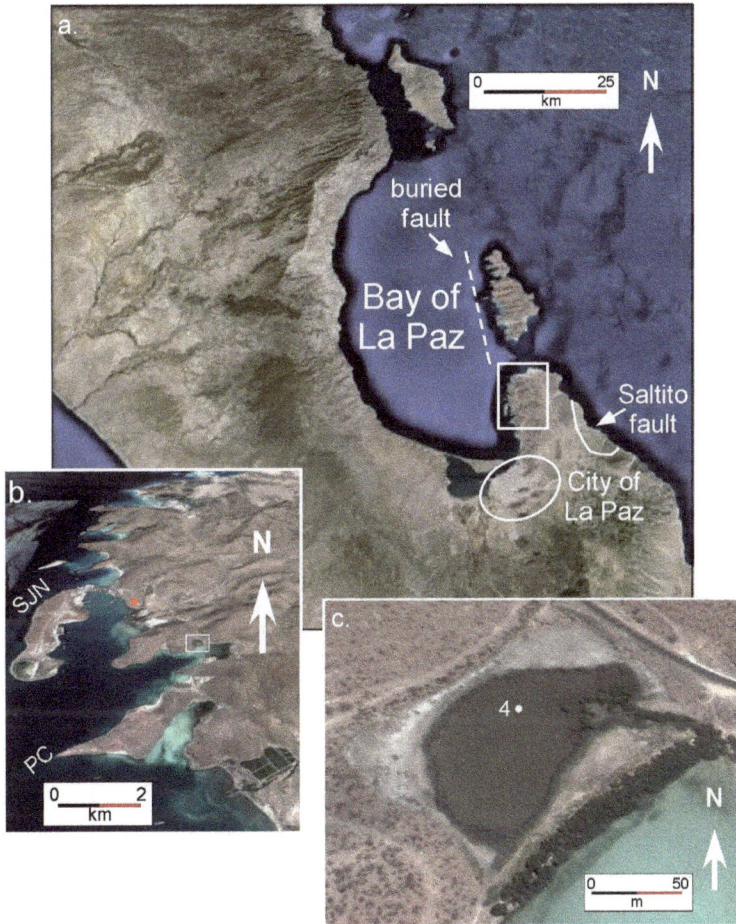

Figure 2. Location of study site. Displayed on Google Earth imagery are (**a**); the Bay of La Paz, the city of La Paz, the Saltito Fault and the buried fault identified by [21,22] as occurring along the western flanks of the islands on the eastern edge of the bay. The area marked by the white square is expanded in (**b**); which displays the location of the two peninsulas surrounding Estero de Bahia Falso, our study site (white square) and the Pichilingue tidal gauge (red square). Estero de Bahia Falso is shown in greater detail in (**c**), with the location of core 4 marked by a white dot.

2.2. Bay of La Paz

Tides in the Bay of La Paz are mixed, mainly semi-diurnal [24], with a maximum tidal range of <2 m [25,26]. The bay is roughly oval with a maximum length of ~75 km N–S and 35 km E–W. The bay opens to the Gulf of California to the east, blocked in the south by several islands bounded on the west by an active fault [21,22], and a peninsula extending into the bay from the south (Figure 2a). The northern section of the bay is relatively deep, with a maximum depth of 420 m in the Alfonso Basin. Water depth in the southern third is <100 m [21,22,27,28]. Surface currents in the northern section are driven by a cyclonic gyre, especially in the fall, while currents are weak in the shallow southern third. Winds are seasonally variable, with weak (⩽5 m/s) southwest winds prevailing during the summer winds, and stronger (⩽12 m/s) northwest winds occur during the rest of the year [29], resulting in a water column that varies from weakly stratified in the winter to highly stratified in the summer [30]. The city of La Paz, the capital of the state of Baja California Sur, occupies the SW corner of the bay, <10 km from our site. The city has a population of >200,000 with an economy based largely on ecotourism and marine activities [31,32].

2.3. Study Site

Our study site is Estero de Bahia Falso, a lake located in a small bay tucked into the convoluted western edge of the southern peninsula, flanked by the 2.5 km long San Juan Nepomuceno (SJN) peninsula to the northwest, and the 1.25 long Punta Colorado (PC) peninsula to the southeast (Figure 2b). The lake is in a small basin surrounded by steep, bare slopes rising to heights of >60 m. The lake is crescent-shaped, ~150 m × 70 m in size, with a maximum depth of <1.0 m, and a surface elevation of ~1.25 m above mean high water (MHW). The lake drains an area of >1 km^2, with larger drainage systems to the northwest and the southeast constraining the catchment area to the immediate vicinity. Salinity was 49 parts per thousand (ppt) at the time of fieldwork in December 2011. The lake is fringed by a thin border of low halophytic shrubs, while red mangroves (*Rhizophora mangle*) form a denser, higher growth along the edge of the bay and black mangroves (*Avicennia germinans*) line the drainage channel that connects the lake to the sea near the lake's eastern end. The lake is separated from the sea by a low, flat ridge of shelly, sandy material which reaches a maximum elevation of ~75 cm above the lake level (~2 m above MWH). This ridge sits on top of a finer-grained beach that slopes to the sea. A salt pan surrounds the lake on the remaining three sides.

The site sits atop a thin alluvial Holocene fill on top of the Comondú Formation that consists of volcanic sandstones and conglomerates, rhyolitic ash-flows, tuffs and andesitic lahars and lava flows [11,31,33], above a Cretaceous basement of metamorphic and plutonic rocks [17]. The short (~2 km) normal north-dipping La Pedrera Fault occurs ~7 km to the northeast, beyond which the Cretaceous basement reaches the surface along the eastern side of the [31,33]. Movement along the Espiritu Santo Fault has resulted in a westward tilting and slow uplift of the peninsula [34]. Estimations of the rate of uplift vary. Calculations based on terrace deposits show a rate of 0.12–0.15 mm yr^{-1} from the last interglacial, while data from an elevated platform suggests little or no uplift since that time. Calculations derived from the formation of notch suggest rates from 0.075 to 0.50 mm yr^{-1} over the last 500–2000 years [34].

The long-term history of human use/disturbance of the area is unknown. However, until the recent development of a tourism-based economy human population levels were low, and the potential for anthropogenic disturbance slight. Material was moved for the construction of a roadway, which passes to the east of the lake, some of which has moved downhill. However the limit of this movement is clear and does not extend into our study areas.

3. Methods

This study focuses on the superficial sediments surrounding the lake, supplemented by data from core 4, taken from a site northeast of the lake's center, approximately 40 m behind the landward edge of the shell-covered beach ridge (Figure 2). The core was extracted from a water depth of ~0.6 m of water by means of a Russian peat borer in seven overlapping sections. All core sections were photographed and described in the field and hermetically sealed before being transported to the Global Change and Coastal Paleoecology Laboratory at Louisiana State University (Baton Rouge, LA, USA), where they were stored in a refrigerated room (4 °C). Surface samples were collected from the lake, the surrounding land surface and the subtidal nearshore area. Elevations and distances were determined by a Nikon range finder (Nikon Corporation, Toyko, Japan) with all locations marked by a hand held Garmin eTrex Venture global positioning system (GPS) unit. Although the listed accuracy of this unit is <15 m, this is the maximum error under poor satellite coverage. Field experience (reoccupying sites) has shown the practical uncertainty to be much less (<3 m). Photos were taken and sketches made of all relevant topographic and geological features. Shells and surface samples were collected from the terrestrial areas surrounding the lake. After opening in the lab, all core sections were photographed and physical properties described. Loss-on-ignition analysis was conducted continuously down the center of each section at 1 cm resolution. Small (~1 cc) samples were weighed wet, dried overnight at 100 °C, then weighed to determine water content (percentage of wet weight), following which they were burned at 550 °C for two hours and weighed to determine organic content (percent dry weight), and finally burned at 1000 °C for one hour and weighed to determine carbonate content (dry weight), following the procedure described by Liu and Fearn [35]. The precise alignment of core segments used to establish final core stratigraphy was based on the matching of the visual, compositional and elemental characteristics of the overlapping sediments. Elemental concentrations were determined by an Innov-X Delta Premium DP-4000 handheld X-ray fluorescence (XRF) unit for all surface samples and at 2 cm resolution down the length of the core (Olympus Corporation, Center Valley, PA, USA). The device analyzes each sample across three frequencies for 30 seconds per frequency producing elemental concentrations in parts per million (ppm) for the following elements: S, Cl, K, Ca, Ti, V, Cr, Mn, Fe, Co, Ni, Cu, Zn, As, Se, Br, Rb, Sr, Zr, Mo, Rh, Pd, Ag, Cd, Sn, Sb, Ba, Pt, Au, Hg, and Pb. We present data for K, Ti, V, Mn, Fe, Co, Zn, Br, and Mo, as representative terrestrial elements. The minimum and maximum error bars (in ppm) for each of these elements for the depth presented (0–35 cm) are: K: 75–140; Ti: 18–34; V: 21–34; Mn: 4–6; Fe: 58–112; Co: 17–26; Zn: 2; Br: 17–23; Rb: 1–2; and Mo: 1–3. The device was calibrated with certified standards NIST 2710a and 2711a.

Tidal amplitude is based on hourly data from a tide gauge located <5 km from our site at Bahia de Pichilingue for the period 15 May 1999 to 3 March 2011. 65,537 of the 103,464 possible hourly readings (63%) are recorded for this period. Due to the missing values we discarded data from months with less than 240 readings (10 days), leaving a set of 95 months. The number of data points per month for the 95 months varied from 252 (once) to 744 (20 times).

Shells and shell fragments are ubiquitous on the shelly ridge fronting the ocean side of the lake, and also sparingly on the back side. Samples of some common types were collected and taken to the laboratory for identification and analysis. Given the immense number of shells encountered and the severely abraded nature of many, identification was performed only on randomly-selected shells, generally the best-preserved (Figure S1). No quantitative assessment of shell assemblages was performed. Shell identification was based on a regional book by Keen [36], plus the assistance of Dr. Emilio Garcia and the Louisiana Malacological Society. Similarly, coral fragments were simply listed as "coral".

Radiocarbon dates were procured from five samples cut from marine shells and from a small subsample (~1 cc) of sediment extracted from a depth of 30 cm in core 4. The latter sample, which was selected in order to date a sedimentary change at the top of the core, was passed through a 63 micron sieve to remove silt and clay. Plant fragments were selected from the remaining material under a dissecting microscope after being washed in de-ionized water. This material was sent, along with the shell samples, to the National Ocean Sciences accelerator mass spectrometry (NOSAMS) laboratory at Woods Hole Oceanographic Institution (Woods Hole, MA, USA), where accelerator mass spectrometry (AMS) radiocarbon dating was performed by means of a 500 kV compact pelletron accelerator. The radiocarbon date for the plant material was calibrated to calendar years using Calib 7.0 and the INTCAL 13 curve, based on the dataset from Reimer and others [37]. The Calib 7.0 program provides a median probability date for each sample, calculated from the probability distribution. This value, which is considered a more reliable calibrated calendar date than the calibrated intercept date [38] is listed in the table and figures. Calibrated shell dates include a correction for the reservoir effect (old marine carbon) through the incorporation of the ΔR value of 253 ± 18 determined for the La Paz area by Frantz and others [39], and the MARINE13 curve.

Due to the extreme spatial heterogeneity of the material comprising the shelly ridge, sediment size was determined visually and reported qualitatively. Both the size of the sediments and the relative percentages of the different size categories (clay, silt, sand, gravel, boulders) varied so dramatically over short distances (both vertically and horizontally) (Figure S2) that quantitative analysis was not deemed as providing useful information.

4. Results

Unlike the uniformly fine-grained sediments of the salt pan which surrounds the eastern, western and northern edges of the lake, the sediments on the ocean side of the lake consist of a mix of material, highly variable both in composition and sediment size. Three areas of particular interest, marked A, B, and C, are highlighted in Figure 3.

Figure 3. Marine deposits along the southern (seaward) edge of the lake. Three areas of interest are marked, as are the locations of the dated shells and core 4.

Figure 4. Area A. Area A is a small section of hillside (**a**) plastered with a shelly mixture of material, including angular boulders and imbricated shells (**b**); The layer sits above a reddish clay (**c**).

Area A is located at the western approach to the small basin that contains the lake. This area is notable for the veneer of mixed material covering a small section of the hillside located above and just landward of a steep (65°–70°) scarp (Figure 4). The sediments are a chaotic jumble of whole shells, disarticulated shells, shell hash and rocks embedded in a predominantly coarse sand matrix (Figure S3). In some sections the shells show evidence of imbrication (Figure 4b and Figure S4). The shells represent species from a spectrum of marine environments. Examples include *Crassotrea* (oysters) that commonly occur in back bays, *Chione californiensis* and *Chionista fluctifraga* that inhabit intertidal and nearshore areas, offshore species, and coral fragments (Table 1, Figure 4). Shell preservation was inconsistent; degree of abrasion ranged from minimal (slight breakage on outside edges of shell, slight erosion of morphological features) to severe breakage. The rocks vary from rounded to extremely angular, indicating large differences in the distance and duration of transport. While the rounded rocks likely originated from the sea, the degree of angularity suggests a terrestrial provenance for some. Sediment varies in size from mud through silt, sand, gravel, cobble, and boulders. The area covered by this mixture is spatially constrained, both laterally and vertically. Maximum lateral extent is ~10 m (~half is shown in Figure 4a), while the vertical extent of the massive shell coverage is ~1.5 m, covering the hillside from ~0.5 m to 2 m above MHW. Less densely packed shells extend the vertical range an additional ~0.5 m upwards. This layer sits above a layer of orange clay (Figure 4c).

Table 1. Name, geographic range and radiocarbon and calibrated dates for identified shells.

Shell Information		Dating Results				
Genus (Subgenus) Species	Geographic Range (Depth)	Lab #	^{14}C Age	Error	2σ Range	Median Probability (Year BP)
Pitar (Pitarella) catharius	West coast of Baja California to Gulf of California (13–80 m)	OS-110978	2020	20	1259–1374	1311
Chione (Chione) californiensis	California to Panama (intertidal to 69 m)	OS-110979	2180	20	1387–1549	1475
Chione (Chionista) fluctifraga	California, Gulf of California, Sonora (intertidal)	OS-110980	2340	25	1560–1765	1660
Glycymeris (Tucetona) strigilata	Gulf of California to Ecuador (13–110 m)	OS-110981	1740	20	951–1130	1032
Barbatia (Cucullaearca) reeveana	Gulf of California to Peru (intertidal to 120 m)	OS-110982	2310	20	1540–1706	1627
Spondylus princeps	Gulf of California, Pacific coast from Baja California Sur to Peru (15–50 m)	NA	NA	NA	NA	NA

There is no apparent sorting, with sediment size exhibiting extreme spatial heterogeneity over very small distances (Figure S2). Boulder-sized rocks are rare or absent (Figure 5c). This ridge reaches a maximum thickness of ~50 cm and sits incongruously on top of a fine-grained surface that slopes down to the bay. The ridge exhibits cliff faces on both the lake and ocean sides (Figure 5a,b). As with the material in Area A, shell preservation was highly variable and identifiable shells represent a variety of marine environments. The top of the ridge is ~2 m above MHW.

Loss-on-ignition analysis indicates that on average the ridge sediments (4.8%) were nearly twice as organic as either the beach (2.9%) or nearshore samples (2.5%), while containing less than half as much carbonate material (11.9%) as either the beach (32.2%) or nearshore samples (26.0%). Elemental concentration values for ridge samples are intermediate between the marine (offshore and beach) and terrestrial (hillslopes) samples for both terrestrial (Ti, V, Mn, Fe, Zn) and marine (Ca, Sr) elements (Figure S5).

Figure 5. Area B. Area B consists of a wedge–shape deposit with a maximum thickness of ~50 cm, consisting of an extremely poorly-sorted mix of marine and terrestrial material. Size varies, although the largest boulders are absent. Clear cliff faces occur on both the seaward and landward edges.

Area B, which extends along the entire front of the lake, is basically a low shelly ridge composed of a mixture of mainly marine sediments. Like the mixture in Area A, the sediments consist of a sand matrix containing a jumble of shells, shell and coral fragments, and rocks, both rounded and angular.

Area C is at the extreme eastern edge of the lake, at the base of a steep (~45°) hillside rising to a height of ~60 m. As in the other two areas, the materials in this location consist of a mixture of shells, shell and coral fragments, and rocks in a sandy matrix (Figure 6). Shells found here include *Glycymeris (Tucentona) strigilata*, whose preferred habitats range from 13 to 110 m [38]. The sediments form a distinct pile, with a sharp cliff face on the northern edge, with a maximum thickness of ~60 cm (Figure 6A). These sediments, marked by a sharp contact, sit on top of a relatively flat surface of fine-grained sediments.

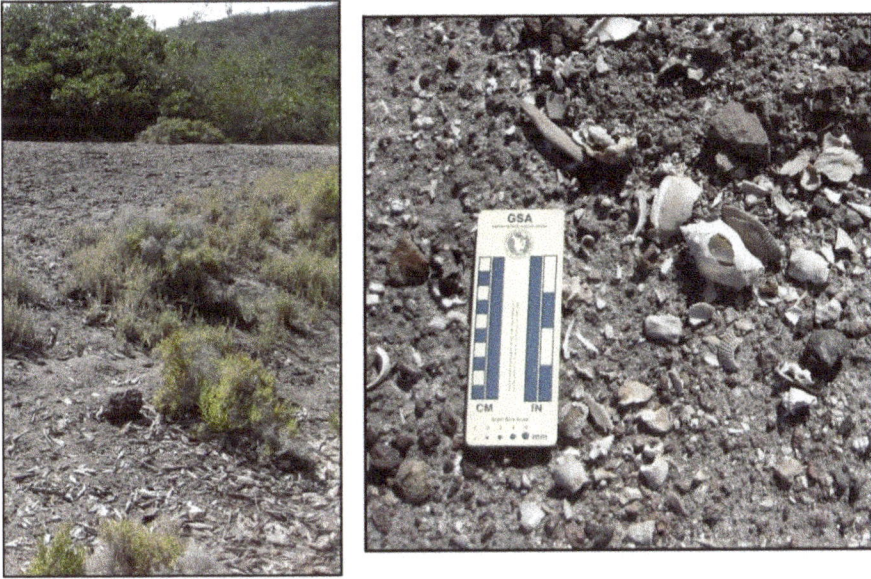

Figure 6. Area C. Area C is a mound of mixed materials, formed at the eastern edge of the lake at the base of a steep slope; thickness is ~60 cm.

Core Stratigraphy

The top of core 4 displays a dramatic stratigraphic change (Figure 7). The top 27 cm consist of a low-organic (6%–10%) brown clay. At 28 cm organic and water contents jump from 9% to 33% and from 47% to 74%, respectively. This marks the stratigraphic change to peat, which dominates the core down to a depth of, ~260 cm (core photo, far left, and the loss-on-ignition graph, center left, Figure 7). The transition at 27 cm (photo, middle) is not only abrupt; it is also marked by irregular deposition with clasts from the underlying peat embedded in the more clastic material (white dashed circle). The geochemical profile also changes at this transition, with the material above 27 cm having distinctly higher concentrations of K, Ti, Mn, Fe, Co, Zn, Rb, and Zr and lower concentrations of Mo and Br than the underlying peat (right), matching the profiles of samples taken from the surrounding hillsides. The plant fragments sampled at 30 cm were dated to 1170 ± 20 ^{14}C year BP, which calibrates to a median date of 1106 cal year BP (Table 2).

Table 2. Radiocarbon and calibrated dates of sediment sample from Core 4.

Name	Type	Lab #	[14]C Age	Error	2σ Range	Relative Probability	Median Probability (Year BP)
CRE 4A 30 cm	Plant/Wood	OS-111567	1170	20	1007–1024	0.07	
					1053–1176	0.93	1106

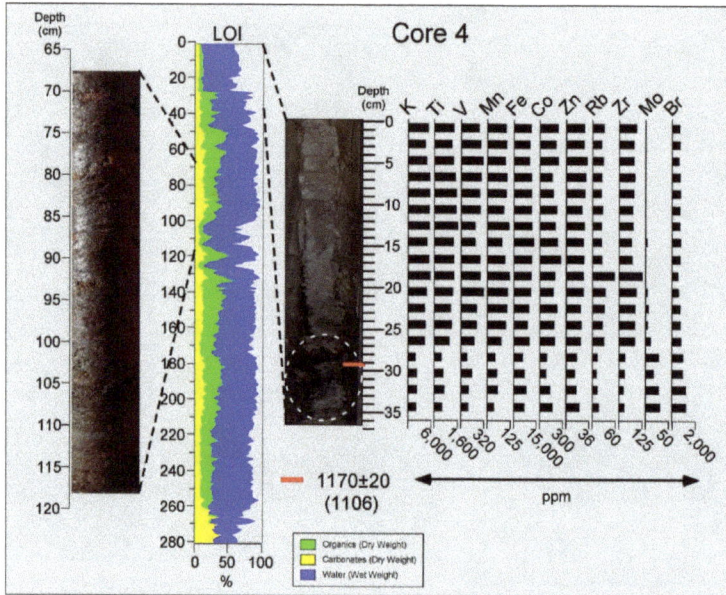

Figure 7. Sediment core 4 from near the center of the lake. Above a calcareous bottom mud from 281 to 262 cm (LOI diagram, center left), the core is predominately organic as shown by the section from 67 to 117 cm (left). Above a boundary at 27 cm, marked by the presence of encapsulated clasts from the underlying material (center right), the sediments abruptly change to a lacustrine mud with an altered geochemical signature (right). The red line marks the depth of the dated sample which returned a median calibrated date of 1106 year BP.

5. Discussion

Areas A, B, and C all contain a mélange of divergent material. Marine elements include shells and corals from the intertidal to deeper offshore water, with the published vertical range of three of the identified shell species beginning at a minimum depth of 13 m (Table 1). The size, angularity, and physical appearance of some rocks occurring within this mélange suggest that the material was derived at least partially from the surrounding hillsides. This is supported by both the geochemical and compositional profiles of the ridge samples, which have elemental concentrations intermediate between the marine and terrestrial samples, and higher organic and lower carbonate percentages than the marine samples (Figure S5).

Changes in either eustatic or relative sea level can potentially explain the presence of this material as being deposited on a beach during a period of higher water. There is no evidence for a former highstand as the single sea level curve for the area [23] indicates that eustatic sea level was never higher than the present.

Recent compilations of uplift-corrected Holocene records from farther north show very similar eustatic records, notably lacking any highstands [40,41].

Relative sea level change is a more likely possibility, given the active tectonic nature of the region. Again, concrete evidence is lacking, as there appears to be no published data concerning seismically-driven vertical motion for the area. In general geologic terms, sudden upward movement at the site does not seem highly likely. The site is situated within the upper edge of the La Paz basin, which being a half graben, is prone to downward movement. The faults in the areas are mainly east-dipping normal faults with vertical motion dominated by downward movement of the hanging walls [17]. Two small faults occur in the vicinity of the study site. Neither is mentioned in a recent, detailed study of the area's faults [17], presumably due to their relative insignificance. The effects of the nearest, the west-dipping Balandra Fault, if felt, would likely result in depression, rather than uplift at the site [33]. The shorter, more distant north-dipping La Pedrera Fault is less likely to result in upward movement at Estero de Bahia Falso. The evidence, derived from studies of adjacent sea notches, for continuing slow uplift suggests that sudden vertical movement has not occurred over the period of notch formation, estimated to have covered the last 500–2000 years [34]. The uplift rates, derived from three different proxies, vary from 0.0 to 0.5 mm year $^{-1}$, producing estimated vertical movement from 0 to 50 cm per 1000 years. The ability of this movement to explain the material in Areas A, B, and C as uplifted marine sediments depends, of course, on the age of the material. This is discussed below.

However, although sudden, seismically-generated vertical movement at the site cannot be definitively eliminated, and slow uplift has likely occurred, internal evidence suggests that if either process resulted in changes in relative sea level, such changes were not sufficient to be responsible for the deposition of the anomalous sediments. With the surface of the lake ~75 cm below the top of the shelly ridge, the 60 cm water depth puts the top of core 4 about 135 cm below the top of the ridge. If the ridge represents approximate sea level at some earlier period the expected sedimentation at core 4, under >1 m of water at the time, would likely be marine clays, not peat. Similarly it seems unlikely that a beach ridge resulting from higher sea level would contain organisms from such a wide range of marine environments.

Another possibility is that the ridge formed as a beach under normal coastal processes and then was abandoned in its present location as the beach prograded seaward. Gradual progradation is, indeed, the process that most likely built the underlying beach, which is a smooth, low-gradient surface composed of well-sorted, uniformly-sized, rounded, fine-grained sediments. All of these features are characteristic of gradual deposition in low-energy environments. However, the shelly ridge sitting on top of this material has distinctly different physical properties. The material is extremely poorly sorted and extremely heterogeneous, both spatially

199

and compositionally, containing, as it does, biological material from across a spectrum of marine environments. The geochemical and loss-on-ignition profiles suggest a significant terrestrial contribution to the material, as does the angularity of many of the larger clastic elements. These characteristics strongly suggest that this ridge did not form under low-energy conditions.

The ridge's formation as part of a chenier plain, which usually forms as a result of shifting sediment supplies and the winnowing of shelly material from the underlying mud flats, seems unlikely. The presence of angular rocks, broken corals, and offshore shell species does not agree with this formation process, nor is there any reason to suggest long-term shifts in either sediment supply or wave energy in this small, protected bay.

The same arguments can be applied to Area A, which contains material from a similar mix of marine environments. Additional arguments against Area A as an uplifted marine environment include the composition of the underlying sediments and the spatial extent of Area A. The clean, shell-free, boulder-embedded orange clay occurring directly below the shell layer (Figure 4C) closely resembles the surrounding hillsides. If the Area A shell layer represents an uplifted sea bed, the underlying material should also be marine. It is difficult to imagine a scenario in which a section of seafloor as small as Area A (~10 m × 1.5 m) could be uplifted and emplaced within an area surrounded by terrestrial environments both vertically and horizontally. Positing that Area A and Area B result from a period of higher relative sea level poses an additional problem in that Area A, representing the sea floor, is at slightly higher elevation than Area B, representing the contemporaneous beach.

Because deposition under normal, fair weather conditions under the current environmental setting does not explain the presence of this material, and neither changes in relative sea level nor vertical movement seem likely, some type of high-energy event becomes the most likely possibility. The piled material in both Area B and Area C is of generally marine origin, and has been deposited as a coherent, spatially continuous unit on top of a distinctly different (and finer-grained) surface, with obvious cliff faces. Identification of this as an event deposit seems the most reasonable.

Evidence from the sediment core suggests that the wetland deepened as a result of the deposition of this event deposit. The upper 27 cm of core 4 consist of a fine dark, low-organic mud with geochemical properties associated with runoff from the surrounding slopes, i.e., a sediment type commonly found on the bottom of shallow coastal lakes in non-carbonate environments. However, apart from a clastic interval (~100–130 cm in core 4), the underlying material, to a depth of ~260 cm is peat. Prior to the transition to lake the area appeared to have been a vegetated wetland of fluctuating water depth, probably in a shallow depression behind a low seaward sill that retarded drainage to the sea. The depositional transition to lake mud indicates a deepening of the flooded wetlands into an open-water lake. This transition, dated to ~1100 year BP in core 4, is marked by turbulent deposition, suggesting a high-energy event.

The physical characteristics of the sedimentary unit covering the small section of hillslope in Area A suggest that this material was deposited under high-energy conditions. The elevation of the site, the presence of coral fragments and deep water marine shells, their imbrication and abraded nature all indicate transport during high water under turbulent conditions.

Occasional shells occur on the hillslopes to the north of the lake, including a *Pitar* (*Pitarella*) *catharius* shell, an offshore species with a depth range from 13 to 80 m [36]. *Glymeris* (*Tucetona*) *strigilata* is another identified offshore species, occupying a similar depth range. Of particular interest is the presence of the upper valve of a *Spondylus princeps* (Pacific thorny oyster), a bivalve species that lives at depths of 15 to 50 m [42] from the Gulf of California and the Pacific coast from Baja California Sur to Peru [43]. *Spondylus* is much more firmly attached to the bottom than other bivalves, typically cementing itself to a solid substrate rather than attaching by byssus threads [44]. This oyster species is also noted for the strength of its ball and socket valve hinge, which is much stronger than the toothed hinge common to other bivalve molluscs. If living at the time of transport, an extremely turbulent force, such as occurs only during a tsunami or intense storm, would have been required to dislodge the upper valve from the firmly-cemented lower valve and transport it from an offshore depth to our study site.

Between the seaward edge of the shelly ridge (Area B) and the bay the ground slopes smoothly, dropping ~1.50 m from the base of the ridge, traversing a barren area of fine-grained material before encountering the landward edge of the mangrove fringe. The surface of the beach and the ground beneath the mangroves are littered with such floatable debris as empty soft drink bottles and styrofoam containers. The line of maximum landward deposition of these objects, presumably corresponding to the reach of the maximum wave height under ordinary conditions (including recent tropical cyclones) is just landward of the mangroves, ~50 cm below the top of the ridge. This supports the view that the event responsible for transporting the materials associated with Areas A, B, and C generated extremely high water levels and/or wave energy. Similar conclusions result from the presence of coral and deep-water shells and the occurrence of rocks apparently transported from the surrounding hillsides. The elevation of both the massive shell coverage in Area A and the top of the ridge in Area B are ~2 m above MHW.

The event most likely occurred around 1100 year BP, although the exact age is somewhat uncertain. Calibrated dates for shells associated with the event range from 1706 year BP to 951 year BP. A wide range in ages is expected as many of the transported shells may have been lying on the sea bottom for centuries before being entrained. However these shells do provide a maximum age of 1130 year BP, the oldest possible date for the youngest dated shell. This is in rough agreement with the sample from the sediment core, with calibrated dates that range from 1176 to 1007 year BP, with a median probability of 1106 year BP. Because the estimated rates of slow uplift can only account for maximum vertical movement of 50 cm over the last 1000 years, this uplift cannot be responsible for the placement of the marine sediments at their present elevations.

Average monthly tidal difference for the 95 months examined was 164 cm. We consider this a conservative estimate of tidal range as data from some months was partial, thereby possibly reducing maximum amplitude due to the missing of the monthly extremes. The range of monthly amplitude varied from 191 cm (January 2006) to 114 cm (March 2000, for which only 11 days of data were recorded). Tide gauge data shows a range of >2 m for the period from 15 May 1999 to 3 March 2011 (Figure 8).

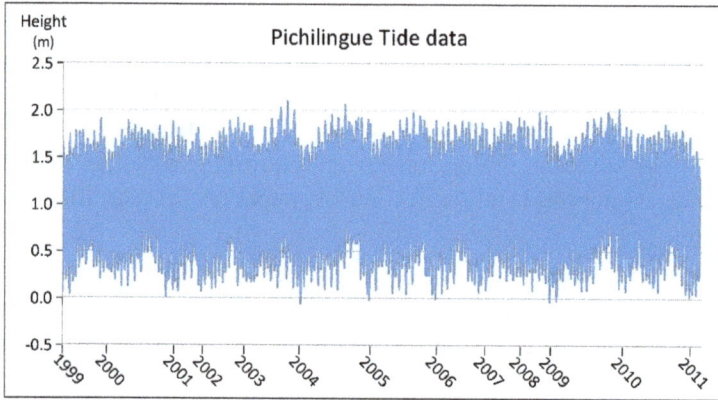

Figure 8. Tide heights at Pichilingue gauge station for the period 15 May 1999 to 3 March 2011.

A tidal amplitude of 1.64 m equates to potential elevation range of ~2–3.6 m above water level for the marine deposits at the time of the event. This should be considered the minimum potential run-up due to the conservative tidal range estimation; in addition, the presence of scattered shells at higher elevations and occasional occurrence on slopes to the north of the lake suggests that the maximum water depth might have been greater.

5.1. Extreme Events

High-energy events for the area are limited to tsunamis and tropical cyclones. Hurricanes are fairly frequent in the area, with 10 hurricanes having passed within

202

50 km of La Paz since 1949, the beginning of the instrumental record (Figure 9) [45], including two category 3 hurricanes, one category 2 hurricane, and seven category 1 hurricanes. Extending the radius to 100 km produces another six hurricanes, all category 1. In the eastern North Pacific, the El Niño-Southern Oscillation (ENSO) exerts an important control over tropical cyclones, as over the instrumental period, tropical cyclones have been more frequent [46–48], and more intense [49,50] during El Niño than neutral and La Niña periods for the region.

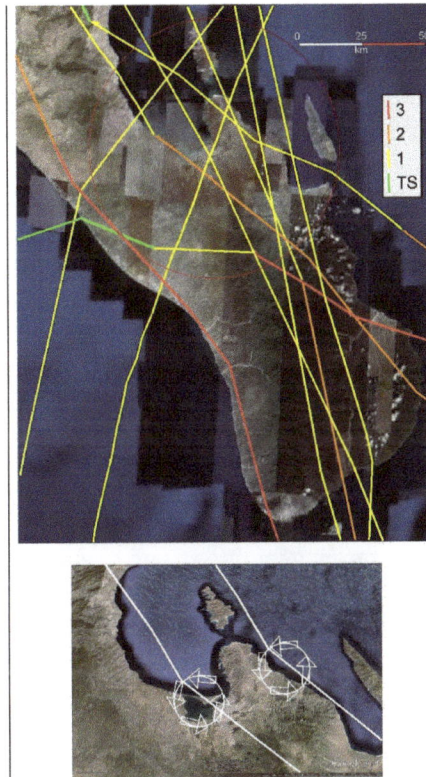

Figure 9. Hurricane history for the La Paz area. Displayed are the tracks of all tropical cyclones that approached within a 50 km radius (red circle) of the city of La Paz at hurricane strength from 1949 to 2014 (top). Five of the ten storms passed to the east of our site, much of their force blocked by the protruding peninsula along the southeastern edge of the bay. Cartoon at bottom shows the cyclonic circulation of hypothetical tropical cyclones with arrows pointing in the direction of wind flow for all four quadrants, demonstrating that storms passing to the west will generally result in an off-shore direction for wind/wave energy, and consequently sediment transport during the passage of the strong right front quadrant. Force of storms passing to the east will be dampened by the intervening peninsula.

The sedimentary signatures of both tsunami [51–57] and tropical cyclone-generated [58–66] deposits have been extensively studied. However, attempts at distinguishing between the two [51,53,55,67] have not been entirely successful. This is due not only to the large overlap of sedimentary features common to both tsunami and tropical cyclone-generated deposits, but also to the highly idiosyncratic depositional patterns of individual events, driven by differences in such parameters as coastal geomorphology, bathymetry, topography, and size and direction of travel [68].

Under current conditions, the deposition at Estero de Bahia Falso seems unlikely to have been generated by tropical cyclones, given the relatively weak tropical cyclones that have impacted the area during the instrumental period (Figure 9), as it is doubtful that storms of this magnitude could generate the surge and wave energy capable of either transporting the large cobbles as far inland as Area C (Figure 6b) as suspended sediment; or depositing the mud/silt/sand/cobbles mixture as bedload.

However, paleorecords show a large increase in ENSO activity for Ecuador [69] staring ~1500 BP and for the Alfonso Basin [70] at ~1000 BP, perhaps corresponding to more frequent and more powerful tropical cyclones near our site during that period. Although paleoclimatic conditions increase the possibility if the passage of an exceptionally powerful tropical cyclone ~1000 years ago, geographical parameters tend to dampen the wave energy of tropical cyclones at the site for two reasons. First, the small fetch acts to limit extreme wave heights, although some potential funneling of wave energy due to coastal morphology is possible under specific wave and wind conditions. Secondly, tropical cyclones, which typically track from south to north in this region, will rarely generate strong onshore winds at the site. Storms passing to the east will result in offshore winds coming from the north and east, from which the lake is well protected by the bulk of the intervening peninsula to the east, which is marked by a rocky spine with elevations >430 m [71]. For storms passing to the west, the lake is subjected to offshore winds and waves during the passage of the strong right front quadrant. It is only after the eye of the storm has passed, during the passage of the weaker right rear quadrant, that the site will be subjected to onshore winds (Figure 9). Furthermore, the intensity of such storms will generally have already been reduced by their transit across the dry peninsula to the south. Additionally, sorting, which is perhaps the characteristic most likely to differentiate tsunami and storm deposits, is extremely poor for the deposits in all areas, especially in Area A where boulders >50 cm in length are mixed with mud, silt, and sand.

The constricted spatial extent of the event deposit, especially the absence of the shelly material along the seaward edge of the beach plain in Area B, is more consistent with the spread of a single, focused surge of water moving up-channel upon reaching the open embayment than with the persistent landward force of waves occurring over the duration of a tropical cyclone. However, this is not a definitive distinguishing feature, as evidenced by the extremely short periods (0.5–1 h) of peak flooding documented for the fast moving Typhoon Haiyan in 2013 [72]. Although a tropical cyclone cannot be eliminated as a candidate process, it would likely have required an unusually large and/or intense storm on a fortuitously-positioned track to generate sufficient energy to transport the anomalous marine material found in Areas A–C.

5.2. Tsunamis

The most likely candidate is a tsunami. Given the protection offered by the Baja California Peninsula, tele-tsunamis from the open Pacific likewise are an unlikely source of such a high-energy event. Large earthquakes occur along the axis of the plate boundary in the central Gulf of California [21,22,73], creating the potential for tsunamis, although the strike/slip nature of the boundary tends to limit vertical movement. However, our study site is protected to a large degree from the effects of such tsunamis by the configuration of the Bay of La Paz, with the eastern islands blocking the entrance to the bay, and especially by the protruding southern peninsula on which our study site is located. These islands are aligned on faults along their western edges, which have been the epicenters of numerous earthquakes throughout the historic period [22], as evidenced by multiple large turbidite layers recorded from the floor of the Alfonso Basin in the northern section of the Bay of La Paz [21,22]. These turbidite layers have been attributed to large, seismically-induced slope failures on the western edges of these islands. The estimated basin-floor volume of the largest turbidite, which is up to 80 cm thick at >10 km offshore, is 10^8 m^3. Not only is this 3–4 orders of magnitude larger than the volume of turbidites associated with tropical cyclones in the same record [22], it is also larger than the total estimated volume (30×10^6 m^3) of the rockslide that produced the 524 m high 1958 tsunami in Lituya Bay, Alaska [6].

It seems reasonable that such slumps could produce large local tsunamis, particularly in the shallow southern section of the bay. Wave focusing resulting from the site's location at the end of a blind bay enclosed within the km-long arms of peninsulas to both the north and south could further increase wave height and energy [74].

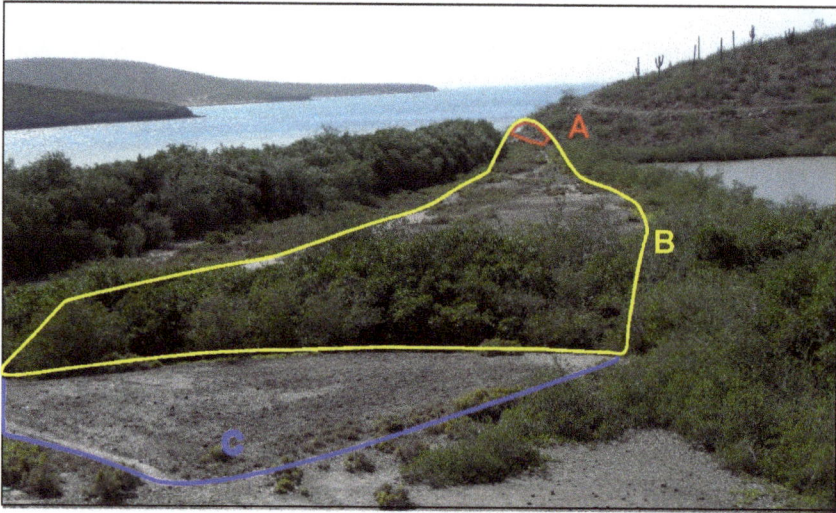

Figure 10. Topography and orientation of the deposited marine material. Looking westward down the inferred path of the marine intrusion. Area A (red) where the largest material was plastered against the hillside is located at the lowest elevation nearest the bay. Mixed sediments, minus the largest terrestrial boulders spread laterally across Areas B (yellow) and C (blue). Steep cliff faces on both sides of the shelly ridge in Area B suggest a narrow path for the wave energy, perhaps due to the angle of approach and the geometry of the projecting hillside. The extension of the reverse of this line leads down the center of the subbay, and eventually towards the city of La Paz.

Such an event would explain the unique geological/geomorphological features observed at the site. When viewed from the eastern end of the lake, the history of the inferred tsunami can be visualized (Figure 10). Area A is located at the closest approach to the lake from the open bay, immediately inside the last projecting hillside. It is likely that the tsunami wave, carrying a load of marine sediments, including corals and shells, living and dead, from a variety of depths, hit the edge of the protruding hillside, scarping the corner and entraining the surface boulders, before slamming into Area A, imbricating shells and plastering the mixed material into the existing hillside. Constrained by the angle of impact the tsunami would have then spread out over the flat terrain (Area B), depositing the thick carpet of material over the foreshore before smashing against the steep hill at the eastern end of the embayment and dropping the entrained sediments (Area C). The ridge of newly deposited material would likely have impounded the marine water, transforming the wetland into a body of open water. Area B and Area C, visually part of the same depositional feature, would, at some point, have been bifurcated by the drainage channel from the lake. This channel, which cuts through the shelly ridge in Area B, might have formed either during the event due to the force of the tsunami back flow, or at a later date during periods of increased precipitation and increased stress on the barrier. The surrounding salt pans, indications of a larger lake, may be related to either the original impoundment or to later variability in precipitation/evaporation levels.

5.3. Local Event History

Gonzalez-Yajimovich and others [22] document three large turbidites over the last 4900 years, the last occurring at ~1500 year BP, all of which they attribute to slumping of the eastern islands. It is possible that the ~1100 year BP event documented at our site corresponds to the 1500 year BP event inferred from the turbidite record. However, it is also possible that our event does not match any major event in the turbidite record, perhaps indicating that even relatively small events can result in tsunamis >2 m in height with devastating impacts at some localities along the Bay of La Paz. This is a risk that needs to be incorporated into local vulnerability studies.

Further research should help to confirm a tsunami as the proximate cause of these anomalous sediments. Examinations of adjacent sites should provide supporting evidence for the occurrence of this event and help determine its spatial extent, while the identification and statistical analyses of a larger number of shells and coral fragments, both at the site and offshore, may produce useful information regarding the path of the wave.

6. Conclusions

Anomalous, mainly allochthonous sediments occur seaward of Estero de Bahía Falso, a shallow lake occupying a small embayment along the western edge of the peninsula extending into the southeastern section of the Bay of La Paz, in Baja California Sur, Mexico. The sediments consist of a mixture of marine and terrestrial material, including biogenic marine material from a wide range of marine environments and depths, including shells of three species with depth ranges >13 m. The material is plastered against the hillside at the western opening of the embayment, beyond which the sediment cover widens and flattens eastward, culminating in a thick, structureless deposit at the base of a steep hillside at the eastern end of the embayment. The deposition of this material coincides both chronologically and stratigraphically with a higher water level within the area occupied by the current lake. We interpret these features as most likely resulting from an extreme wave event with a minimum runup of ~2–3.6 m above MHW, which we suggest can be attributed to a paleotsunami occurring ~1100 year BP.

This probable tsunami was likely generated by the seismically-induced slumping of the western edge of one or more of the islands along the eastern edge of the bay. Because such events, which Gonzalez-Yajimovich and others [22] suggest have occurred repeatedly in the past, are associated with an active fault, they are likely to reoccur in the future. Such events present a significant societal risk to the nearby city of La Paz, a metropolitan area with marine tourism-based economy [31,32]. Therefore, we suggest that the potential for devastating tsunamis generated by relatively small events in enclosed waters is a geological risk that needs to be fully incorporated into coastal vulnerability studies. This applies specifically to La Paz, but also more broadly to other tectonically-active areas with steep topography surrounding confined bodies of water.

Acknowledgments: Funding was provided by IAI grant SPG-CRA-2050 to Kam-biu Liu. We wish to thank Luis Farfan (CICESE) and Graciela Raga (UNAM) for logistical support and Emilio Garcia for his help in identifying shells.

Author Contributions: K.-B.L., T.A.M., and T.A.B. conceived and conducted the fieldwork involved in this project. T.A.B. and T.A.M. conducted the laboratory analysis along with K.-B.L. T.A.M. primarily wrote the article which was edited and revised by T.A.B. and K.-B.L.

Conflicts of Interest: The authors declare no conflict of interest.

References

1. Lay, T.; Kanamori, H.; Ammon, C.J.; Nettles, M.; Ward, S.N.; Aster, R.C.; Beck, S.L.; Bilek, S.L.; Brudzinski, M.R.; Butler, R.; *et al.* The Great Sumatra-Andaman Earthquake of 26 December 2004. *Science* **2005**, *38*, 1127–1133. PubMed]

2. Paris, R.; Lavigne, F.; Wassmer, P.; Sartohadi, J. Coastal sedimentation associated with the December 26, 2004 tsunami in Lhok Nga, west Banda Aceh (Sumatra, Indonesia). *Mar. Geol.* **2007**, *238*, 93–106.

3. Fujii, Y.; Satake, K.; Sakai, S.; Shinohara, M.; Kanazawa, T. Tsunami source of the 2011 off the Pacific coast of Tohoku Earthquake. *Earth Planets Space* **2011**, *63*, 815–820.

4. Mori, N.; Takahashi, T.; Yasuda, T.; Yanagisawa, H. Survey of 2011 Tohoku earthquake tsunami inundation and run-up. *Geophys. Res. Lett.* **2011**, *38*, L00G14.

5. Heidarzadeh, M.; Satake, K. Waveform and spectral analyses of the 2011 Japan tsunami records on tide gauge and DART stations across the Pacific Ocean. *Pure Appl. Geophys.* **2013**, *170*, 1275–1293.

6. Weiss, R.; Fritz, H.M.; Wünnemann, K. Hybrid modeling of the mega-tsunami runup in Lituya Bay after half a century. *Geophys. Res. Lett.* **2009**, *36*, L09602.

7. Miller, D.J. *Giant Waves in Lituya Bay, Alaska*; USGS Professional Paper 354-C. United States Government Printing Office: Washington, DC, USA, 1960.

8. Voight, B.; Janda, R.J.; Glicken, H.; Douglass, P.M. Nature and mechanics of the Mount St Helens rockslide-avalanche of 18 May 1980. *Géotechnique* **1983**, *33*, 243–273.

9. Müller-Salzburg, L. The Vajont catastrophe-A personal review. *Eng. Geol.* **1987**, *24*, 423–444.

10. Müller-Salzburg, L. The rock slide in the Vajont Valley. *Rock Mech. Eng. Geol.* **1964**, *2*, 148–212.

11. Hausback, B.P. Cenozoic volcanic and tectonic evolution of Baja California Sur, Mexico. In *Geology of the Baja California Peninsula*; Pacific Section Special Paper 39; Frizzell, V.A., Jr., Ed.; Society of Economic Paleontologists and Mineralogists: Tulsa, OK, USA, 1984; pp. 219–236.

12. Atwater, T. Implications of plate tectonics for the Cenozoic tectonics of western North America. *Geol. Soc. Am. Bull.* **1970**, *81*, 125–133.

13. Stock, J.M.; Hodges, K.V. Pre-Pliocene extension around the Gulf of California and the transfer of Baja California to the Pacific Plate. *Tectonics* **1989**, *8*, 99–115.

14. Mayer, L.; Vincent, K.R. Active tectonics of the Loreto area, Baja California Sur, Mexico. *Geomorphology* **1999**, *27*, 243–255.

15. Sumy, D.F.; Gaherty, J.B.; Kim, W.-Y.; Diehl, T.; Collins, J.A. The mechanisms of earthquakes and faulting in the southern Gulf of Mexico. *Bull. Seismol. Soc. Am.* **2013**, *103*, 487–506.

16. Fletcher, J.M.; Kohn, B.P.; Foster, D.A.; Gleadow, J.W. Heterogeneous Neogene cooling and exhumation of the Los Cabos block, southern Baja California: Evidence from fission-track thermochronology. *Geology* **2000**, *28*, 107–110.

17. Coyan, M.M.; Arrowsmith, J.R.; Umhoefer, P.; Coyan, J.; Kent, G.; Martínez Gutíerrez, G.; Driscoll, N. Geometry and Quaternary slip behavior of the San Juan de los Planes and Saltito fault zones, Baja California Sur, Mexico: Characterization of rift-normal faults. *Geosphere* **2013**, *9*, 426–443.

18. Ortlieb, L. Quaternary shorelines along the northeastern Gulf of California: Geochronological data and neotectonic implications. In *Studies of Sonoran Geology*; Special Paper 254; Pérez-Segura, E., Jacques-Ayala, C., Eds.; Geological Society of America: Boulder, CO, USA, 1991; pp. 95–120.

19. Ortlieb, L. Quaternary vertical movements along the coasts of Baja California and Sonora. In *The Gulf and Peninsular Province of the Californias*; Dauphin, J.P., Simoneit, B.R.T., Eds.; American Association of Petroleum Geologists: Tulsa, OK, USA, 1991; Memoir 47; pp. 447–480.

20. Welles, D.L.; Coppersmith, K.J. New empirical relationships among the magnitude, rupture length, rupture width, rupture area, and surface displacement. *Bull. Seismol. Soc. Am.* **1994**, *84*, 974–1002.

21. Gorsline, D.S.; de Diego, T.; Nava-Sanchez, E.H. Seismically triggered turbidites in small margin basins: Alfonso Basin, Western Gulf of California and Santa Monica Basin, California Borderland. *Sediment. Geol.* **2000**, *135*, 21–35.

22. Gonzalez-Yajimovich, O.E.; Gorsline, D.S.; Douglas, R.G. Frequency and sources of basin floor turbidites in alfonso basin, Gulf of California, Mexico: Products of slope failures. *Sediment. Geol.* **2007**, *199*, 91–105.

23. Curray, F.; Emmel, J.; Crampton, P.J. Coastal Lagoons: A Symposium. In *Holocene History of a Strand Plain Lagoonal Coast, Nayarit, Mexico*; Ayala-Castanares, A., Ed.; Universidad Nacional Autonoma de Mexico Press: Mexico, Mexico, 1969; pp. 63–10.

24. Gómez-Valdés, J.; Delgado, J.A.; Dworak, J.A. Overtides, compound tides, and tidal-residual current in Ensenada de la Paz lagoon, Baja California Sur, Mexico. *Geofis. Int.* **2003**, *42*, 623–634.

25. Servicio Mareográfico Nacional Universidad Nacional Autónoma de México, Instituto de Geofísica. Available online: http://www.mareografico.unam.mx/portal/ (accessed on 15 November 2015).

26. Zavala, J.; The Head of the Servicio Mareográfico Nacional Universidad Nacional Autónoma de México, Instituto de Geofísica. Personal communication, 2015.

27. Nava-Sanchez, E.H. Modern fan deltas of the west coast of the Gulf of California, Mexico. Ph.D. Thesis, University of Southern California, Los Angeles, CA, USA, 1997.

28. Pérez-Cruz, L.; Urrutia-Fucugauchi, J. Magnetic mineral study of Holocene marine sediments from the Alfonso Basin, Gulf of Mexico-implications for depositional environment and sediment sources. *Geofis. Int.* **2009**, *48*, 305–318.

29. Sánchez-Velasco, L.; Beier, E.; Avalos-García, C.; Lavín, M.F. Larval fish assemblages and geostrophic circulation in Bahía de La Paz and the surrounding southwestern region of the Gulf of California. *J. Plankton Res.* **2006**, *28*, 1081–1098.

30. Oleg Zaytsev, O.; Rabinovich, A.B.; Thompson, R.E.; Silverberg, N. Intense diurnal surface currents in the Bay of LaPaz, Mexico. *Cont. Shelf Res.* **2010**, *30*, 608–619.

31. De Los Monteros, R.L.-E. Evaluating ecotourism in natural protected areas of La Paz Bay, Baja California Sur, Mexico: Ecotourism or nature-based tourism? *Biodivers. Conserv.* **2002**, *11*, 1539–1550.

32. Barr, R.F.; Mourato, S. Investigating the potential for marine resource protection environmental service markets: An exploratory study from La Paz, Mexico. *Ocean Coast. Manag.* **2009**, *52*, 568–577.

33. Servicio, G.M. *Carta Geologic–Minera Coyote G12-D73 Baja California Sur*; Servicio Geológico Mexicano: Pachuca, Mexico, 2008.

34. Trenhaile, A.S.; Porter, N.I.; Prestanski, K. Shore platform and cliff notch transitions along the La Paz Peninsula, southern Baja, Mexico. *Geol. Acta* **2015**, *13*, 167–180.

35. Liu, K.B.; Fearn, M.L. Reconstruction of prehistoric landfall frequencies of catastrophic hurricanes in northwestern Florida from lake sediment records. *Quat. Res.* **2000**, *54*, 238–245.

36. Keen, A.M. *Sea Shells of Tropical West America: Marine Mollusks from Baja California to Peru*, 2nd ed.; Stanford University Press: Stanford, CA, USA, 1971.

37. Reimer, P.J.; Bard, E.; Bayliss, A.; Beck, J.W.; Blackwell, P.G.; Bronk Ramsey, C.; Buck, C.E.; Cheng, H.; Edwards, R.L.; Friedrich, M.; *et al.* IntCal13 and MARINE13 radiocarbon age calibration curves 0–50000 years cal BP. *Radiocarbon* **2013**, *55*, 1869–1887.

38. Telford, R.J.; Heegaard, E.; Birks, H.J.B. The intercept is a poor estimate of a calibrated radiocarbon age. *Holocene* **2004**, *14*, 296–298.

39. Frantz, B.R.; Kashgarian, M.; Coale, K.H.; Foster, M.S. Growth rate and potential climate record from a rhodolith using 14C accelerator mass spectrometry. *Limnol. Oceanogr.* **2000**, *45*, 1773–1777.

40. Engelhart, S.E.; Vacchi, M.; Horton, B.P.; Nelson, A.R.; Kopp, R.E. A sea-level database for the Pacific coast of central North America. *Quat. Sci. Rev.* **2015**, *113*, 78–92.

41. Reynolds, L.C.; Alexander, R.; Simms, A.R. Late Quaternary relative sea level in Southern California and Monterey Bay. *Quat. Sci. Rev.* **2015**, *126*, 57–66.

42. Pillsbury, J. The thorny oyster and the origins of empire: Implications of recently uncovered Spondylus imagery from Chan Chan, Peru. *Lat. Am. Antiq.* **1996**, *7*, 313–340.

43. Moore, E.J. *Tertiary Marine Pelecypods of California and Baja California: Nuculidae through Malleidae*; Geological Survey Professional Paper-1228-A. United States Government Printing Office: Washington, DC, USA, 1983.

44. Instituto de Investigaciones Marinas. Available online: http://institutonazca.org/bivalve-spondylus/ (accessed on 18 July 2015).

45. NOAA Coastal Services Center. Available online: http://coast.noaa.gov/hurricanes/ (accessed on 9 April 2015).

46. Jien, J.Y.; Gough, W.A.; Butler, K. The influence of El Niño-Southern Oscillation on tropical cyclone activity in the Eastern North Pacific Basin. *J. Clim.* **2015**, *28*, 2459–2474.

47. Jáuregui, E. Climatology of landfalling hurricanes and tropical storms in Mexico. *Atmósfera* **2003**, *16*, 193–204.

48. Rodgers, E.B.; Adler, R.F.; Pierce, H.F. Contribution of tropical cyclones to the North Pacific climatological rainfall as observed from satellites. *J. Appl. Meteorol.* **2000**, *39*, 1658–1678.

49. Romero-Vadillo, E.; Zaytsev, O.; Morales-Pérez, R. Tropical cyclone statistics in the Northeastern Pacific. *Atmósfera* **2007**, *20*, 197–213.

50. Jin, F.F.; Bouchare, J.; Lin, I.I. Eastern Pacific tropical cyclones intensified by El Niño delivery of subsurface ocean heat. *Nature* **2014**, *516*, 82–85. PubMed]

51. Nanayama, F.; Shigeno, K.; Satake, K.; Shimokawa, K.; Koitabashi, S.; Miyasaka, S.; Ishii, M. Sedimentary differences between the 1993 Hokkaido-nansei-oki tsunami and the 1959 Miyakojima typhoon at Taisei, southwestern Hokkaido, northern Japan. *Sediment. Geol.* **2000**, *135*, 255–264.

52. Bussert, R.; Aberhan, M. Storms and tsunamis: Evidence of event sedimentation in the Late Jurassic Tendaguru Beds of southeastern Tanzania. *J. Afr. Earth Sci.* **2004**, *39*, 549–555.

53. Goff, J.; McFadgen, B.G.; Chagué-Goff, C. Sedimentary differences between the 2002 Easter storm and the 15th-century Okoropunga tsunami, southeastern North Island, New Zealand. *Mar. Geol.* **2004**, *204*, 235–250.

54. Dawson, S. Diatom biostratigraphy of tsunami deposits: Examples from the 1998 Papua New Guinea tsunami. *Sediment. Geol.* **2007**, *200*, 328–335.

55. Morton, R.A.; Gelfenbaum, G.; Jaffe, B.E. Physical criteria for distinguishing sandy tsunami and storm deposits using modem examples. *Sediment. Geol.* **2007**, *200*, 184–207.

56. Goff, J.; Chagué-Goff, C.; Nichol, S.; Jaffe, B.; Dominey-Howes, D. Progress in paleotsunami research. *Sediment. Geol.* **2012**, *243–244*, 70–88.

57. Cuven, S.; Paris, R.; Falvard, S.; Miot-Niorault, E.; Benbakkar, M.; Schneider, J.-L.; Billy, I. High-resolution analysis of a tsunami deposit: Case-study from the 1755 Lisbon tsunami in southwestern Spain. *Mar. Geol.* **2013**, *337*, 98–111.

58. Donnelly, J.P.; Bryant, S.S.; Butler, J.; Dowling, J.; Fan, L.; Hausmann, N.; Newby, P.; Shuman, B.; Stern, J.; Westover, K.; *et al.* 700 yr sedimentary record of intense hurricane landfalls in southern New England. *Geol. Soc. Am. Bull.* **2001**, *113*, 714–727.

59. Scott, D.B.; Collins, E.S.; Gayes, P.T.; Wright, E. Records of prehistoric hurricanes on the South Carolina coast based on micropaleontological and sedimentological evidence, with comparison to other Atlantic Coast records. *Geol. Soc. Am. Bull.* **2003**, *115*, 1027–1039.

60. Murnane, R.J., Liu, K.-B., Eds.; *Hurricanes and Typhoons: Past, Present and Future*; Columbia University Press: New York, NY, USA, 2004.

61. Hippensteel, S.P. Limiting the limits of bioturbation, or at least focusing on the positive. *Palaios* **2005**, *20*, 319–320.

62. Williams, H.F.L. Stratigraphy, sedimentology and microfossil content of Hurricane Rita storm surge deposits in Southwest Louisiana. *J. Coast. Res.* **2009**, *254*, 1041–1051.

63. Williams, H.F.L. Storm surge deposition by Hurricane Ike on the Mcfaddin National Wildlife Refuge, Texas: Implications for paleotempestology studies. *J. Foramin. Res.* **2010**, *40*, 210–219.

64. Hippensteel, S.P. Spatio-lateral continuity of storm overwash deposits in back barrier marshes. *Geol. Soc. Am. Bull.* **2011**, *123*, 2277–2294.

65. Liu, K.-B.; Li, C.; McCloskey, T.A.; Yao, Q.; Weeks, E. Storm deposition in a coastal backbarrier lake in Louisiana caused by hurricanes Gustav and Ike. *J. Coast. Res.* **2011**, *64*, 1866–1870.

66. McCloskey, T.A.; Liu, K.B. A sedimentary-based history of hurricane strikes on the southern Caribbean coast of Nicaragua. *Quat. Res.* **2012**, *78*, 454–464.

67. Kortekaas, S.; Dawson, A.G. Distinguishing tsunami and storm deposits: An example from martinhal, SW Portugal. *Sediment. Geol.* **2007**, *200*, 208–221.

68. Phantuwongraj, S.; Choowong, M.; Nanayama, F.; Hisada, K.-I.; Charusiri, P.; Chutakositkanon, V.; Pailoplee, S.; Chabangbon, A. Coastal geomorphic conditions and styles of storm surge washover deposits from Southern Thailand. *Geomorphology* **2013**, *192*, 43–58.

69. Moy, C.M.; Seltzer, G.O.; Rodbell, D.T.; Anderson, D.M. Variability of El Nino/Southern Oscillation activity at millennial timescales during the Holocene epoch. *Nature* **2002**, *420*, 162–165. PubMed]

70. Staines-Urias, F.; Gonzalez-Yajimovich, O.; Beaufort, L. Reconstruction of past climate variability and ENSO-like fluctuations in the southern Gulf of California (Alfonso Basin) since the last glacial maximum. *Quat. Res.* **2015**, *83*, 488–501.

71. Jankaew, K.; Atwater, B.F.; Sawai, Y.; Choowang, M.; Charoentitirat, T.; Prendergast, A.; Martin, M.E. Medieval forewarning of the 2004 Indian Ocean tsunami in Thailand. *Nature* **2008**, *455*, 1228–1231.

72. Soria, J.; Switzer, A.; Villanoy, C.; Fritz, H.; Bilgera, P.; Cabrera, O.; Siringan, F.; Sta Maria, Y.; Ramos, R.; Fernandez, I. Repeat storm surge disasters of Typhoon Haiyan and its 1897 predecessor in the Philippines. *Bull. Am. Meteor. Soc.* **2015**. in press.

73. U.S. National Earthquake Data Center. Available online: http://earthquake.usgs.gov/earthquakes/search (accessed on 10 July 2015).

74. DeLange, W.P.; Moon, V.G. Tsunami washover deposits, Tawharanui, New Zealand. *Sediment. Geol.* **2007**, *200*, 232–247.

A Numerical Modelling Study on the Potential Role of Tsunamis in the Biblical Exodus

José M. Abril and Raúl Periáñez

Abstract: The reliability of the narrative of the Biblical Exodus has been subject of heated debate for decades. Recent archaeological studies seem to provide new insight of the exodus path, and although with a still controversial chronology, the effects of the Minoan Santorini eruption have been proposed as a likely explanation of the biblical plagues. Particularly, it has been suggested that flooding by the associated tsunamis could explain the first plague and the sea parting. Recent modelling studies have shown that Santorini's tsunami effects were negligible in the eastern Nile Delta, but the released tectonic stress could have triggered local tsunamigenic sources in sequence. This paper is aimed to a quantitative assessment of the potential role of tsunamis in the biblical parting of the sea. Several "best case" scenarios are tested through the application of a numerical model for tsunami propagation that has been previously validated. The former paleogeographic conditions of the eastern Nile Delta have been implemented based upon recent geological studies; and several feasible local sources for tsunamis are proposed. Tsunamis triggered by submarine landslides of 10–30 km^3 could have severely impacted the northern Sinai and southern Levantine coasts but with weak effects in the eastern Nile Delta coastline. The lack of noticeable flooding in this area under the most favorable conditions for tsunamis, along with the time sequence of water elevations, make difficult to accept them as a plausible and literally explanation of the first plague and of the drowning of the Egyptian army in the surroundings of the former *Shi-Hor* Lagoon.

Reprinted from *J. Mar. Sci. Eng.* Cite as: Abril, J.M.; Periáñez, R. A Numerical Modelling Study on the Potential Role of Tsunamis in the Biblical Exodus. *J. Mar. Sci. Eng.* **2015**, *3*, 745–771.

1. Introduction

The Minoan eruption of Thera (Santorini), dated around 1613 BC, was one of the largest Plinian eruptions on earth in the past 10,000 years [1]. Stanley and Sheng (1986) [2] reported the evidence for the presence of ash ejected from this eruption in sediment cores from the eastern Nile Delta, and they suggested that it could be associated with the Exodus plague of darkness. The connection between this extreme volcanic event and the biblical plagues preceding the Exodus, has been largely argued in the scientific literature [3–7].

214

Recent archaeological findings [8,9] provide new insight on the Exodus route and on the location of the sea crossing. Thus, the site of Migdol (Exodus 14:2; see Figure 1) would equate one of the military fortresses in the *Horus Way*, placed in the shoreline of the *Shi-Hor*, a paleo-lagoon opened to the Mediterranean Sea. The water-body that the Israelites crossed when leaving Egypt, called *yam suph*, the Sea of Reeds, should have been a large body of water in the area of Migdol. Nevertheless, the above scenario would correspond to the context of two centuries after the Minoan Santorini eruption. Indeed, the chronology of the biblical Exodus has been a matter of controversial [10] (see for example the review by Sivertsen, 2009, [4]). This last author proposes that there are two distinct episodes that merged in a single narrative: the exodus-flight (a group of western Semites leaving the *Wadi Tumilat* area slightly before 1600 BC, during the Hyksos rule in Egypt), and the exodus-expulsion (of Israelite slaves during the co-rulership of Tuthmossis III and Amenophis II, about 1450 BC).

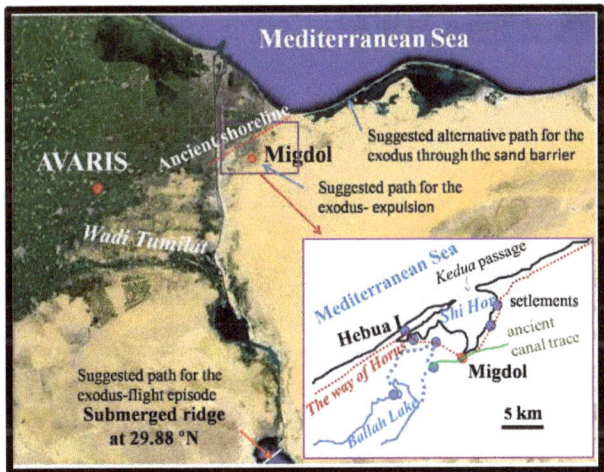

Figure 1. Map of the eastern Nile Delta and the Suez Canal, including the main geographical references cited in the text. The zoon-box shows the palaeogeographical reconstruction of the former shoreline at the second millennium BC around the Migdol site and the *Shi Hor* lagoon (after Hoffmeier, 2005, [9]).

Sivertsen (2009) [4] had suggested that the first plague would have been the likely result of the Minoan eruption and its associated tsunamis. According to this author, in the exodus-flight episode persisting strong wind conditions would have dried any of the two submerged ridges lying south of Lake Timshah, allowing for the sea crossing, while in the exodus-expulsion, a volcanic eruption in the area of Yali and Nisyros (eastern Aegean Sea) would have produced the tsunami that drowned the Egyptian army at the *Shi-Hor* Lagoon.

The modern reef at 29.88° N extends about 10 km under the Gulf of Suez, and it is not of uniform depth. Voltzinger and Androsov (2003) [11] published a numerical model for the drying of the reef by strong winds using a simplified bathymetry for the Gulf and a shallow reef with a uniform depth of 3 m, getting an exposure time of 4 h under winds of 33 m/s. More recently, Drews and Han (2010) [12] pointed out the strong effect of a non-uniform depth. Thus, they showed that a 3 m deep notch (1 km length) within a reef of 2 m depth would have never been exposed under a wind of 33 m/s. Furthermore, these modelling exercises look for steady-state solutions with typical simulation times of the order of one day; but they omit the tidal oscillations, which are up to 1.5 m range in this area [13,14] and a major factor in the involved physical problem. Thus, this scenario for the sea crossing in the exodus-flight seems quite unlikely.

Drews and Han (2010) [12] studied an alternative scenario for a sea parting by strong winds: the Lake of Tanis, northeast of the *Shi-Hor* palaeo-lagoon. They used the paleogeographic reconstructions by Hoffmeier (2005) [9] in which the Lake of Tanis did not exist and the mouth of the *Shi-Hor* lagoon (the *Kedua* passage) was connected to the open sea. Instead, in their modified topography a sand bank delimited the northern extension of Lake Tanis, opened to the sea through the narrow Pelusium mouth. This geometry critically controls the water flux to allow a partial drying under persisting strong winds. Despite of these hypothetical and critical topographic settings, this scenario does not fit well with the exodus path and the location of the Egyptian military fortresses [9].

The idea of a tsunami producing the sea parting in the Exodus has been suggested by several authors [4,15,16], although not supported by quantitative analysis. It is known that the Minoan Santorini eruption generated tsunamis by the entry of pyroclastic flows and by caldera collapse. They left their fingerprints, identified through sedimentary deposits, in most of the coastlines of the Aegean Sea, SW Turkey and Crete [17], Cyprus [18], the coastal area of Israel and Gaza [19–21] and eastern Sicily [22,23]. While model results for these tsunamis are able to reasonably account the estimated runups in the Aegean Sea coasts, they fail to explain the observed effects in other areas (e.g., eastern Sicily and Israel and Gaza coasts); and, particularly, they predict negligible effects in the eastern Nile Delta [17,24,25]. In fact, the Aegean Sea is a semi-enclosed water body, and the islands of the Hellenic arc comprise a physical barrier that dissipates most of the tsunami energy. Moreover, the outer border of the Nile Delta, where a sharp drop in water depth occurs, produces partial reflection and energy dissipation in the tsunami waves (as it can be seen in the dispersion pattern of the Cyprus 1222 earthquake, modelled by Periáñez and Abril, 2014a, [25]).

To explain the whole set of tsunamigenic deposits from the Santorini event, Periáñez and Abril (2014a) [25] suggested a scenario of sequential tsunamis linked to intense tectonic stress release. Particularly, a tsunami triggered by a submarine landslide in the eastern area of the Nile Delta would have been able to generate the isochronous tsunamigenic deposits found in Israel and Gaza. Nevertheless, this does not necessarily support the hypothesis of tsunamis flooding large areas in the eastern Nile Delta, as required in the Sivertsen hypothesis for the first plague, or in the Jacobovici and Cameron interpretation of the sea crossing. A quantitative analysis is needed, taking into account the former paleogeographic conditions and the potential tsunamigenic sources compatible with the known geological settings of the eastern Nile Delta. This is the aim of the present paper, which tries to improve our insight on whether these extraordinary natural disasters could had been only a source of inspiration for the Exodus narrative or they could have conformed reliable scenarios for the sea parting and the drowning of the Egyptian army (e.g., the heated debate between minimalist and maximalist viewpoints).

Going further beyond the previous work [25], the present study includes a more detailed paleogeographical reconstruction of the former coastline, including the target area of the *Shi-Hor* Lagoon [9]. As potential tsunamigenic sources (Section 2) this work considers submarine landslides in the eastern edge of the stable Nile Delta shelf, defined as large as possible within the known constrains stated by geological studies. A hypothetical and sudden colmation of a branch of the former Damietta canyon has been also considered. In the tree cases sensitivity tests have been conducted for the total volume and the prescribed motion of the landslide, up to conform a set of 14 simulations. Two tsunamigenic scenarios triggered by geological faults are also considered along the Pelussium Mega-Shear line, and the combination of two simultaneous sources. This variety of sources, selected as the "best cases" for the drowning of the Egyptian army in the *Shi-Hor* area, reveal that due to the physical settings of the former eastern Nile Delta, most of the tsunami energy dissipates within the stable shelf or it travels out. The inner shoreline remains only weakly affected by flooding (Section 3).

2. Materials and Methods

2.1. Model Description

Numerical models for tsunami propagation are a relatively well established methodology which has been widely validated against recorded data from historical events over the world [26–29]. For the present study, the selected tsunami propagation model is based on the 2D depth-averaged nonlinear barotropic shallow water equations (see for instance Kowalik and Murty, 1993, [30]):

$$\frac{\partial \zeta}{\partial t} + \frac{\partial}{\partial x}(Hu) + \frac{\partial}{\partial y}(Hv) = 0 \tag{1}$$

$$\frac{\partial u}{\partial t} + u\frac{\partial u}{\partial x} + v\frac{\partial u}{\partial y} + g\frac{\partial \zeta}{\partial x} - \Omega v + \frac{\tau_u}{\rho H} = A\left(\frac{\partial^2 u}{\partial x^2} + \frac{\partial^2 u}{\partial y^2}\right) \tag{2}$$

$$\frac{\partial v}{\partial t} + u\frac{\partial v}{\partial x} + v\frac{\partial v}{\partial y} + g\frac{\partial \zeta}{\partial y} + \Omega u + \frac{\tau_v}{\rho H} = A\left(\frac{\partial^2 v}{\partial x^2} + \frac{\partial^2 v}{\partial y^2}\right) \tag{3}$$

where u and v are the depth averaged water velocities along the x and y axes, h is the depth of water below the mean sea level, ζ is the displacement of the water surface above the mean sea level measured upwards, $H = h + \zeta$ is the total water depth, Ω is the Coriolis parameter ($\Omega = 2w \cdot \sin\lambda$, where w is the Earth rotational angular velocity and λ is latitude), g is acceleration due to gravity, ρ is a mean value of water density and A is the horizontal eddy viscosity. τ_u and τ_v are friction stresses which have been written in terms of a quadratic law:

$$\tau_u = k\rho\, u\sqrt{u^2 + v^2}; \tau_v = k\rho\, v\sqrt{u^2 + v^2} \tag{4}$$

where k is the bed friction coefficient. Essentially, these equations express mass and momentum conservation. They have been written in Cartesian coordinates given the relatively small model domain.

Horizontal viscosity has been set as $A = 10$ m^2/s and the bed friction coefficient as $k = 0.0025$. Good model results when estimating wave amplitudes and runups have been obtained with these values. Actually, model results (amplitudes, runups and wave arrival times) have been compared with observations in previous works [28]. Consequently, these values have been retained in the present application. Moreover, the model sensitivity to the bed friction coefficient has been studied in Periáñez and Abril (2014b) [31].

Still waters are used as initial conditions. As boundary conditions, water flow towards a dry grid cell is not allowed, and a gravity wave radiation condition is imposed at the open boundaries [32], which is implemented in an implicit form. Due to the CFL stability condition [30] time step for model integration was fixed as $\Delta t = 2$ s.

A flood/dry algorithm is required since when the tsunami reaches the coast new wet or dry grid cells may be generated due to run-up or rundown. The numerical scheme described in Kampf (2009) [33] and in Periáñez and Abril (2013) [28] has been adopted. Wet grid cells are defined as those with a total water depth larger than a threshold value typically set as a few centimeters. Dry cells are defined as cells where depth is smaller than the threshold one. Flooding and drying is implemented in the code via the calculation of the water velocity normal to the interface between wet and dry cells. The calculation is performed when the pressure gradient force is directed towards the dry cell. Otherwise velocity is set to zero at this point. In the case of a non-zero velocity, water level in the dry cell will increase and the cell turns into a wet one once the water depth is larger than the threshold depth, which has been set as

10 cm [28]. All the equations are solved using explicit finite difference schemes [30] with second order accuracy. In particular, the MSOU (Monotonic Second Order Upstream) is used for the advective non-linear terms in the momentum equations.

For tsunamis produced by earthquakes in geological faults, the vertical sea-floor deformation is considered as the initial condition for the tsunami calculation, and it is computed using the classical Okada formulae [34]. Inputs for this equation are fault plane strike, rake, dip, slip, location, length and width, as well as seismic moment and rigidity. Validation and applications of this model can be found in Periáñez and Abril (2013 [28], 2014a [25], 2014b [31]), and Abril *et al.* (2013) [35].

The methodology of Harbitz (1992) [36] and Cecioni and Bellotti (2010) [37] has been adopted to simulate the generation of tsunamis by submarine landslides, along with the modification provided by Periáñez and Abril (2014a) [25] to describe varying slopes and asymmetric velocity profiles. In the first stage, the slide can be described as a solid body whose downslope movement locally modifies the bathymetry, resulting in an almost instantaneous and equal change in the level of the overlapping waters, which propagates as gravity waves. The adopted geometry is a box of length L, width B and maximum height Δh, and with an exponential smoothing over a distance S in the front and rear and $B/2$ on the flanks (see scheme in Figure 2). The resulting volume is $V = 0.90B \cdot \Delta h(L + 0.90S)$ [36]. As a practical approach, the former slide at its initial position will be superposed onto the present-day bathymetry of the source area. The greatest part of the energy transfer to the water column takes place during the first stage of the slide displacement (approximately until the time of its maximum velocity) [38]. Details of the subsequent history (including deformation and breaking up of the slide, and the accurate update of the seafloor bathymetry) will contribute only as second order corrections. Thus, corrections of the present day bathymetry in the depositional areas have been omitted in this modelling approach. Away from the coastal area, changes in bathymetry during the last 4 ka are negligible in this context, and the former paleogeography conditions of the eastern coastline of the Nile Delta have been re-created for this work based upon available geological studies, as explained further below.

The motion of the slide can be known after solving the governing dynamic equations [39,40], or it can be defined by imposing a prescribed motion based upon a maximum velocity, U_{max}, and the displacement, R [36]. Here we adopt the approach by Harbitz (1992) [36], in which U_{max} is estimated as a function of the slope angle, α, the average thickness of the slide, $\overline{h}$, its density, $\overline{\rho}(\sim 1.7 \times 10^3 \text{ kg} \cdot \text{m}^{-3})$, the density of turbidity currents, ρ_t $\overline{\rho}(\sim 1.1 \times 10^3 \text{ kg} \cdot \text{m}^{-3})$, and the friction ($\mu$) and drag ($C_D^u$) coefficients:

$$U_{max} \cong \sqrt{\frac{(\overline{\rho} - \rho_t)g\overline{h}(\sin\alpha - \mu\cos\alpha)}{\frac{1}{2}C_D^u\rho_t}} \tag{5}$$

with C_D^u, the drag coefficient along the upper surface of the slide, being estimated from the roughness length parameter, k (in the range of 0.01 to 0.1 m):

$$C_D^u = \left[1.89 + 1.62\log\frac{(L+S)}{k}\right]^{-\frac{5}{2}} \tag{6}$$

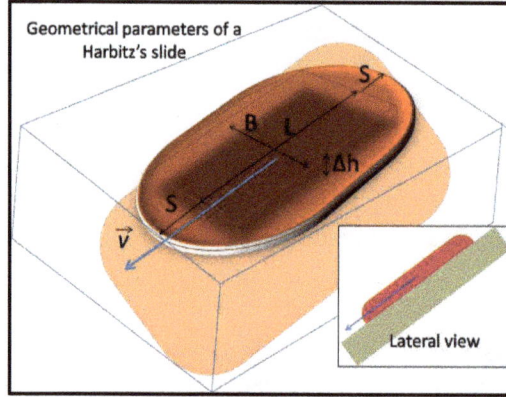

Figure 2. Sketch with the geometrical parameters defining the submarine landslide (following the formalism proposed by Harbitz, 1992, [36]). The adopted geometry is a box of length L, width B and maximum height Δh, and with an exponential smoothing over a distance S in the front and rear, and $B/2$ on the flanks.

The value of U_{max} strongly depends on the estimation of the Coulomb friction coefficient, μ, within an acceptable range (being its upper limit $\mu_{st} = \tan\alpha$), and U_{max} must remain within the range of the reported values in scientific literature [39,41–43].

In many cases a first and large slope angle is involved in the triggering mechanism. After a displacement R_1, the slope angle decreases, but the moving masses still complete a second displacement R_2. For each slope angle the maximum velocity $U_{max,1}$ and $U_{max,2}$ are estimated as commented above, and the following function of time is imposed for the slide velocity, v_s (Periáñez and Abril, 2014a, [25]):

$$v_s = U_{max,1}\sin(\frac{\pi t}{T_1}) \qquad 0 \leqslant t \leqslant t_{trans} \tag{7}$$

$$v_s = U_{max,2}\sin[\frac{\pi}{T_2}(t - t_{trans}) + \frac{\pi}{2}] \qquad t_{trans} < t \leqslant t_{trans} + \frac{T_2}{2} \tag{8}$$

with

$$T_1 = \frac{\pi}{2}\frac{R_1}{U_{max,1}}$$

$$T_2 = \frac{\pi[R_1 + R_2 - S(t_{\text{trans}})]}{U_{\text{max},2}}$$

$$t_{\text{trans}} = \frac{T_1}{\pi} \arcsin \left(\frac{U_{\text{max},2}}{U_{\text{max},1}} \right)$$

and $S(t)$ is the instantaneous position of the slide front at time t.

The "two slope angle" kinematics is a model choice, being more general than the single sinus function used by Harbitz (1992) [36], but containing it as a particular case, and it allows generating asymmetric velocities profiles (as the ones used by Lastras *et al.*, 2005, [39]). Applications of this model can be found in Periáñez and Abril (2014a) [25] and in Abril and Periáñez (2015) [38]. Friction stresses over the moving slide are formulated in terms of relative speed, as in Harbitz (1992) [36], in such a way that a slice moving faster than the water column can transfer energy to it.

The slide model requires four input (but not free) parameters: maximum velocities (governed by μ) and displacements. The candidate source areas will be characterized by depth profiles along their respective transects, what allows the identification of slope angles. The displacements can be partially estimated from the previous profiles, or introduced as plausible values. These displacements have to be understood as effective run-out distances (displacement of the sliding block), over which the transfer of energy to the tsunami takes place. More details will be provided along with the source definitions.

A further validation for the performance of the numerical model with its flood-drying algorithm can be found in Periáñez and Abril (2015) [44]. The submarine landslide submodel has been tested against independent modelling works and subject to extensive sensitivity tests [38,44]. The effects of model resolution and sensitivity to the friction coefficient have been studied in Periáñez and Abril (2014b) [31].

2.2. Model Domain and Bathymetric Map of the Former Eastern Nile Delta

The computational domain used for the numerical modelling of tsunami propagation comprises from 27.0° E to 36.0° E in longitude and from 30.5° N to 37.3° N in latitude (see partial views in Figure 3). Water depths have been obtained from the GEBCO08 digital atlas, available on-line, with a resolution of 30 s of arc both in longitude and latitude. The emerged lands appear with a minimum elevation of 1 m, the minimum depth for waters is 1 m, and both are provided with a resolution of 1 m.

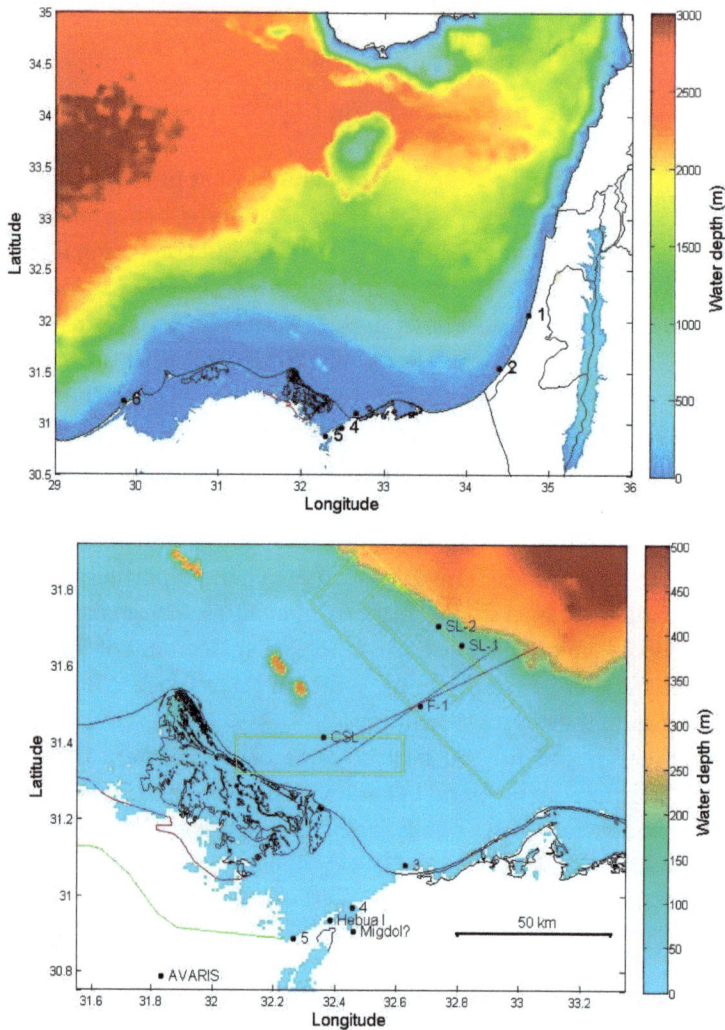

Figure 3. Details of the domain used for the tsunami propagation model: General view of the former Nile Delta and the Levantine coastlines (up); detailed view of the eastern Nile Delta and the *Shi-Hor* lagoon (bottom). Water depths (m) with 30 s of arc resolution from GEBCO08 bathymetry and paleogeographic reconstructions (see text) are drawn. Black line is the present coastline, the red and green ones are, respectively, the former coastline and the marshland limit at 3500 year BP after Coutellier and Stanly (1987) [45]. The blue line delimits wetlands at northern Ballah Lake. *Tjaru* (Hebua I) and Migdol are Egyptian military fortresses around the former *Shi-Hor* paleolagoon [9]. Dots 1 to 6 are synthetic gauges. The locations of tsunamigenic sources (Tables 1 and 2) are also depicted: yellow boxes for submarine landslides and red lines for faults.

During the last 5 ka, with a stable sea level, the major changes in the coastline of the Nile Delta have been driven by fluvial deposits, coastal erosion and land subsidence. The last is estimated as 2.0 mm/year in the area of Alexandria [46]. Coutellier and Stanley (1987) [45] estimated mean Holocene sedimentation rates of 5.0 m/ka in the eastern Nile Delta. As a result, the coastal margin in the studied area migrated northward some 50 km during the past 5 ka. As a practical and crude approach for the reconstruction of the former coastline in the entire Nile Delta at ca 3500 years BP, we followed the methodology described by Periáñez and Abril (2014a) [25]. Briefly, we applied a correction to the present depths/elevations, linearly decreasing from the modern to the former coastline (this last as defined by Coutellier and Stanley, 1987, [45]). The small-scale details of the coastline will not significantly affect the general patterns of tsunami propagation, but they are of capital importance for assessing local run-ups. The detailed paelogeographical reconstruction published by Hoffmeier (2005) [9] for the area of the *Shi-Hor* Lagoon has been digitalized and merged with the previous corrected bathymetry, applying then a filter for smoothing the bathymetry among adjacent grid-cells. Results are shown in Figure 3.

2.3. Tsunamingenic Sources in the Eastern Nile Delta

Works and reviews on the geodynamics of the eastern Mediterranean can be found, among others, in Barka and Reilinger (1997) [47], Mascle *et al.*, (2000) [48], or Yolsal and Taymaz (2012) [29]. Its complex tectonic is responsible for intense earthquake and volcanism activity, triggering large tsunamis in the past [17,29,49,50]. The study by Papadopoulos *et al.*, (2007) [51] on tsunami hazards in the Eastern Mediterranean concluded that the mean recurrence of strong tsunamis (most of them were caused by earthquakes) is likely equal to about 142 years. The assessment of the probability for landslide tsunami scenarios is a quite difficult task, due to the insufficient statistics from the recorded events, as discussed in detail by Harbitz *et al.*, (2014) [52]. Such probabilities of occurrence must not be confused with the probability of that a given coastal area is affected by a tsunami of certain level (e.g., with a maximum water height).

The first map of the Nile deep-sea fan has not been completed until recent years [53,54]. The Messinian salinity crisis led to the deposition of salt and anhydrite throughout the Mediterranean basin, while the proto-Nile river excavated deep canyons and transported offshore large volumes of terrigenous sediments. Thick Plio-Quaternary sediments covered then the ductile evaporitic layers, what triggered some giant gravity-driven salt tectonics [54–58]. A series of fault trends have been identified across the stable shelf of the Nile Delta, e.g., Rosetta, Baltim and Temsah fault trends, and the Pelusium Mega-Shear Fault system [59–61]. The Nile Delta is a major gas and condensate province with several mud volcanoes and geodynamics associated to fluid seepage [62,63].

A general overview on multi-scale slope instabilities along the Nile deep-sea fan can be found in Loncke *et al.*, (2009) [57], and in Urgeles and Camerlenghi (2013) [64]. Garziglia *et al.*, (2008) [60] identified and dated seven mass-transport deposits on the Western province, northern to the Rosetta Canyon, with volumes ranging from 3 to 500 km^3, mean thickness from 11 to 77 m and run-out distances from 18 to 150 km, being the youngest deposit older than 8940 ± 30 cal. year BP. Ducassou *et al.*, (2009), based upon 42 sediment cores collected across the entire Nile deep sea fan, identified several slump deposits and turbidities in the last 2000 ka BP [55], but none in the Mid and Late Holocene (ca 5 ka to present). Thus, any candidate source area for submarine landslides must be compatible with the "empty spaces" within this cloud of cores. Recently, Ducassou *et al.*, (2013) [56] reported four highly mobile debris flows in the Nile deep-sea fan system, with a chronology confined between 5599 and 6915 cal. years BP for the most recent event, in the Rosetta province.

Results from numerical models have proved that the effects of the Minoan Santorini tsunamis (triggered by entry of pyroclastic flows and caldera collapse) were negligible in north-eastern Egypt and Levantine coasts [17,25], as already commented. Thus, for any less energetic event in the inner Aegean Sea (as the volcanic eruption in the area of Nisyros and Yale suggested by Sivertsen for the sea parting in the exodus-expulsion) a similar behaviour would be expected. Even if potential tsunami directionality is invoked, the Isle of Rhodes prevents any direct pathway for energy transfer from these sources towards the eastern Nile Delta. Moreover, for the Cyprus AD 1222 earthquake tsunami, with source area in southern Cyprus, model results (Periáñez and Abril, 2014a, [25]) show that the outer shelf of the Nile Delta acts as a natural barrier and slows waves at these shallow depths (<500 m), thus increasing wave amplitude in this area, but with negligible impacts in the inner shoreline.

Recent mass-wasting events can be recognized in bathymetric maps by the presence of head and footwall scars [57,64]. Periáñez and Abril (2014a) [25] suggested several hypothetical candidate sites in the Nile Delta for submarine landslides of 9–10 km^3, accomplishing for high slopes and being distant enough from the already studied areas, where any mass-wasting deposits in the last 5 ka can be discarded. They showed that tsunamis generated by sources in the western and northern Nile Delta did not significantly affect the coastal zones of Israel and Gaza, and their effects on the eastern area of the delta were equally negligible. Thus, the source area within the Nile Delta able to produce tsunamis potentially linked to the Exodus has to be confined to its eastern zone.

In this work the geometry of the moving boxes and their displacement will be defined with the criteria of being generously large, but compatible with the "empty areas" defined by the studied sediment cores. The value of μ will be subject to sensitivity tests; and 50 m/s has been adopted as the upper limit for

U_{max}, according to Harbitz (1992) [36]. The tsunamigenic sources selected for this study are summarized in Tables 1 and 2 and briefly discussed further below.

Landslide SL-1 is the one discussed and modelled by Periáñez and Abril (2014a) [25], but applied here along with the more detailed paleogeography. The slide front is placed at the eastern border of the stable Nile delta, facing a down-slope of 2.8 degrees. The second displacement prevents reaching the sites of cores studied by Ducassou *et al.* (2009) [55], although its numerical value has a minor effect in the tsunamigenic potential of the slide. The slide extends over a large area (B = 80 km; L = 20 km), but with a moderate value for its maximum height (6.0 m), accounting for a total volume of 9.80 km^3. The value of μ has been fixed as 0.1, 0.3, 0.5, 0.8 and 0.9 of its maximum (static, μ_{st}) limit in model runs R1, R2, R3, R4 and R5, respectively. A second version of this slide uses a larger value for its maximum height (20.0 m), leading to a total volume of 32.7 km^3. The larger height increases the slide speed, which reaches values up to 49.5 m/s for $\mu = 0.5\mu_{st}$ (run R6), and of 31.3 m/s for $\mu = 0.8\mu_{st}$ (run R7). The displacement of the slide over the second slope, with a smaller angle, has a minor contribution to its tsunamigenic potential; thus, and for the sake of simplicity, a value of $\mu = 0.75\mu_{st}$ has been adopted for runs R1 to R7.

Table 1. Source parameters for submarine landslides [#].

Landslide	Run	Geometrical Parameters					Front Position		Direction [$]			Kinematics [¶]				
		L (km)	S (km)	B (km)	h_m (m)	V (km^3)	$\lambda_E°$	$\Phi_N°$	$\theta°$	R_1 (km)	α_1 (°)	μ/μ_{st}	$U_{max,1}$ (m/s)	R_2 (km)	α_2 (°)	$U_{max,2}$ (m/s)
SL-1	R1	20.0	3.0	80.0	6.0	9.80	32.800	31.658	45	6.26	2.8	0.1	38.6	7.73	0.16	4.6
	R2											0.3	32.1			
	R3											0.5	27.1			
	R4											0.8	17.1			
	R5											0.9	12.1			
	R6				20.0	32.7						0.5	49.5			8.4
	R7				20.0	32.7						0.8	31.3			8.4
SL-2	R1	8.0	3.0	40.0	26.0	10.0	32.725	31.708	45	4.36	2.8	0.6	47.7	11.73	0.6	17.5
	R2											0.7	41.3			
	R3											0.8	33.7			
	R4											0.9	23.9			
CSL [*]	R1	10.0	1.0	50.0	20.4	10.0	32.350	31.417	90	3.25	3.0	0.5	48.9	3.25	3.0	48.9
	R2											0.7	37.9			37.9
	R3											0.9	21.9			21.9

[#] Defined as in Harbitz (1992) [36]. Slide volume $V = 0.9Bh_m(L + 0.9S)$; B, width; L, length; S, smoothing distance; h_m, average height; [$] From the positive X direction (West to East); [¶] R_1 and R_2 are down-slope displacements, with slope angles α_1 and α_2 and maximum speeds $U_{max,1}$ and $U_{max,2}$, respectively. For SL-1 and SL-2 slides, μ/μ_{st} = 0.75 for the second slope; while a single slope scheme is adopted for CSL slides; [*] Implemented along with a change in bathymetry defining an hypothetical canyon lying E-W, 62 km length, 6.5 km wide and 35 m deep, excavated on the reconstructed 3500 year BP bathymetry, and centred at 31.27° E, 31.45° N.

Slide SL-2 is also extracted from the present bathymetry. Its front (Table 1), at the south-eastern border of the stable shelf in the Nile Delta, is initially at 71 m water depth and displaces 4.3 km down a slope with angle 2.8°, and then 11.7 km along a slope of 0.6°. The displacement ends at a plateau. The area covered by this hypothetical mass-wasting does not disagree with the geological surveys above referred. With a maximum height of 26.0 m and a total volume of 10 km^3, the moving slide reaches a maximum speed of 47.7 m/s for $\mu = 0.6\mu_{st}$ (SL-2, run R1). Runs R2 to R4 for this slide use increasing values of μ in the first displacement, while the same criteria than for SL-1 has been adopted for the second displacement (Table 1).

Table 2. Fault parameters used in the simulations. Geographical coordinates correspond to the fault center. Rake is 90° in all cases.

Tsunami	$\lambda_E°$	$\Phi_N°$	Length (km)	Width (km)	Slip (m)	Strike (degree)	Dip (degree)	Potential Energy [#] (J)
F-1	32.667	31.500	60.0	20.0	8.0	55.0	45.0	4.6×10^{13}
F-2	32.667	31.500	80.0	24.0	12.0	65.0	35.0	1.8×10^{14}
F-2 + SL-2 R2 [*]	32.667	31.500	80.0	24.0	12.0	65.0	35.0	1.8×10^{14}

[#] Initial potential energy linked to the Okada's deformation. [*] Double source, composed by a fault earthquake and a simultaneous submarine landslide (SL-2 R2 in Table 1).

Loncke *et al.*, (2006) [54] reported (their Figure 9) the paths of the Messinian canyons in the Nile Delta. Particularly, at the Damietta branch, a former canyon of some 60 km length and 6 km wide ran eastward. In the GEBCO08 bathymetry, the stable shelf area shows some holes of several hundred meters deep and few km wide (dark spots in Figure 3), likely remains of the former canyons. Due to its proximity to the former *Shi-Hor* Lagoon, and as a potential and extreme case, the source CSL (Table 1) simulates the effect of a large submarine landslide running into a still partially colmated canyon. The box geometry has been selected as large as possible, and its displacement has to be confined within the width of the former canyon. A maximum height of 20.4 m has been adopted to account for a total volume of 10.0 km^3. For simplicity a uniform slope angle of 3.0 degrees has been selected, with three options for μ values, leading to maximum speeds of 48.9, 37.9 and 21.9 m/s in runs CSL R1, R2 and R3, respectively.

Gamal (2013) [59] provided a detailed study of the Pelusium Mega-Shear Fault system. The Pelusium line runs north-westwards crossing the south-eastern Nile Delta. Hypothetical earthquakes triggered by this fault with epicentres in this area of the Nile Delta could have been tsunamigenic sources whose potential effects on the nearby former shoreline of the *Shi-Hor* Lagoon will be studied here. Thus, sources F-1 and F-2 (Table 2) are defined with tentative values for length, width and slip, and angular parameters inspired in the studies of Badawy and Abdel Fattath (2001) [65], and Gamal (2013) [59]. As the tsunamis triggered by geological faults are less energetic events when compared with the studied submarine landslides, the application of sensitivity tests has been discarded in this case.

Finally, the case of a fault-earthquake that triggers an almost simultaneous submarine landslide is also considered by combining the two most energetic sources, F-2 and SL-2 R2.

3. Results and Discussion

The numerical model has been run for all the tsunamigenic sources (Tables 1 and 2) with a typical simulation time of 5 h and a time step of 2 s. Figure A1 (in Electronic Supplementary Material) shows the time series of slide velocities for some of the studied submarine landslides (Table 1).

Figure 4 shows the computed maximum amplitudes for the submarine landslides SL-1 R6, SL-2 R1 and CSL R1 (Table 1), selected as the extreme cases of each tsunamigenic source. Sources SL-1 and SL-2 are both located at the outer edge of the stable Nile Delta, and they show strong directionality towards the coastline of Israel, were wave amplitudes exceeds 20 m in some areas. For source CSL, most of the tsunami energy propagates northwards. In the three cases, the former Egyptian coastline is much less impacted by tsunami waves. An example of tsunami propagation can be seen in Supplementary material, which shows an animation with some snapshots of the propagation of the tsunami generated by the submarine landslide SL-2 R2.

The energy of the tsunami can be evaluated at any time by integration over the whole domain of the potential and kinetic energies of the water column at each grid-cell. Submarine landslides transfer potential and kinetic energy to the water column, competing with energy dissipation by frictional stress at the seafloor (Equations (2)–(4)). Thus, the peak energy usually appears around the time of the maximum velocity for the moving slide. In the scenario of landslide SL-1, with maximum thickness $h = 6.0$ m, the tsunami peak energy increases with $U_{max,1}$ from 5.3×10^{14} J in R5 up to 4.2×10^{15} J in R1 (see Table 1). When a maximum thickness $h = 20.0$ m is adopted, the tsunami peak energy reaches 8.4×10^{15} J in R6 (the maximum computed wave amplitudes for this extreme event are shown in Figure 4). In the scenario of landslide SL-2, with maximum thickness $h = 26.0$ m, the tsunami peak energy only slightly increases with $U_{max,1}$, from 2.6×10^{15} J in R4 up to 3.7×10^{15} J in R1 (see Table 1, and Figure 4 for the maximum wave amplitudes computed for R1). Similarly, for landslide CSL the tsunami peak energy increases from 2.0×10^{15} J (R3) up to 3.9×10^{15} J (R1).

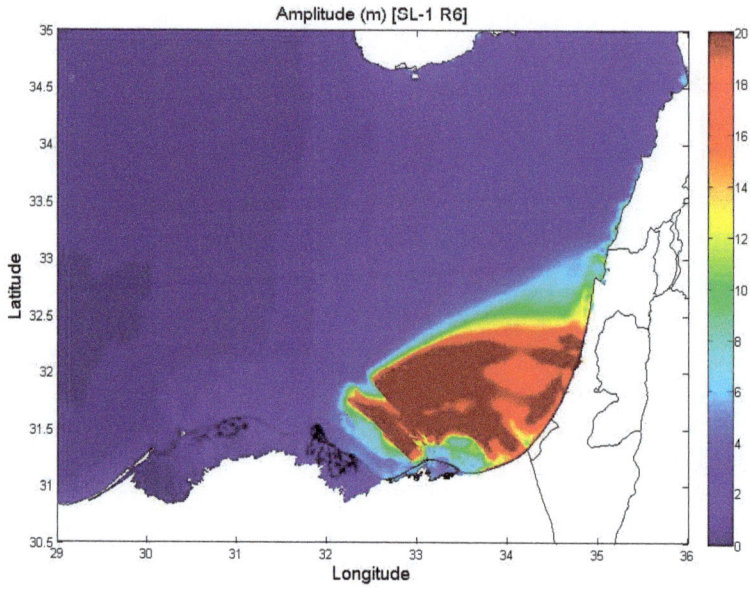

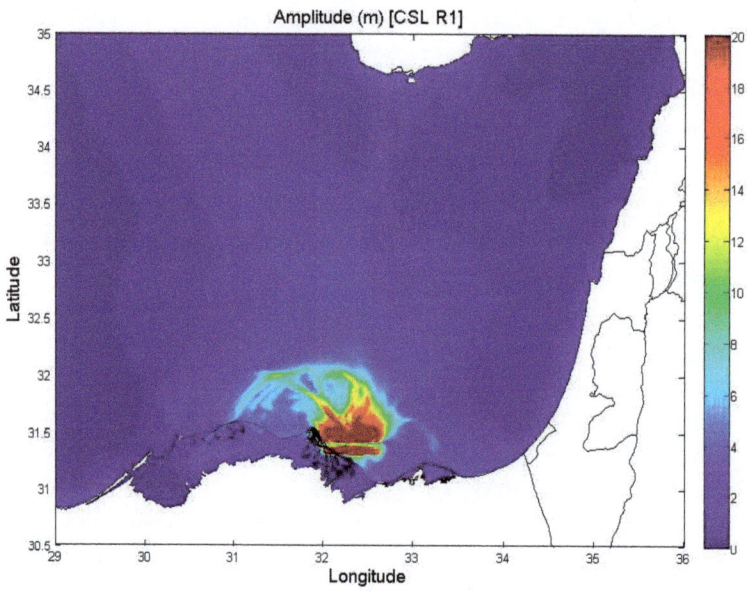

Figure 4. *Cont.*

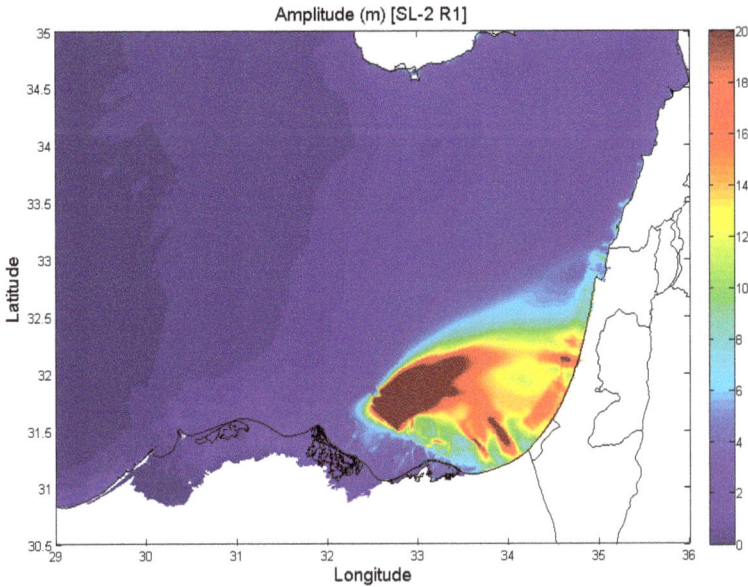

Figure 4. Computed maximum amplitude for water elevations (m) due to submarine landslides SL-1 R6, SL-2 R1 and CSL R1. Source parameters are presented in Table 1. Simulation time is 5 h.

Figures 5 and 6 show the computed time series of water elevations for all the landslide tsunamis (Table 1) at three of the synthetic tidal gauges TG-3, TG-4 and TG-5 (see Figure 3). TG-4, at the *Shi-Hor* Lagoon, is the candidate scenario for the drowning of the Egyptian army, as mentioned in the introduction. For all the runs with SL-1 landslide, a weak withdrawal of the sea of few tens cm takes place, and lasting about one hour. After that, water level suddenly rises about 0.5 m, and then continues increasing for 40–50 min, and reaching levels around or below 1 m for Runs R1 to R5, and below 1.5 m for R5 and R6. For runs R1 to R5, differences in the first maximum water height are of the order of 10–20 cm, what can be seen as a model sensitivity test for the Coulombian friction coefficient (and then for $U_{max,1}$). These differences are only of few cm for landslides SL-2 and CSL (see Figure 6). Concerning the evolution of water level at TG-4 for the series of SL-2 tsunamis (Figure 6), it is similar to the one commented for SL-1. For CSL landslides, the withdrawal of the sea lasts about two hours, but interfered by the arrival of a weak wave of few tens cm. At TG-5, in the inner shoreline (Figure 3), the tsunami signal arrives latter, and with a similar although smoother pattern than in TG-4; but the maximum water heights are amplified about 0.5 m for SL-1 and 0.25 m for SL-2 and CSL submarine landslides. At TG-3, landslides SL-1 closely follow the same pattern, with a withdrawal of the sea of about 4 m that lasts one hour, followed by a sudden rise of 2–3 m, and then a continuous increase over 50–60 min up to reach water levels of 3–5 m (this last for the 32.7 km^3 landslide).

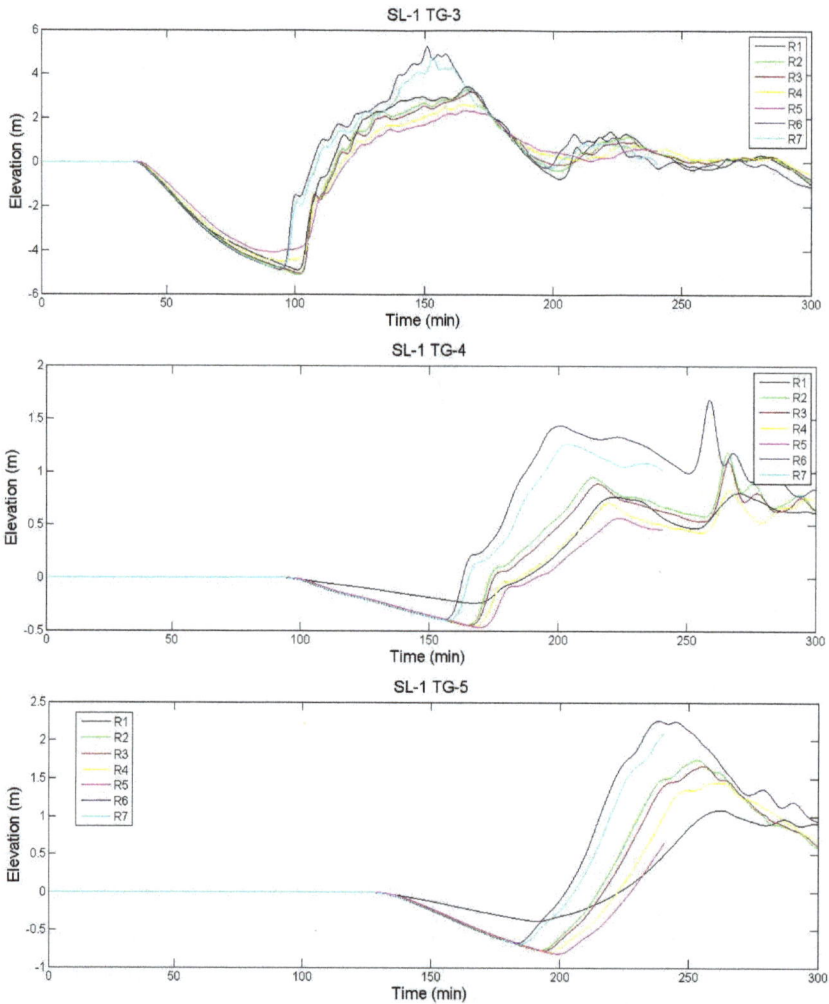

Figure 5. Computed time series of water elevations at tidal gauges TG-3, TG-4 and TG-5 for the submarine landslide SL-1, runs R1 to R7 (Table 1). See Figure 3 for the location of the synthetic gauges. Initial water depths were 9.0, 2.1 and 8.0 m for TG-3, TG-4 and TG-5, respectively.

For tsunamis triggered by geological faults the maximum energy corresponds to the initial potential energy due to the deformation of the free water surface. It was 4.6×10^{13} J for F-1and 1.8×10^{14} J for F-2. The two studied sources are less energetic events than the submarine landslides. Figure 7 shows the maximum computed wave amplitude for tsunamis F-1, F-2, and for the simultaneous source F-2+SL-2 R2 (a landslide triggered in sequence with the earthquake). In this last case, the landslide component dominates the propagation pattern. In tsunamis F-1 and F-2, the highest amplitudes remains confided within the stable shelf of the Nile Delta, and they show strong directionality towards the coastline delimiting the Bardawil Lake, but with weak effects around the *Shi-Hor* lagoon, as shown in the computed time series of Figure 8.

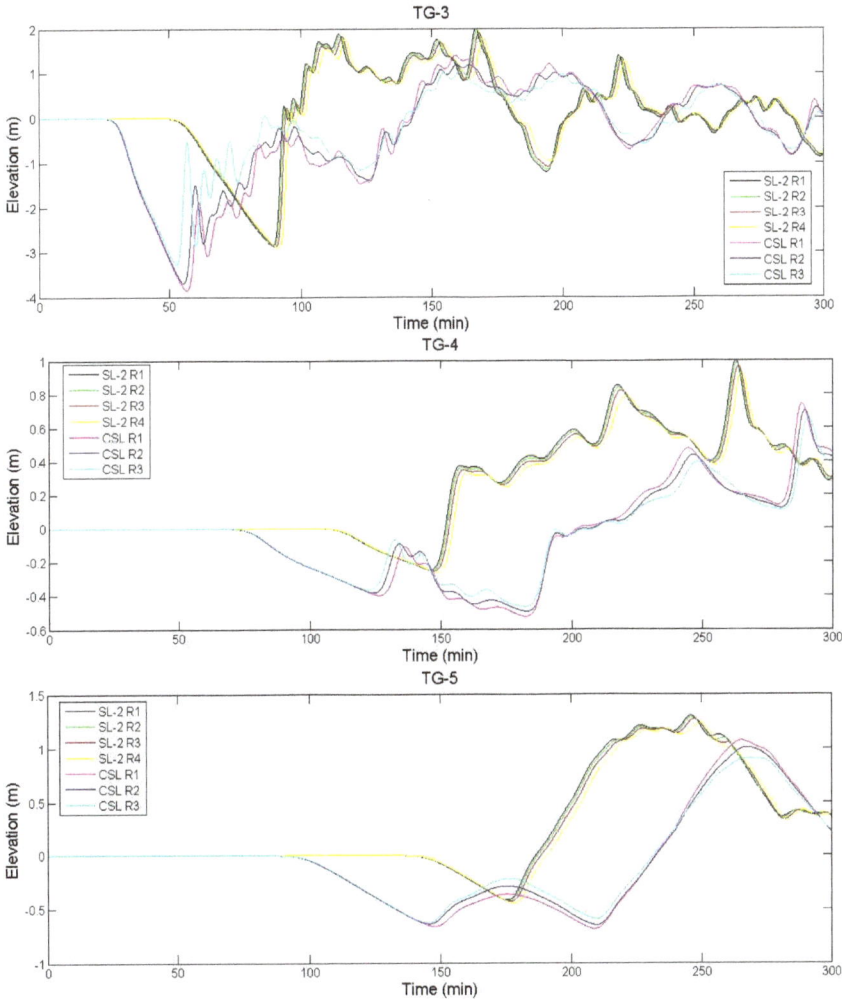

Figure 6. Computed time series of water elevations at tidal gauges TG-3, TG-4 and TG-5 for the submarine landslide SL-2, with runs R1 to R4; and CSL, with runs R1 to R3 (Table 1). See Figure 3 for the location of the synthetic gauges. Initial water depths were 9.0, 2.1 and 8.0 m for TG-3, TG-4 and TG-5, respectively.

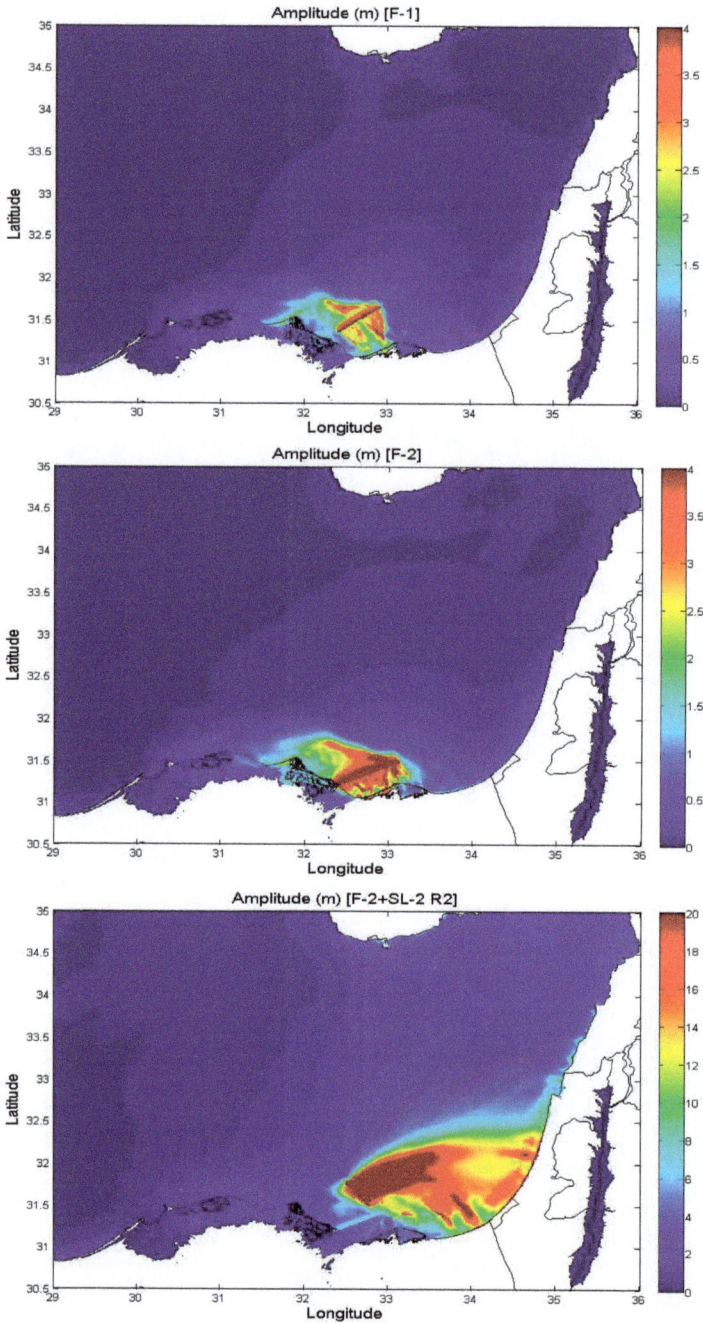

Figure 7. Computed maximum amplitude for water elevations (m) due to geological faults F-1, F-2 and the double source F-2 + SL-2 R2. Source parameters are presented in Table 1. Simulation time is 5 h.

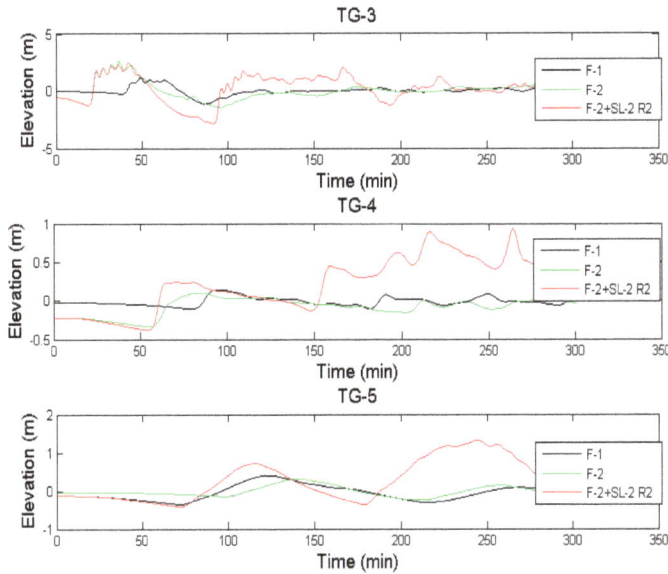

Figure 8. Computed time series of water elevations at tidal gauges TG-3, TG-4 and TG-5 for the geological faults F-1. F-2, and the double source F-2 + SL-2 R2 (Table 1). See Figure 3 for the location of the synthetic gauges. Initial water depths were 9.0, 2.1 and 8.0 m for TG-3, TG-4 and TG-5, respectively.

Figure A2 shows the computed time series of water elevations at the synthetic gauges TG-1, TG-2 and TG-6 (see Figure 3) for a sub-set of tsunamis. Tsunamis triggered by geological faults (F-1 and F-2) have negligible effects in the Levantine coasts and in Alexandria (TG-6), in the western Nile Delta. The most energetic submarine landslides impact the Levantine coasts with waves above 15 m height, while their effects in the western Nile Delta are much weaker, with wave amplitudes of 0.5 m.

For the studied landslide tsunamis, water currents exceed 5 to 10 m/s along the eastern delta shelf edge and the southern Levantine coastline (Figure 9 shows some examples). These are the areas where the major energy dissipation (Equations (2)–(4)) and, consequently, the greatest impacts occur.

In all the cases, wave amplitudes only slightly exceed 1–1.5 m for the most energetic tsunamis at the *Shi-Hor* Lagoon (synthetic gauges 4 in Figures 5 and 6). The paleo-lagoon never dries out, nor the *Kedua* passage. Thus, around *Shi-Hor* Lagoon, only the coastal areas less than 1.0–1.5 m above sea level would have had some risk of being flooded. These dynamical tsunami scenarios (Figures 5 and 6), and particularly the time sequence of water elevations, perhaps could produce some scene of fear and chaos in a military group camped at the *Tjaru*'s seaside, but in our opinion they hardly can explain a sudden drowning of the Egyptian army.

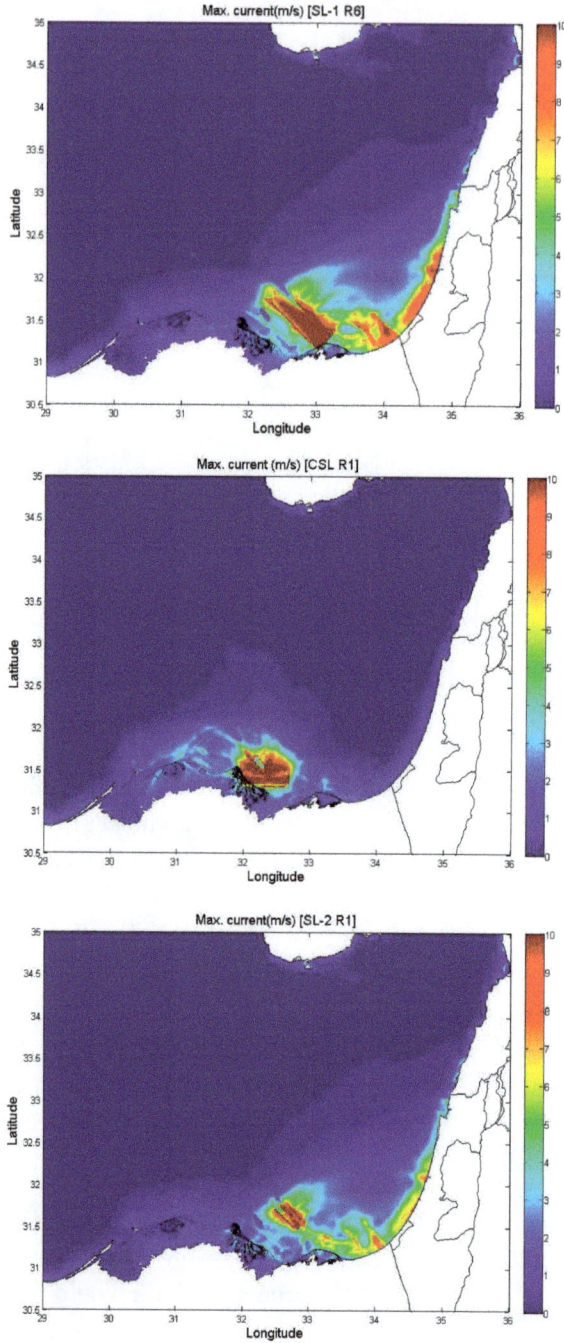

Figure 9. Computed maximum water currents (m/s) for tsunamis SL-1 R6, SL-2 R1 and CSL R1. Source parameters are presented in Table 1. Simulation time is 5 h.

234

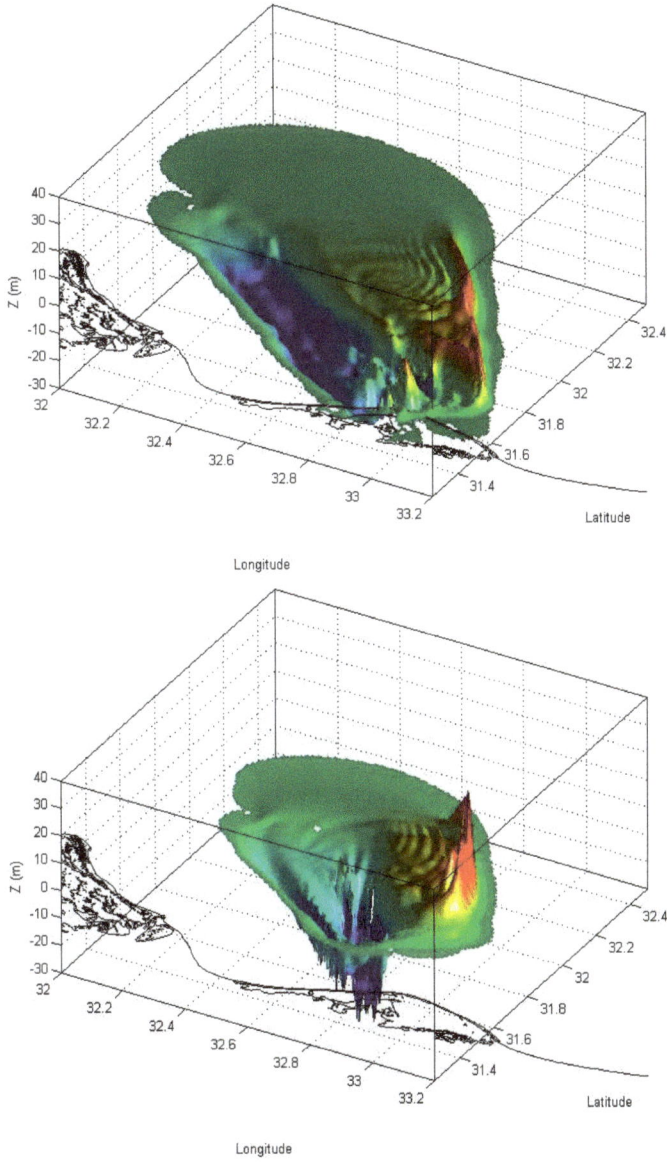

Figure 10. 3D-view of the free water surface deformation, computed after 10 min of the beginning of landslides SL-1 R1 (up) and SL-2 R2 (bottom).

The above results are a consequence of the particular geological setting of the studied marine system. A large area of the Nile Delta shelf extends in front of the Shi Hor Lagoon. In these shallow waters, tsunami waves travels at low speed, while the bed friction is high (it is inversely proportional to water depth). The implemented formulation for friction (Equations (2)–(4)) has been widely used and validated in many oceanographic studies, and particularly in models for tidal and tsunami propagation (see for instance Periáñez and Abril, 2014b, [31]). Figure A3 shows a sensitivity test for the friction coefficient, applied to the most energetic tsunami from our set of simulations (SL-1 R6). When the nominal value of the friction coefficient ($k = 0.0025$) is increased / decreased by a 50%, water elevations at the gauges TG-3 to TG-5 slightly decrease/increase, without affecting in the essential to the main conclusions of this study.

Scenarios for severe flooding can be found along the northern Sinai coasts, as shown in Figure 4. Nevertheless, they do not match with the plausible paths for the exodus. It is worth noting the great exposure to most of the tsunamis of the sand barrier delimiting the present Bardawil Lagoon. It is known that it was a military path in Persian times, and it has been proposed by some authors as the exodus path, although the most recent studies concluded that this sand barrier was not a passable path at the second millennium BC [9].

Tsunamis triggered by submarine landslides are characterized by a depression of the free water surface over the line of rupture while a huge wave appears at the front of the moving slice. For some exceptional witnesses on the northern Sinai shoreline or in cabotage sailing in this area, a submarine landslide along the eastern Nile Delta shelf edge could have been seen as a parting of the sea (as shown in the 3D-view of Figure 10).

The fingerprint of noticeable tsunami impacts in coastal areas can be potentially found in their characteristic sedimentary deposits, although they are not always well preserved; and huge submarine landslides can be detected and dated through their associated turbidite deposits. The selected tsunamigenic sources in this paper are compatible with the main geological features of the Nile Delta, and they are not in contradiction with our present knowledge from field studies. Obviously, this is not a proof for their feasibility, what can only be provided by empirical evidences. They represent extreme and near-field cases in an attempt to provide the best scenarios for the maximalist hypothesis of tsunamis behind the biblical Exodus. But, even for these favourable scenarios, the lack of noticeable flooding of the area, along with the time sequence of water elevations, make difficult to accept a tsunami as a reliable hypothesis neither for the first plague nor for the drowning of the Egyptian army in the surroundings of the *Shi-Hor* Lagoon.

The tsunamigenic deposits from the Santorini event in the coastal area of Israel and Gaza [19] can be explained by the occurrence of tsunamis with source in the eastern Mediterranean, likely triggered by submarine landslides. In our opinion this could have been a source of inspiration for the Exodus narrative, which has to be understood as the likely result of the merging of several stories from different sources [4].

4. Conclusions

A numerical model for tsunami propagation, based on the 2D shallow water equations and previously applied and validated in a wide set of scenarios, has been adapted, from available geological and archaeological studies, to the former paleogeographic conditions in the eastern Nile Delta.

Seventeen potential tsunamigenic sources in the eastern Nile Delta have been suggested as possible candidates within the Minoan tsunami sequence, comprising faults along the Pelusium Mega-shear system and several submarine landslides at the shelf edge and along the hypothetical remains of the Messinian Damieta canyon. They have been defined as large as possible within the known constrains stated by geological studies, and comprise sensitivity tests for the total volume and the prescribed motion of the landslides.

Tsunamis triggered by submarine landslides could have severely impacted the northern Sinai and southern Levantine coasts with waves over 10–15 m height and water currents over 5–10 m/s, but with weak effects in the surroundings of the *Shi-Hor* Lagoon, the proposed scenario for the Biblical sea parting.

The real occurrence of tsunamis can only be proved by empirical evidences; but as the proposed ones are representative of the most favorable scenarios for the hypothesis to be tested, the lack of any noticeable flooding, and the particular time sequence of water elevations, make difficult to accept them as plausible and literally explanation for the first plague and for the drowning of the Egyptian army in the surroundings of the *Shi-Hor* Lagoon, although they might have been a source of inspiration for the Biblical narrative.

Supplementary Materials: Snapshots for the propagation of the tsunami SL-2 R2. Scale in the color bar is given in m. Source parameters are presented in Table 1. Supplementary materials can be accessed at: http://www.mdpi.com/2077-1312/3/3/745/s1.

Author Contributions: Conceived and designed the experiments: JMA, RP. Performed the experiments: JMA, RP. Analyzed the data: JMA, RP. Wrote the paper: JMA, RP.

Conflicts of Interest: The authors declare no conflict of interest.

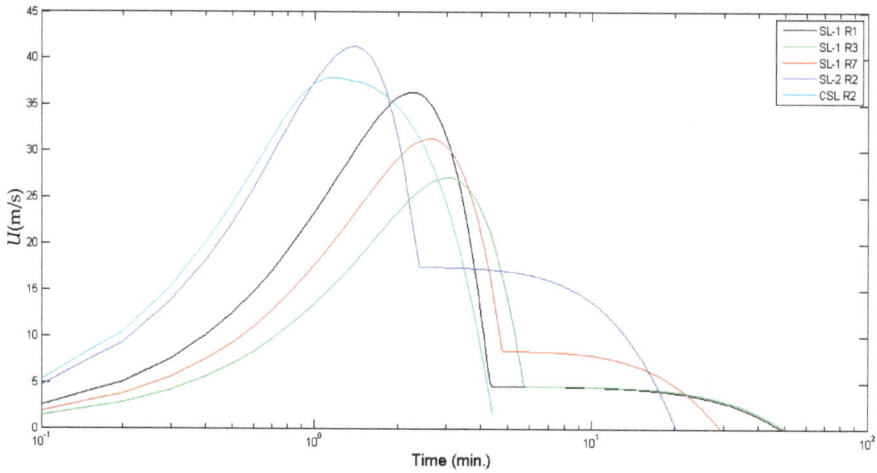

Figure A1. Prescribed time series for the slice velocity (Equations (7)–(8)) for some of the submarine landslide scenarios defined in Table 1.

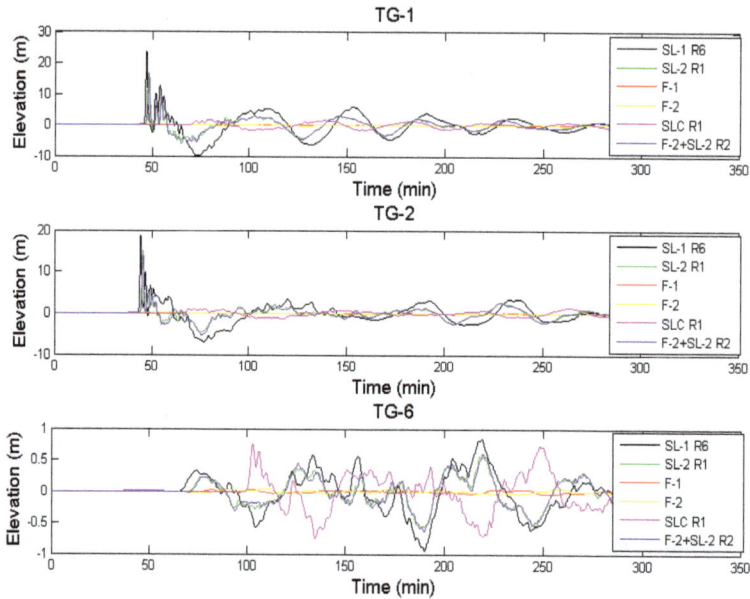

Figure A2. Computed time series of water elevations at tidal gauges TG-1, TG-2 and TG-6 for six of the tsunamigenic sources given in Tables 1 and 2. See Figure 3 for the location of the synthetic gauges. Initial water depths were 36, 33 and 36 m for TG-1, TG-2 and TG-6, respectively.

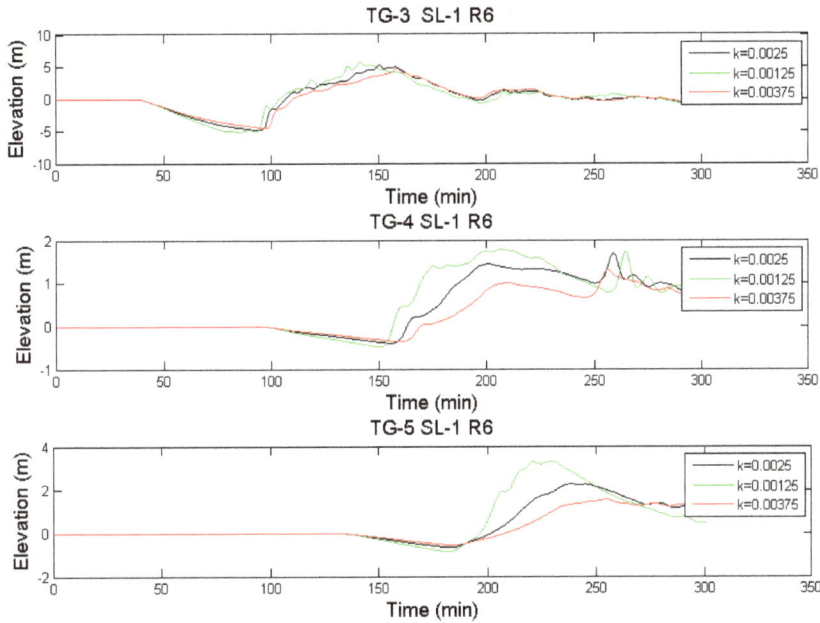

Figure A3. Computed time series of water elevations at tidal gauges TG-3, TG-4 and TG-5 for tsunami SL-1 R6 by increasing/decreasing by a 50% the nominal value of the friction coefficient (Equations (2)–(4)). See Figure 3 for the location of the synthetic gauges. Initial water depths were 9.0, 2.1 and 8.0 m for TG-3, TG-4 and TG-5, respectively.

References

1. Friedrich, W.L. The Minoan Eruption of Santorini around 1613 B.C. and its consequences. Tagungen des Landesmuseums für Vorgeschichte Halle. *Band* **2013**, *9*, 37–48.
2. Stanley, J.D.; Sheng, H. Volcanic shards from Santorini (Upper Minoan ash) in the Nile Delta, Egypt. *Nature* **1986**, *320*, 733–735.
3. Bruins, H.J.; van der Plicht, J. The Exodus enigma. *Nature* **1996**, *382*, 213–214.
4. Sivertsen, B.J. *The Parting of the Sea. How volcanoes, earthquakes, and plagues shaped the story of Exodus*; Princeton University Press: Princeton, NJ, USA, 2009.
5. Trevisanato, S.I. Treatments for burns in the London Medical Papyrus show the first seven biblical plagues of Egypt are coherent with Santorini's volcanic fallout. *Med. Hypotheses* **2006**, *66*, 193–196.
6. Trevisanato, S.I. Six medical papyri describe the effects of Santorini's volcanic ash, and provide Egyptian parallels to the so-called biblical plagues. *Med. Hypotheses* **2006**, *67*, 187–190.

7.   Trevisanato, S.I. Medical papyri describe the effects of the Santorini eruption on human health, and date the eruption to August 1603–March 1601 BC. *Med. Hypotheses* **2007**, *68*, 446–449.

8.   Byers, G.A. New evidence from Egypt on the location of the Exodus Sea crossing, part 1. *Bible Spade* **2006**, *19*, 14–22.

9.   Hoffmeier, J.K. *Ancient Israel in Sinai. The evidence for the authenticity of the wilderness tradition*; Oxford University Press Inc.: New York, NY, USA, 2005.

10. Wood, B.G. Debunking "The Exodus Decoded". ABR, 20 September 2006. Available online: Http://www.biblearchaeology.org/post/2006/09/20/Debunking-The-Exodus-Decoded.aspx (accessed on 15 May 2015).

11. Voltzinger, N.; Androsov, A. Modeling the hydrodynamic situation of the Exodus. *Izv. Atmos. Ocean. Phys.* **2003**, *39*, 482–496.

12. Drews, C.; Han, W. Dynamics of wind setdown at Suez and the Eastern Nile Delta. *PLoS ONE* **2010**, *5*, e12481.

13. Abril, J.M.; Abdel-Aal, M.M. Marine radioactivity studies in the Suez Canal, Part I: Hydrodynamics and transit times. *Estuar. Coast. Shelf Sci.* **2000**, *50*, 489–502.

14. Abril, J.M.; Abdel-Aal, M.M. Marine radioactivity studies in the Suez Canal, Part II: Field experiments and a modelling study of dispersion. *Estuar. Coast. Shelf Sci.* **2000**, *50*, 503–514.

15. Jacobovici, S.; Cameron, J. *The Exodus decoded*; Documentary Film: Toronto, ON, Canada; 19; April; 2006.

16. Salzman, R. *Mega-Tsunami: The true story of the Hebrew Exodus from Egypt*; iUniverse: Lincoln, NE, USA, 2005.

17. Novikova, T.; Papadopoulos, G.A.; McCoy, F.W. Modelling of tsunami generated by the giant Late Bronze Age eruption of Thera, South Aegean Sea, Greece. *Geophys. J. Int.* **2011**, *186*, 665–680.

18. Mészáros, S. Some words on the Minoan tsunami of Santorini. In *Thera and the Aegean World II*; Doumas, C., Ed.; The Thera Foundation: London, UK, 1978; pp. 257–262.

19. Goodman-Tchernov, B.N.; Dey, H.W.; Reinhardt, E.G.; McCoy, F.; Mart, Y. Tsunami waves generated by the Santorini eruption reached Eastern Mediterranean shores. *Geology* **2009**, *37*, 943–946.

20. Neev, D.; Balker, N.; Emery, K.O. *Mediterranean Coast of Israel and Sinai, Holocene Tectonics from Geology and Geophysics and Archaeology*; Taylor and Francis: New York, NY, USA, 1987; p. 130.

21. Pfannenstiel, M. Erläuterungen zu den bathymetrischen Karten des östlichen Mittelmeeres. *Bull. Inst. Oceanogr.* **1960**, *57*, 60.

22. De Martini, P.M.; Barbano, M.S.; Smedile, A.; Gerardi, F.; Pantosti, D.; del Carlo, P.; Pirrotta, C. A unique 4000 year long geological record of multiple tsunami inundations in the Augusta Bay (eastern Sicily, Italy). *Mar. Geol.* **2010**, *276*, 42–57.

23. De Martini, P.M.; Barbano, M.S.; Pantosti, D.; Smedile, A.; Pirrotta, C.; del Carlo, P.; Pinzi, S. Geological evidence for paleotsunamis along eastern Sicily (Italy): An overview. *Nat. Hazards Earth Syst. Sci.* **2012**, *12*, 2569–2580.

24. Ozel, N.M.; Ocal, N.; Cevdet, Y.A.; Dogan, K.; Mustafa, E. Tsunami hazard in the Eastern Mediterranean and its connected seas: Towards a tsunami warning center in Turkey. *Soil Dyn. Earthq. Eng.* **2011**, *31*, 598–610.

25. Periáñez, R.; Abril, J.M. Modelling tsunamis in the Eastern Mediterranean Sea. Application to the Minoan Santorini tsunami sequence as a potential scenario for the biblical Exodus. *J. Mar. Syst.* **2014**, *139*, 91–102.

26. Choi, B.H.; Pelinovsky, E.; Kim, K.O.; Lee, J.S. Simulation of the trans-oceanic tsunami propagation due to the 1883 Krakatau volcanic eruption. *Nat. Hazards Earth Syst. Sci.* **2003**, *3*, 321–332.

27. Kowalik, Z.; Knight, W.; Logan, T.; Whitmore, P. Numerical modelling of the Indian Ocean tsunami. In *The Indian Ocean Tsunami*; Murty, T.S., Aswathanarayana, U., Nipurama, N., Eds.; Taylor and Francis Group: London, UK, 2007; pp. 97–122.

28. Periáñez, R.; Abril, J.M. Modeling tsunami propagation in the Iberia–Africa plate boundary: Historical events, regional exposure and the case-study of the former Gulf of Tartessos. *J. Mar. Syst.* **2013**, *111–112*, 223–234.

29. Yolsal-Cevikbilen, S.; Taymaz, T. Earthquake source parameters along the Hellenic subduction zone and numerical simulations of historical tsunamis in the Eastern Mediterranean. *Tectonophysics* **2012**, *536–537*, 61–100.

30. Kowalik, Z.; Murty, T.S. *Numerical Modelling of Ocean Dynamics*; World Scientific: Singapore, 1993.

31. Periáñez, R.; Abril, J.M. A numerical modeling study on oceanographic conditions in the former Gulf of Tartessos (SW Iberia): Tides and tsunami propagation. *J. Mar. Syst.* **2014**, *139*, 68–78.

32. Herzfeld, M.; Schmidt, M.; Griffies, S.M.; Liang, Z. Realistic test cases for limited area ocean modelling. *Ocean Model.* **2011**, *37*, 1–34.

33. Kampf, J. *Ocean Modelling for Beginners*; Springer-Verlag: Heidelberg, Germany, 2009.

34. Okada, Y. Surface deformation due to shear and tensile faults in a half-space. *Bull. Seismol. Soc. Am.* **1985**, *75*, 1135–1154.

35. Abril, J.M.; Periáñez, R.; Escacena, J.L. Modeling tides and tsunami propagation in the former Gulf of Tartessos, as a tool for Archaeological Science. *J. Archaeol. Sci.* **2013**, *40*, 4499–4508.

36. Harbitz, C.B. Model simulations of tsunamis generated by the Storegga slides. *Mar. Geol.* **1992**, *105*, 1–21.

37. Cecioni, C.; Bellotti, G. Modeling tsunamis generated by submerged landslides using depth integrated equations. *Appl. Ocean Res.* **2010**, *32*, 343–350.

38. Abril, J.M.; Periáñez, R. A modelling approach to Late Pleistocene mega-tsunamis likely triggered by giant submarine landslides in the Mediterranean. *Mar. Geol.* **2015**, under review.

39. Lastras, G.; de Blasio, F.V.; Canals, M.; Elverhøi, A. Conceptual and numerical modeling of the BIG'95 debris flow, western Mediterranean Sea. *J. Sediment. Res.* **2005**, *75*, 784–797.

40. Ruff, L.J. Some aspects of energy balance and tsunami generation by earth-quakes and landslides. *Pure Appl. Geophys.* **2003**, *160*, 2155–2176.

41. De Blasio, F.; Elverhøi, A.; Issler, D.; Harbitz, C.B.; Bryn, P.; Lien, R. On the dynamics of subaqueous clay rich gravity mass flows -the giant Storegga slide, Norway. *Mar. Pet. Geol.* **2005**, *22*, 179–186.

42. Fine, I.V.; Rabinovich, A.B.; Bornhold, B.D.; Thomson, R.E.; Kulikov, E.A. The Grand Banks landslide-generated tsunami of November 18, 1929: Preliminary analysis and numerical modeling. *Mar. Geol.* **2005**, *215*, 45–57.

43. Tappin, D.R.; Watts, P.; Grilli, S.T. The Papua New Guinea tsunami of 17 July 1998: Anatomy of a catastrophic event. *Nat. Hazards Earth Syst. Sci.* **2008**, *8*, 243–266.

44. Periáñez, R.; Abril, J.M. Computational fluid dynamics simulations of the Zanclean catastrophic flood of the Mediterranean (5.33 Ma). *Palaeogeogr. Palaeoclimatol. Palaeoecol.* **2015**, *424*, 49–60.

45. Coutellier, V.; Stanley, D.J. Late quaternary stratigraphy and paleogeography of the eastern Nile Delta, Egypt. *Mar. Geol.* **1987**, *77*, 257–275.

46. El-Raey, M. Vulnerability assessment of the coastal zone of the Nile Delta of Egypt, to the impacts of sea level rise. *Ocean Coast. Manag.* **1997**, *37*, 29–40.

47. Barka, A.; Reilinger, R. Active tectonics on the Eastern Mediterranean region: Deduced from GPS, neotectonics and seismicity data. *Ann. Geofis.* **1997**, *40*, 587–610.

48. Mascle, J.; Benkhelil, J.; Bellaiche, G.; Zitter, T.; Woodside, J.; Loncke, L.; Prismed II Scientific Party. Marine geologic evidence for a Levantine-Sinai Plate; a new piece of the Mediterranean puzzle. *Geology (Boulder)* **2000**, *28*, 779–782.

49. Hamouda, A.Z. Numerical computations of 1303 tsunamigenic propagation towards Alexandria, Egyptian Coast. *J. Afr. Earth Sci.* **2006**, *44*, 37–44.

50. Stiros, S.C. The 8.5+ magnitude AD365 earthquake in Crete: Coastal uplift, topography changes, archaeological and historical signature. *Quat. Int.* **2010**, *216*, 54–63.

51. Papadopoulos, G.A.; Daskalaki, E.; Fokaefs, A.; Giraleas, N. Tsunami hazards in the Eastern Mediterranean: Strong earthquakes and tsunamis in the East Hellenic Arc and Trench system. *Nat. Hazards Earth Syst. Sci.* **2007**, *7*, 57–64.

52. Harbitz, C.B.; Løvholt, F.; Bungum, H. Submarine landslide tsunamis: How extreme and how likely? *Nat. Hazards* **2014**, *72*, 1341–1374.

53. Gaullier, V.; Mart, Y.; Bellaiche, G.; Macle, J.; Vendeville, B.; Zitter, T.; Prismed II scientific parties. Salt tectonics in and around the Nile Deep-Sea Fan: Insights from the PRISMED II cruise: From the Artic to the Mediterranean: Salt, shale and igneous diapirs in and around Europe. *Geol. Soc.* **2000**, *174*, 111–129.

54. Loncke, L.; Gaullier, V.; Macle, J.; Vendeville, B.; Camera, L. The Nile deep-sea fan: An example of interacting sedimentation, salt tectonics, and inherited subsalt paleotopographic features. *Mar. Pet. Geol.* **2006**, *23*, 297–315.

55. Ducassou, E.; Migeon, S.; Mulder, T.; Murat, A.; Capotondi, L.; Bernasconi, S.M.; Macle, J. Evolution of the Nile deep-sea turbidite system during the Late Quaternary: Influence of climate change on fan sedimentation. *Sedimentology* **2009**, *56*, 2061–2090.

56. Ducassou, E.; Migeon, S.; Capotondi, L.; Mascle, J. Run-out distance and erosion debris-flows in the Nile deep-sea fan system: Evidence from lithofacies and micropaleontological analyses. *Mar. Pet. Geol.* **2013**, *39*, 102–123.

57. Loncke, L.; Gaullier, V.; Droz, L.; Ducassou, E.; Migeon, S.; Mascle, J. Multi-scale slope instabilities along the Nile deep-sea fan, Egyptian margin: A general overview. *Mar. Pet. Geol.* **2009**, *26*, 633–646.

58. Loncke, L.; Vendeville, B.C.; Gaullier, V.; Mascle, J. Respective contributions of tectonic and gravity-driven processes on the structural pattern in the Eastern Nile deep-sea fan: Insights from physical experiments. *Basin Res.* **2010**, *22*, 765–782.

59. Gamal, M.A. Truthfulness of the existence of the Pelusium Megashear Fault System, East of Cairo, Egypt. *Int. J. Geosci.* **2013**, *4*, 212–227.

60. Garziglia, S.; Migeon, S.; Ducassou, E.; Loncke, L.; Macle, J. Mass-transport deposits on the Rosetta province (NW Nile deep-sea turbidite system, Egyptianmargin): Characteristics, distribution, and potential causal processes. *Mar. Geol.* **2008**, *250*, 180–198.

61. Khaled, K.A.; Attia, G.M.; Metwalli, F.I.; Fagelnour, M.S. Subsurface geology and petroleum system in the eastern offshore area, Nile Delta, Egypt. *J. Appl. Sci. Res.* **2014**, *10*, 254–270.

62. Bayon, G.; Loncke, L.; Dupré, S.; Caprais, J.C.; Ducassou, E.; Duperron, S.; Etoubleau, J.; Foucher, J.P.; Fouquet, Y.; Gontharet, S.; *et al.* Multi-disciplinary investigation of fluid seepage on an unstable margin: The case of the Central Nile deep sea fan. *Mar. Geol.* **2009**, *261*, 92–104.

63. Prinzhofer, A.; Deville, E. Origins of hydrocarbon gas seeping out from offshore mud volcanoes in the Nile Delta. *Tectonophysics* **2013**, *591*, 52–61.

64. Urgeles, R.; Camerlenghi, A. Submarine landslides of the Mediterranean Sea: Trigger mechanisms, dynamics and frequency-magnitude distribution. *J. Geophys. Res. Earth Surf.* **2013**, *118*, 2600–2618.

65. Badawy, A.; Abdel Fattah, A.-K. Source parameters and fault plane determinations of the 28 December 1999 northeastern Cairo earthquakes. *Tectonophysics* **2001**, *343*, 63–77.

Ionospheric Electron Density Perturbations Driven by Seismic Tsunami-Excited Gravity Waves: Effect of Dynamo Electric Field

John Z. G. Ma, Michael P. Hickey and Attila Komjathy

Abstract: The effect of an ionospheric dynamo electric field on the electron density and total electron content (TEC) perturbations in the F layer (150–600 km altitudes) is investigated at two arbitrarily selected locations (noted as 29° N and 60° N in latitudes) in the presence of seismic tsunami-excited gravity waves propagating in a stratified, nondissipative atmosphere where vertical gradients of atmospheric properties are taken into consideration. Generalized ion momentum and continuity equations are solved, followed by an analysis of the dynamo electric field (E). The E-strength is within several mV/m, determined by the zonal neutral wind and meridional geomagnetic field. It is found that, at the mid-latitude location, n'_e is dominated by the atmospheric meridional wind when $E = 0$, while it is determined by the zonal wind when $E \neq 0$. The perturbed TEC over its unperturbed magnitude lies in around 10% at all altitudes for $E = 0$, while it keeps the same percentage at most altitudes for $E \neq 0$, except a jump to >25% in the F2-peak layer (300–340 km in height). By contrast, at the low-latitude location, the TEC jump is eliminated by the locally enhanced background electron density.

Reprinted from *J. Mar. Sci. Eng.* Cite as: Ma, J.Z.G.; Hickey, M.P.; Komjathy, A. Ionospheric Electron Density Perturbations Driven by Seismic Tsunami-Excited Gravity Waves: Effect of Dynamo Electric Field. *J. Mar. Sci. Eng.* **2015**, *3*, 1194–1226.

1. Introduction

Tsunamis have been a significant threat to humans living in coastal regions throughout recorded history. The Sumatra tsunami of 2004 took the largest toll of human life on record, with approximately 228,000 casualties attributed to the tsunami waves [1]. It is therefore necessary to provide reliable tsunami forecast and warning system which provides estimates of tsunami properties before the tsunami itself arrives at a given shore. From as early as the 1970s, modeling results have demonstrated that the ionospheric signature of an ocean tsunami can potentially be detected as traveling ionospheric disturbances (TIDs) produced by internal gravity waves propagating upward in the upper atmosphere in periods of 10–30 min, horizontal wavelengths and phase speeds of several hundreds of km and ~200 m/s, respectively, and vertical speeds of the order of 50 m/s (e.g., [2–18]), and these tsunami-driven TIDs have been identified in ionospheric total electron content (TEC) data, the vertical integration of the ionospheric electron density (n_e), as measured from ground-based GPS radio signals (e.g., [19,20]) or satellite altimeter radar [21].

During and after 2004–2007 earthquakes over Sumatra, more and more tsunami-related ionospheric perturbations have been studied. For example, data analyses of the Jason-1 and Topex/Poseidon satellite altimeters demostrated that, between 250 and 350 km of altitude, the effect of neutral-plasma coupling was verified to be maximum, with 10% TEC perturbations, and the electron density, n_e, up to $\sim 5 \times 10^{12}$ m^{-3} [21]. In addition, Occhipinti *et al.* demonstrated that when tsunami is in the northern hemisphere the perturbed electron density, n_e'/n_{e0} (where n_{e0} is the local density), does not exceed 10% in both E and F regions; however, when it travels south, n_e'/n_{e0} can reach to 80% for a simple tsunami model of wave package with a principal period of 20 min and wavelength of 230 km [8]. Example TEC variation curves were also produced when the tsunami is located at $-10°$ N, $0°$ N, $10°$ N, and $20°$ N of latitude for a northward propagating tsunami, and $10°$ N, $0°$ N, $-10°$ N, and $-20°$ N for a southward one, respectively. In the northward case, the authors showed the latitudinal dependence of the local electron density perturbations: in the GPS-favored F-region, gravity wave-driven perturbation is strongly dependent on magnetic inclination angle (I): within $I = [-40°, 20°]$ with $I = 0°$ near $10°$ N, the perturbation appears the most intensive, while outside the range the perturbation turns to zero.

Moreover, TEC measurements reported 30% and 40%–70% changes, respectively, in 40 min to two hours of the wave propagation [22]. The most recent extensive study examined events from Sumatra 2004 to Tohoku-Oki 2011 as measured by local networks SEAMERGES (30 stations), CTO/SUGAR (up to 32 stations), CTO/CENTRAL-ANDES (10 stations), SAMOANET (13 stations), and GEONET (1000 stations) [23]. The study explored the perturbations close to the epicenter of several events with different network (note that only Tohoku 2011 use GEONET). The amplitude of TEC perturbations exceed ± 0.25 TECU (but within ± 0.5 TECU; 1TECU = 10^{16} m^{-2}) in the three different stations measured on a same day (Figure 6 of the paper). However, this amplitude varies for different locations and times. For example, by comparing with the data of the previous day of another station to ensure earthquake-related oscillations, Rolland *et al.* reported that the perturbation was nearly 2 TECU [24]: upper right panel, Figure 3.

Relative to the measured $\leq 20\%$ and 10%–70% variations in n_e and TEC, respectively, either earlier or later predictions from theoretical modeling of gravity waves overestimated n_e-fluctuations and diversified TEC-modulations in response to the wave propagation through an assumed adiabatic or more realistic atmosphere. A standard perturbation treatment was firstly established by Hooke to obtain the magnitude of the effects of gravity waves on the creation of ionospheric irregularities [25]. However, a thorough investigation by following the procedure acquired a peak n_e-fluctuation of 20%–40%, an estimation the authors considered to be exaggerated for realistic perturbations [4]. After Occhipinti *et al.*'s first development and application of a full-wave model (FWM) to compare a numerical modeling results with real data [21], Hickey *et al.* [11] continued their early work on FWM (e.g., [27–30]) and employed MacLeod's ion dynamics [26] to expose $\delta n_e \sim 50\%$ down to 5% from adiabatic to nonadiabatic situations; while both the electron number density and TEC can reach 100% under quasi-adiabatic conditions with neutral wind perturbations of several 100 m/s, the corresponding responses of δn_e in realistic viscous atmosphere is an order of magnitude smaller than before, and the TEC responses were only a factor of ~ 3 smaller than before. By contrast, a case study on the 2004 Sumatra tsunami by employing Hooke's model provided $\delta n_e \sim 15\%$–40%, with $n'_e \sim (1–6) \times 10^{11} / m^3$, from dissipative to loss-free models at 100–450 km altitudes, and in the lossless case, $n'_e / n_{e0} > 100\%$ above 255 km; the peak perturbation of electron density, $\sim 2 \times 10^{11} / m^3$, appears at 238 km with $\delta n_e = 37\%$; at the 303 km height of maximum n_{e0}, $\delta n_e = 14\%$; while the TEC perturbation is within $\pm 7.5\%$ of the unperturbed magnitude, consistent with measurements of <10% variations [12].

Contributions by previous authors, e.g., Hooke, Occhipinti *et al.*, Hickey *et al.*, especially the most recent study by Meng *et al.* [31], are classical to elucidate the long-lasting issues of travelling ionospheric disturbances (TIDs) in a plasma environment, and of gravity wave propagations under realistic atmospheric conditions (e.g., [32–34]), respectively. The above mentioned overestimation in electron density perturbation might originate from some terms which, though necessary to be included in the set of equations to describe realistic atmospheric/ionospheric processes, were omitted either intentionally for convenience of mathematical manipulations, or unintentionally due to a lack of sufficient experimental support at early time to show their importance to be involved. After a revisit to these pioneer work, we find the missing terms are related to ionospheric dynamo electric field (**E**) which may play an appreciable role in both Hooke's electron density equation and MacLeod's ion dynamic equation. Luckily, Occhipinti *et al.* included Earth's electric field terms [8,21]. However, the modeling argued that the effect of the electric field polarization is negligible on the resulting perturbation of ion velocities, and ion's and electron's densities under the effect of gravity waves.

As a matter of fact, from the early 1970s the **E**-influence on atmospheric phenomena drew much attention (see pioneer work by Cole in [35,36]). Albeit the fact, it is understandable that the electric field effect was neglected in early models. This is because only after abundant spaceborne data became available from the 1980s (e.g., [37–45]), has the **E**-effect on charge kinetics, linear wave excitation

246

and propagation, and nonlinear plasma dynamics been eventually recognized unambiguously, investigated attentively, and developed consistently in the last 35 years (e.g., [46–69]). In auroral ionosphere, for example, Cole's model exposed a magnitude of $|\mathbf{E}| \sim$ a few mV/m drives ions away from Maxwellian substantially due to the $\mathbf{E} \times \mathbf{B}$ drift (where $\mathbf{B} = B\mathbf{b}$ is the local geomagnetic field and $\mathbf{b}$ is the unit vector along magnetic field lines), the magnitude of which is larger than the local thermal speeds of neutrals. These abundant theoretical and experimental studies demonstrated that the electric field effect is more conspicuous at high latitude (seldom relevant to tsunami-related applications); however, whenever $\mathbf{E}$ exists anywhere, including regions at mid- or low latitudes, plasma properties are modulated accordingly. Therefore, it is necessary for us to include the electric field effect in accommodating ionospheric perturbations driven by tsunami-excited atmospheric waves at mid- or low latitudes. Our studies to be introduced in this paper will validate this argument.

Ground-based GPS receivers are an efficient tool to detect tsunami-driven TEC perturbations. They not only provide data with highly localized coverage in particular regions, but also detect tsunami-generated signals far from landmasses where ground-based GPS coverage is nonexistent. The primary objective of this series of papers is to investigate whether ionospheric radio occultation measurements can be used to detect tsunami wave fronts while providing increased coverage and data density by comparing modeling results and actual data [70]. In this first paper, we extend the procedure described in [11] to obtain electron density perturbation by employing (1) the classical ionospheric electrodynamics to replace MacLeod's ion momentum equation; and (2) Kendall and Pickering's generalized perturbation theory [71] to directly get the electron density perturbation equation, instead of the traditionally used indirect ion perturbation equation in Hooke's model (which was then used to obtain the electron perturbation indirectly by assuming a charge neutrality condition). More importantly, we will consider the influence of the ionospheric dynamo electric field on the electron density and TEC perturbations. The purpose of the paper is to provide an extended data-fit model in the presence of ionospheric dynamo electric field which is able to grant data analysis and case study of space measurements with an algorithm of less errors, so as to estimate the tsunami wave front and subsequently help to confirm and image tsunamis by comparing both space-borne and ground-based GPS measurements with our theoretical results, thus establish a more effective and efficient tsunami warning and alarming system in future work.

The rest of the paper is as follows: Section 2 exhibits tsunami-driven disturbance at the sea surface and its upward propagation in atmosphere; Section 3 discusses ionospheric plasma properties in the upper atmosphere; Section 4 introduces the ionospheric dynamo electric field and its effects on plasma momentum and continuity equations; Section 5 estimates the electron density and TEC perturbations driven by tsunami-excited gravity waves in both the absence and presence of the dynamo electric field. At last, Section 6 outlines the results and gives a quick summary. The concerned altitudes are from 150 km up to 600 km, the ionospheric F2 region where GPS signals of electron density and TEC perturbations are measured for tsunami applications. It is worth mentioning here that this first paper employs a WKB approximation model which assumes linear wavelike solutions in time and 2D horizontal coordinates, but not in the vertical direction only along which the mean-field properties are supposed to vary, while keeping their homogeneities in the horizontal plane [72]. A 1D vertical Taylor-Goldstein equation (or, equivalently, a quadratic equation) can thus be derived in the presence of the height-varying temperature and wind shears to describe the vertical propagation of tsunami-excited gravity waves. This basic study will be expanded in a sister paper to a 3D backward ray-tracking algorithm to account for the detection of this kind of gravity waves with the radio-occultation data which are measured in situations where the atmospheric mean-field properties are also nonuniform in the 2D horizontal plane.

2. Tsunami-Driven Disturbance at the Sea Surface and Its Upward Propagation in Atmosphere

2.1. Tsunami Displacement and Its Vertical Speed Amplitude at $z = 0$

The surface waveforms to characterize the coupling (or modeling) of the tsunami with the overlying atmosphere at the sea level, $z = 0$ (where z is the altitude in atmosphere), were described in details from the late 1960s [2,3,21,73–75]. Based on these studies, a concise model was presented by Hickey *et al.* [11] (or Hickey [76]) to model the propagation of a tsunami-generated gravity wave packet. We continue to use this model in the present paper, with the initial tsunami displacement $Z(x)$ determined by the Airy function, Ai, in the horizontal plane at $t = 0$ and $z = 0$:

$$Z(x,0,0) = A \left[Ai(1-x)\frac{x}{2}e^{1-x/2} \right] \tag{1}$$

where x is the horizontal distance at the sea surface in units of 100 km, and $A \sim 0.5$ m is the amplitude of the forcing in meters [21] for a dominant horizontal-scale size of $\lambda_h = 400$ km. Let ω and $k_h = 2\pi/\lambda_h$ be the wave frequency and horizontal wavenumber, respectively. The k_h-spectrum of the forcing can be obtained from the Fourier transform of Equation (1):

$$\hat{Z}(k_h,0,0) = \frac{1}{2\pi} \int_{-\infty}^{\infty} Z(x,0,0)e^{ikx}dx \tag{2}$$

which gives the vertical speed spectrum, $w(k_h)$, as follows:

$$w(k_h, 0, 0) = i\omega\hat{Z}(k_h, 0, 0) \tag{3}$$

For typical values of ω and k_h, the shallow-water phase speed is $c_{ph} = \sqrt{gh} \approx$ 200 m/s for an ocean depth of about 4 km, where g is the gravitational acceleration and h is the ocean depth [74]. For a monochromatic wave of such a period, the amplitude of $w(k_h)$ is 1.57×10^{-3} m/s. However, because the displacement is due to the sum of all the waves in the the bandwidth of spectrum, the final value of $w(k_h)$ turns out to be 1.17×10^{-4} m/s [11].

2.2. Wave Amplification during Upward Propagation

The vertical displacement of the tsunami-driven disturbance at the sea surface acts like a moving corrugation at the base of the atmosphere. Tsunamis were therefore firstly postulated [2] and then demonstrated [3] to be capable of triggering atmospheric gravity waves that could subsequently propagate to high altitudes due to the fact that the tsunami speeds, wavelengths and periods lie well within the range of those of the gravity waves, featured by the motions of air parcels which are dominantly influenced by gravity and buoyancy.

In order to understand the impact of seismic tsunami-excited gravity waves on ionospheric electron density perturbations, and, based on this knowledge, to develop suitable approaches for the solution of more realistic problems through a series of incremental steps in following sister papers, this first paper deals with a stably stratified atmosphere featured by homogeneous density, pressure, temperature, and zonal and meridional winds in the horizontal plane, however, with vertical gradients, under nondissipative conditions, *i.e.*, in the absence of eddy process, molecular viscosity and thermal conduction, ion-drag, and Coriolis effect. It is worth mentioning here that previous studies argued that the nondissipative assumption can be valid from the sea level up to about 200 km altitude [77–80].

Let p_1, ρ_1, and T_1 be the perturbations of the atmospheric mean-field pressure p_0, mass density ρ_0, and temperature T_0; and, (u, v, w) be the perturbed components of the mean-field wind velocity $(U, V, 0)$. The upward propagating gravity waves incarnated from the tsunamis surface waves can well be formulated by the full-wave model (e.g., [11,13,30–32,77,81–86]). After a straightforward, but tedious, simplification to the set of FWM differential equations (see, e.g., Appendix in [83]), we obtain the governing Taylor-Goldstein equation for $\tilde{w}(z)$ [or, equivalently, a quadratic equation for $w(z) = \tilde{w}e^{-\frac{1}{2}\int f(z)dz}$] and the full solutions for the rest atmospheric perturbations as follows (*cf.* [87]):

$$\frac{\partial^2 \tilde{w}}{\partial z^2} + q^2(z)\tilde{w} = 0, \text{ or, } \frac{\partial^2 w}{\partial z^2} + f(z)\frac{\partial w}{\partial z} + g(z)w = 0$$

$$i\Omega\frac{p_1}{p_0} = -(\beta - 1)\left[\gamma\frac{\partial w}{\partial z} + \left(\gamma\frac{V_{k1}}{\Omega} + \frac{k_p}{k_h}\right)k_h w\right]$$

$$i\Omega\frac{\rho_1}{\rho_0} = -(\beta - 1)\left[\frac{\partial w}{\partial z} + \left(\frac{V_{k1}}{\Omega} + \frac{k_p}{k_h} + \frac{\beta}{\beta-1}\frac{k_p - \gamma k_\rho}{\gamma k_h}\right)k_h w\right]$$

$$i\Omega\frac{T_1}{T_0} = -(\beta - 1)\left\{(\gamma - 1)\frac{\partial w}{\partial z} + \left[(\gamma - 1)\frac{V_{k1}}{\Omega} + \frac{k_T}{k_h} - \frac{\beta}{\beta-1}\frac{k_p - \gamma k_\rho}{\gamma k_h}\right]k_h w\right\}$$

$$iku = -\beta\frac{k^2}{k_h^2}\frac{\partial w}{\partial z} + \left[\frac{k}{\Omega}\frac{\partial U}{\partial z} - \beta\frac{k^2}{k_h^2}\left(k_h\frac{V_{k1}}{\Omega} - \frac{1}{\gamma H}\right)\right]w$$

$$ilv = -\beta\frac{l^2}{k_h^2}\frac{\partial w}{\partial z} + \left[\frac{l}{\Omega}\frac{\partial V}{\partial z} - \beta\frac{l^2}{k_h^2}\left(k_h\frac{V_{k1}}{\Omega} - \frac{1}{\gamma H}\right)\right]w$$

$$\left. \right\} \qquad (4)$$

where

$$f(z) = -\frac{1}{H} + \beta k_T + 2(\beta - 1)\frac{V_{k1}}{C_{ph}}$$

$$g(z) = \frac{1}{C_d^2}\left(\frac{1}{\beta-1}\omega_b^2 - 2V_{k1}^2\right) - \frac{1}{\beta}k_h^2 + \frac{\beta^2}{\beta-1}\frac{k_T}{\gamma H} + \frac{1}{C_{ph}}\left\{V_{k2} - \left[1 + \beta\left(k_T H - \frac{2}{\gamma}\right)\right]\frac{V_{k1}}{H}\right\}$$

$$q^2(z) = g(z) - \frac{1}{4}f^2(z) - \frac{1}{2}\frac{df}{dz}$$

$$\left. \right\} \qquad (5)$$

In the above, following notations are used:

$$k_\rho = \frac{d(\ln\rho_0)}{dz}, \; k_p = \frac{d(\ln p_0)}{dz}, \; k_T = \frac{d(\ln T_0)}{dz}, \; k_h^2 = k^2 + l^2, \; H = -\frac{1}{k_p}, \; \omega_b^2 = \frac{\gamma-1}{\gamma}\frac{g}{H}$$

$$\Omega = \omega - (kU + lV), \; \omega = c_{ph}k_h, \; C_{ph} = c_{ph} - V_k, \; C_d^2 = C^2 - C_{ph}^2, \; \beta = \frac{C^2}{C_d^2}$$

$$V_k = \frac{k}{k_h}U + \frac{l}{k_h}V, \; V_{k1} = \frac{k}{k_h}\frac{dU}{dz} + \frac{l}{k_h}\frac{dV}{dz}, \; V_{k2} = \frac{k}{k_h}\frac{d^2U}{dz^2} + \frac{l}{k_h}\frac{d^2V}{dz^2}$$

$$\left. \right\} \qquad (6)$$

in which k and l are the wave vector components along x and y in the horizontal plane, respectively; H is scale height; ω_b is the Brunt-Väisälä buoyancy frequency; Ω is the intrinsic (or, Doppler-shifted) frequency [accordingly ω is called the extrinsic (ground-based) frequency]; C_{ph} and c_{ph} are the intrinsic and extrinsic phase speeds, respectively; C_d is the complementary phase speed; $C = \sqrt{\gamma g H}$ is the sound speed; and γ is the ratio of specific heats; (k_ρ, k_p, k_T) are the three scale numbers in density, pressure, and temperature, respectively.

The vertical profiles up to 600 km height of the undisturbed mean-field parameters (ρ_0, p_0, T_0, U, V), as well as related (k_ρ, k_p, k_T) are calculated by employing the two empirical neutral atmospheric model, NRLMSISE-00 [88] and horizontal wind model, HWM93 [89]. We arbitrarily choose a position at 60° latitude and −70° longitude for a local apparent solar time of 1600 h on the 172th day of a year, with supposed daily solar $F_{10.7}$ flux index and its 81-day average of 150. The daily geomagnetic index is assumed 4. Figure 1 demonstrates the mean-field profiles of the neutral atmosphere. The upper left panel gives ρ_0 (pink) and p_0 (blue). Density ρ_0 decreases all the way up from 1.225 kg/m^3 (or, 2.55×10^{25} 1/m^3) at the sea level to only 2.44×10^{-13} kg/m^3 at 600 km altitude. Pressure p_0 has a similar tendency to ρ_0. It reduces from 10^5 Pa at the sea level to 1.56×10^{-7} Pa finally. The upper right panel presents T_0 (pink) and C (blue). Temperature T_0 is 281 °K at the sea level. It decreases linearly to 224 °K at 13 km, and then returns to 281 °K at 47 km, followed by a reduction again to 146 °K at 88 km. Above this height, the temperature goes up continuously and reaches a stable exospheric value of ~1250 °K above 400 km altitude. At 194 km it is 1000 °K. Sound speed C follows roughly the variation of $\sqrt{T_0}$. At the sea level, it is 336 m/s; at 600 km altitude, it is around 1.2 km/s. The lower left panel dipicts k_ρ (pink), k_p (blue), and k_T (black). Up to 200 km altitude, $k_\rho \neq k_p$ keeps alive, thus, the isothermal condition $k_T = 0$ is broken in atmosphere, except at three heights: 13.1 km, 47.2 km, and 87.9 km where $k_T = 0$. However, above 100 km altitude, k_T eventually keeps its positive polarization after two times of adjustment from negative to positive values. Above 200 km altitude, $k_T = 0$ can be considered valid. Note that the scale height H is equal to $-1/k_p$. At the sea level, H is calculated as 8.44 km. It soars to as high as 75.6 km when approaching to about 200 km altitude and beyond. The lower right panel illustrates U (blue) and V (pink). Both of them oscillate twice dramatically in altitude within ±51 m/s in amplitude below 200 km altitude. Above this height, their magnitudes grow monotonously with height to 75 m/s and 23 m/s, respectively.

Compared with the vertical profiles of atmospheric properties, NRLMSISE-00 and HWM93 also provide the horizontal gradients of ρ_0, T_0, p_0, U, and V. These inhomogeneities are always at least $10^{2\sim3}$ smaller than the vertical gradients. It is reasonable to assume, as most authors did, that the mean-field parameters are uniform and stratified in the horizontal plane, free of any inhomogeneities compared to that in the vertical direction, i.e., $\partial/\partial x \simeq 0$, $\partial/\partial y \simeq 0$ and $\nabla \cong (\partial/\partial z)\hat{e}_z$.

Under the initial condition, $w_0 = 1.17 \times 10^{-4}$ m/s, as given in Subsection 2.1, and assuming $dw_0/dz = 0$ at the sea level, we use an adaptive-step, 4th-order Runge-Kutta method to solve the Taylor-Goldstein equation, Equation (4). Note that the initial conditions of the six perturbations are all determined by w_0 and dw_0/dz. This is an alternative approach to the amplification of upward propagating tsunami-driven surface waves in addition to the traditionally used plane-wave linearization method. In order to show the features of tsunami waves of different horizontal scale sizes but with the same horizontal wavenumber $k_h \sim 2\pi/400$ km^{-1}, we take into account a couple of cases: one has specific scale sizes of 2000 km and 400 km in the x and y directions, respectively; the other has 400 km and 2000 km, respectively. Clearly, the k/l-ratios in the two cases are 1/5 and 5/1, respectively. The profiles of all the perturbations p_1, ρ_1, T_1 and u, v, w in the F layer (150–600 km altitudes) are demonstrated in Figure 2 where the curves of the two cases are in blue and pink, respectively.

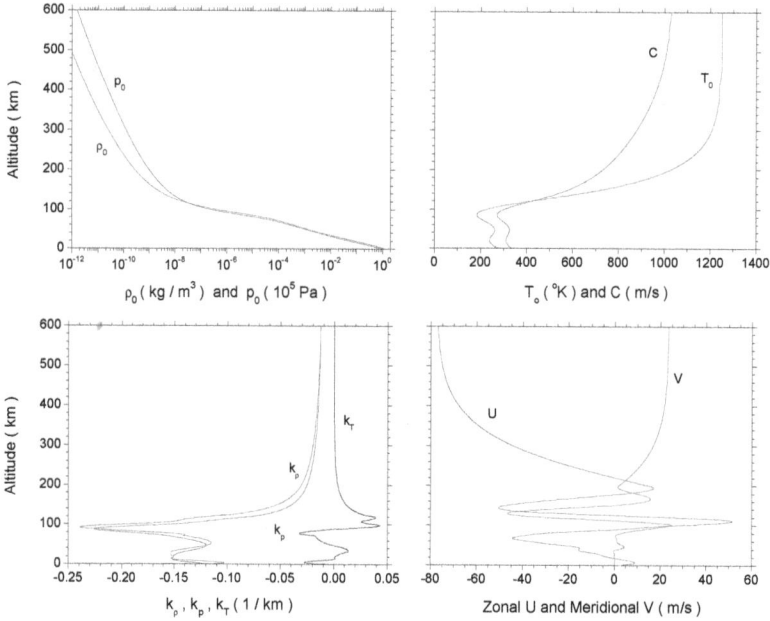

Figure 1. Vertical profiles of the atmospheric mean-field properties from NRLMSISE-00 and HWM93. Upper left: mass density ρ_0 (pink) and pressure p_0 (blue); upper right: temperature T_0 (pink) and sound speed C (blue); lower left: density scale number k_ρ (pink), pressure scale number k_p (blue), and temperature scale number k_T (black); lower right: zonal (eastward) wind U (blue) and meridional (northward) wind V (pink).

First of all, among all the profiles in either case, there are no two ones which exhibit the same phase and the amplitude growth. For example, the blue curves in all the panels expose that the maximal amplitude of p_1/p_0 is smaller than that of both ρ_1/ρ_0 and T_1/T_0, while the phases of the last two are opposite; similarly, the velocity components of u, v, w evolve differently from each other, either in amplitude or phase. This feature is in a sharp contrast with that obtained with the traditional plane-wave perturbation modeling where all the fluctuations follow a same rule of thumb, $\sim e^{i(\mathbf{k}\cdot\mathbf{r}-\omega t)}$. Secondly, below 150 km altitude (not shown due to outside the F region), the amplitudes of all the perturbations are always relatively smaller than those at higher altitudes and can be reasonably neglected. See the upper right panel as an example: at 150 km, the blue T_1/T_0 is no more than 0.07; above this height, its amplitude grows to about 1.5 at 560 km. Thirdly, at any altitudes, the three perturbations in pressure, density, and temperature in the upper three panels, respectively, satisfy the perturbed equation of state, $p_1/p_0 = \rho_1/\rho_0 + T_1/T_0$, which can be used to check the validity of the simulations. Fourthly, the perturbed zonal

252

wind, u in blue (or pink), in the lower left panel has a smaller (or larger) amplitude than that of the perturbed meridional wind, v in blue (or pink), in the lower middle panel for the small (or large) k/l. This is expressed by the last two equations in Equation (4): the ratio of the two amplitudes is approximately proportional to k/l. Finally, the amplitude of all the perturbations grow monotonically in altitude, regardless of the mean-field profiles which can be either increase all the way up in, e.g., the panels of pressure, density, and temperature, or, oscillate in altitudes in, e.g., the wind panel.

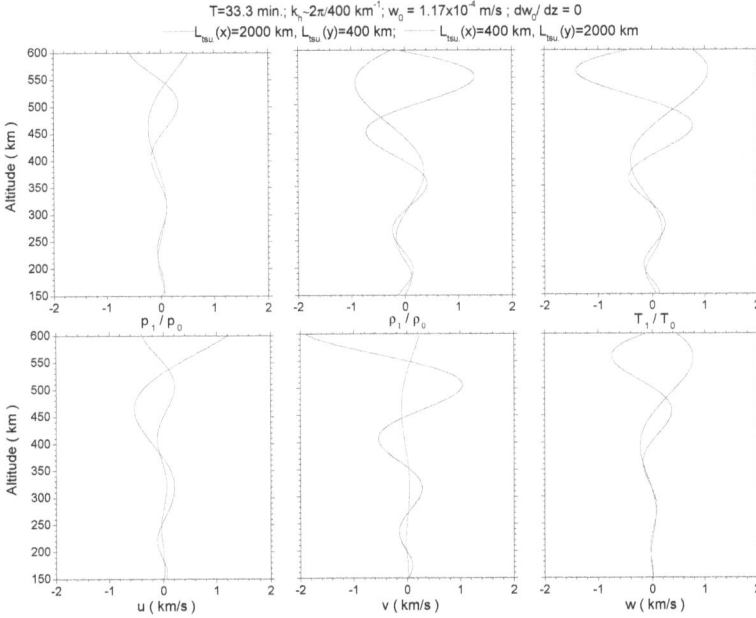

Figure 2. Amplification of upward propagating tsunami-driven surface waves in the form of atmospheric gravity waves in ionospheric F layer (150–600 km). The perturbations are calculated from Equation (4) under initial conditions at the sea level of $w_0 = 1.17 \times 10^{-4}$ m/s (given in Subsection 2.1) and assumed $dw_0/dz = 0$. Wave period, $2\pi/\omega$, is 33.3 min, andHickey *et al.*'s horizontal wave number, $\lambda_h \sim 400$ km, is considered [11] in two cases of specific horizontal scale sizes: $(x,y) = (2000,400)$ km (in blue) and $(x,y) = (400,2000)$ km (in pink).

3. Ionospheric Plasma Properties in the Upper Atmosphere

Above 80~85 km altitude, the atmosphere becomes weakly ionized to form the ionosphere. In this plasma system, in addition to the atmospheric neutral particles of density n_n and mass m_n, there are two extra types of charged particles: electrons of charge $q_e = -e$, density n_e, and mass m_e; ions of charge q_i, density n_i, and mass m_i. For simplicity but without loss of generality, we use a single mean ion component for multiple charges like NO^+, N_2^+, and O_2^+; and use a single neutral gas with the same mean atomic mass for multiple neutrals in the same region where $m_e \ll m_i \approx m_n$ and $n_e \approx n_i \ll n_n$.

Classical ionospheric electrodynamics (e.g., [90–92]) demonstrates that the behavior of such a system is determined on the whole by its massive and dense neutral component, and by the strong coupling between charged particles and the neutrals via electron-neutral and ion-neutral collisions in frequencies ν_{en} and ν_{in}, respectively, which behave as two drag forces for the neutral gas, namely, the electron-drag f_{en} and ion-drag f_{in} defined as follows:

$$f_{en} = n_e m_e \nu_{en}(\mathbf{v}_e - \mathbf{v}_n), \quad f_{in} = n_i m_i \nu_{in}(\mathbf{v}_i - \mathbf{v}_n) \tag{7}$$

where $\mathbf{v}_e$, $\mathbf{v}_i$, and $\mathbf{v}_n$ are the electron, ion, and neutral velocities, respectively, and

$$\nu_{in} = 2.6 \times 10^{-15}\,(n_n + n_i)\,\frac{1}{\sqrt{A}}, \quad \nu_{en} = 5.4 \times 10^{-16} n_n \sqrt{T_e} \tag{8}$$

in which $A = 28.97$ is the mean molecular weight (in amu) of either the neutrals or the ions; ν_{en} is the electron-neutral collision frequency [93]; and ν_{in} is the ion-neutral collision frequency [94]. The units of n_n, n_e, and n_i are in m^{-3}.

3.1. Generalized Ion Momentum Equation

Using Equation (7) in the momentum equations of charged particles yields

$$n_e m_e \frac{D\mathbf{v}_e}{Dt} = -\nabla p_e + n_e m_e \mathbf{g} - e n_e (\mathbf{E} + \mathbf{v}_e \times \mathbf{B}) - n_e m_e \nu_{en}(\mathbf{v}_e - \mathbf{v}_n) \tag{9}$$

$$n_i m_i \frac{D\mathbf{v}_i}{Dt} = -\nabla p_i + n_i m_i \mathbf{g} + q_i n_i (\mathbf{E} + \mathbf{v}_i \times \mathbf{B}) - n_i m_i \nu_{in}(\mathbf{v}_i - \mathbf{v}_n) \tag{10}$$

in which p_e and p_i are the electron and ion pressure, respectively; and, $\mathbf{g} = -g\hat{\mathbf{e}}_z$ is gravity acceleration. After neglecting the acceleration (or inertia) terms on the LHS of the above two equations, because the response time scale in the order of wave periods (e.g., tens of minutes for gravity waves) is much longer than both the gyration periods (e.g., $\Omega_e^{-1} \sim 0.1$ μs and $\Omega_i^{-1} \sim 3$ ms in F layer, where $\Omega_e = eB/m_e$ and $\Omega_i = q_i B/m_i$ are the electron and ion gyro-frequencies, respectively, in which $B = |\mathbf{B}|$ is the magnitude of the local geomagnetic field) and collision relaxation time scales (e.g., $\nu_{en}^{-1} \sim 10$ ms and $\nu_{in}^{-1} \sim (0.1–1)$ s in the F layer), Equations (9) and (10) provide

$$\mathbf{v}_e = \mathbf{v}_n - \frac{\nu_{en}\Omega_e}{\nu_{en}^2 + \Omega_e^2}\frac{\mathbf{E}'_{e\perp}}{B} + \frac{\Omega_e^2}{\nu_{en}^2 + \Omega_e^2}\frac{\mathbf{E}'_{e\perp} \times \mathbf{B}}{B^2} - \frac{\Omega_e}{\nu_{en}}\frac{\mathbf{E}'_{e\parallel}}{B} \tag{11}$$

$$\mathbf{v}_i = \mathbf{v}_n + \frac{\nu_{in}\Omega_i}{\nu_{in}^2 + \Omega_i^2}\frac{\mathbf{E}'_{i\perp}}{B} + \frac{\Omega_i^2}{\nu_{in}^2 + \Omega_i^2}\frac{\mathbf{E}'_{i\perp} \times \mathbf{B}}{B^2} + \frac{\Omega_i}{\nu_{in}}\frac{\mathbf{E}'_{i\parallel}}{B} \tag{12}$$

where subscripts "$\perp$" and "$\parallel$" denote the components perpendicular and parallel to **b**, respectively; and,

$$\frac{\mathbf{E}'_{e\perp}}{B} = \frac{\mathbf{E}_\perp + \mathbf{v}_n \times \mathbf{B}}{B} + \frac{\frac{\nabla_\perp p_e}{n_e m_e} - \mathbf{g}_\perp}{\Omega_e}, \quad \frac{\mathbf{E}'_{e\parallel}}{B} = \frac{\mathbf{E}_\parallel}{B} + \frac{\frac{\nabla_\parallel p_e}{n_e m_e} - \mathbf{g}_\parallel}{\Omega_e} \tag{13}$$

$$\frac{\mathbf{E}'_{i\perp}}{B} = \frac{\mathbf{E}_\perp + \mathbf{v}_n \times \mathbf{B}}{B} - \frac{\frac{\nabla_\perp p_i}{n_i m_i} - \mathbf{g}_\perp}{\Omega_i}, \quad \frac{\mathbf{E}'_{i\parallel}}{B} = \frac{\mathbf{E}_\parallel}{B} - \frac{\frac{\nabla_\parallel p_i}{n_i m_i} - \mathbf{g}_\parallel}{\Omega_i} \tag{14}$$

In the above equations, the relative importance of the terms related to pressure gradient and gravity can be estimated as follows. Considering $\Omega_i \sim 10^2$ rad/s and $\Omega_e \sim 10^5$ rad/s, and, $|\mathbf{v}_n| \sim$ tens of m/s in ionosphere, we have $\Omega_{e,i}|\mathbf{v}_n \times \mathbf{b}| \sim (10^4 - 10^7) \gg g$, or, the **g**-terms in both of the $\mathbf{v}_e$ and $\mathbf{v}_i$ equations contribute to an additional drift of no more than several cm/s, much less than $|\mathbf{v}_n|$. Thus, the **g**-terms can be reasonably omitted. Similarly, for the pressure gradient terms, the thermal speed v_T of charged particles is in the order of $v_T = \sqrt{k_b T_0 / m_{e,i}} \sim 1$–242 km/s. Because $|\nabla p/(m_e, n_0)| = \frac{v_T^2}{H_0} \ll \Omega_{e,i}|\mathbf{v}_n \times \mathbf{b}| \sim (10^4 - 10^7)$, where H_0 is the scale height more than 8 km, we obtain

$$\frac{\left|\frac{\nabla p_{e,i}}{e n_{e,i}}\right|}{|\mathbf{v}_n \times \mathbf{B}|} = \frac{\left|\frac{\nabla p_{e,i}}{m_{e,i} n_{e,i}}\right|}{\Omega_{e,i}|\mathbf{v}_n \times \mathbf{b}|} \ll 1 \tag{15}$$

which reveals that the pressure-gradient terms can be reasonably omitted. As a result, the final expresses of Equations (11) and (12) are as follows:

$$\mathbf{v}_e = \frac{-v_{en}\Omega_e \frac{\mathbf{E}_{*1}}{B} + \Omega_e^2 \frac{\mathbf{E}_{*2}}{B} + v_{en}^2 \mathbf{v}_n}{v_{en}^2 + \Omega_e^2} - \frac{\Omega_e}{v_{en}} \frac{\Omega_e^2}{v_{en}^2 + \Omega_e^2} \frac{\mathbf{E}_\parallel}{B} \tag{16}$$

$$\mathbf{v}_i = \frac{v_{in}\Omega_i \frac{\mathbf{E}_{*1}}{B} + \Omega_i^2 \frac{\mathbf{E}_{*2}}{B} + v_{in}^2 \mathbf{v}_n}{v_{in}^2 + \Omega_i^2} + \frac{\Omega_i}{v_{in}} \frac{\Omega_i^2}{v_{in}^2 + \Omega_i^2} \frac{\mathbf{E}_\parallel}{B} \tag{17}$$

in which

$$\frac{\mathbf{E}_{*1}}{B} = \frac{\mathbf{E}}{B} + \mathbf{v}_n \times \mathbf{b}, \quad \frac{\mathbf{E}_{*2}}{B} = \frac{\mathbf{E}}{B} \times \mathbf{b} + v_{n\parallel}\mathbf{b} \tag{18}$$

Note that it is **E**, rather than $\mathbf{E}_\perp$ is used in the above equations. Equation (17) extends MacLeod's result [26] by including the extra terms contributed by the ionospheric electric field, **E**.

Equations (16) and (17) contribute to an ionospheric current, **j**:

$$\mathbf{j} = e(n_i\mathbf{v}_i - n_e\mathbf{v}_e) = \sigma_0 \left(\sigma_\parallel^* \mathbf{E}_\parallel + \sigma_P^* \mathbf{E}^* + \sigma_H^* \mathbf{b} \times \mathbf{E}^*\right) \tag{19}$$

where the quasi-neutrality condition, $n_e \approx n_i = n_0$, is applied; and

$$\sigma_0 = \frac{e n_0}{B}, \ \sigma_\parallel^* = \frac{\Omega_e}{\nu_{en}} \frac{\Omega_e^2}{\nu_{en}^2 + \Omega_e^2} + \frac{\Omega_i}{\nu_{in}} \frac{\Omega_i^2}{\nu_{in}^2 + \Omega_i^2}, \ \sigma_P^* = \frac{\nu_{en}\Omega_e}{\nu_{en}^2 + \Omega_e^2} + \frac{\nu_{in}\Omega_i}{\nu_{in}^2 + \Omega_i^2}, \ \sigma_H^* = \frac{\Omega_e^2}{\nu_{en}^2 + \Omega_e^2} - \frac{\Omega_i^2}{\nu_{in}^2 + \Omega_i^2} \quad (20)$$

in which $\sigma_\parallel^*$, σ_P^*, and σ_H^* are the three classical ionospheric conductivities: parallel conductivity, Pedersen conductivity, and Hall conductivity, respectively (e.g., [94]). Note that they are dimension-free. In Equation (19), $\mathbf{E}$ has been Lorentz-transformed to $\mathbf{E}^*$ in the frame of reference of the atmosphere which is moving at a velocity $\mathbf{v}_n$: $\mathbf{E}^* = \mathbf{E} + \mathbf{v}_n \times \mathbf{B}$.

Either the theoretical Chapman profile or measurements from GPS/ionosonde demonstrates that the F2 layer (220–600 km altitude; peak plasma density $\sim 10^{12}$ m^{-3}) provides primary plasma contents (more than 90%), and dominates ionospheric perturbation in electron density or TEC (e.g., [92]). We therefore concentrate on his region. In addition, as pointed out in, e.g., [28,95]), that the mean-field parameters of ionospheric properties have much smaller horizontal derivatives that those in the vertical direction, while their horizontal scales are of $\sim$1000 km, appreciably exceeding the variation scale in the vertical direction, we neglect the horizontal profiles of both neutral and charged particles in evaluating plasma perturbations.

The two upper panels in Figure 3 illustrate the vertical profiles of the ionospheric F electron and ion densities n_e and $n_i \approx [\mathrm{O}^+]$ (upper left), and their temperatures T_e and T_i (upper right), respectively. They are calculated by employing the IRI-2012 empirical model [96] at two different locations: one is the previous 60° latitude one (labelled hereafter as 60° N) used to exhibit globally stratified atmospheric properties for reference; and the other is assumed at 29° latitude and 81° longitude, UT 19:30 on the 108th day (labelled hereafter as 29° N). Clearly, the atmospheric stratified assumption is not applicable for ionosphere: in the horizontal plane, the plasma properties demonstrate significant variations at different locations. For instance, the maximal plasma densities at around 300 km altitude increase about three times when moving equatorward from the 60° N-location to the 29° N location. Note that the electron temperature is always higher than that of ions, and the temperatures tend to decrease in the equatorward direction.

The lower left panel of Figure 3 illustrates the vertical profiles of ν_{in} and ν_{en}. Due to $n_n \gg n_i$, ν_{in} depends only on the neutral density n_n, and independent of plasma properties. By contrast, ν_{en} is also related to electron temperature T_e. However, T_e appears exerting an inappreciable influence. For example, the two vertical profiles of ν_{en} nearly superimpose upon each other with different T_e values of two locations, 29° N and 60° N, respectively. Besides, ν_{en} is roughly 2 orders higher than ν_{in}. The lower right panel in the figure presents the three vertical profiles of $\sigma_\parallel^*$, σ_P^*, and σ_H^* in the F layer, respectively. They are independent of geographic locations and universal time. Obviously, $\sigma_H^* \ll \sigma_P^* \ll \sigma_\parallel^*$.

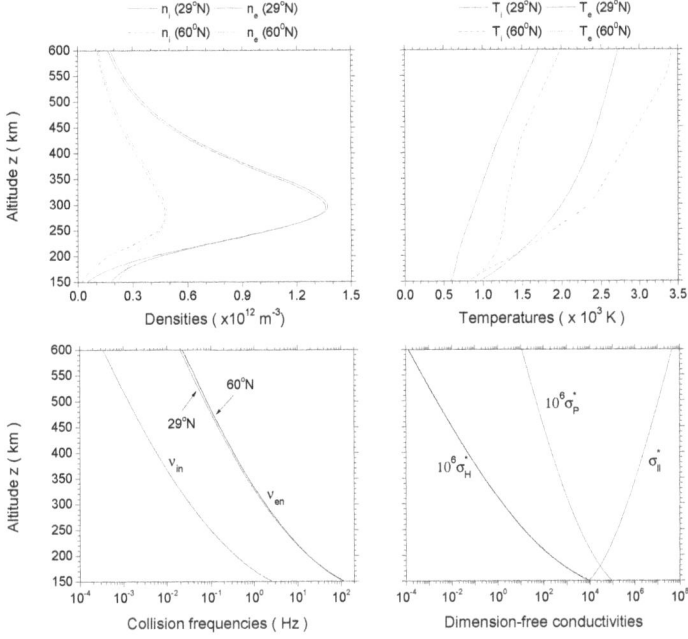

Figure 3. Vertical profiles of ion and electron densities n_i and n_e (upper left panel) and temperatures T_i and T_e (upper right panel) from the IRI-2012 empirical model [96] at two different locations; corresponding ion-neutral and electron-neutral collision frequencies ν_{in} and ν_{en} (lower left panel) from Equation (8); and, three dimension-free conventional conductivities $\sigma_{\parallel}^*$, σ_P^*, and σ_H^* from Equation (20) (lower right panel), which are independent of geographic locations and universal time.

3.2. Generalized Electron Continuity Equation

In the frame of reference of the atmosphere, the equation of ionospheric plasma motion is determined by the summation of Equations (9) and (10):

$$n_p m_e \nu_{en}(\mathbf{v}_e - \mathbf{v}_n) + n_p m_i \nu_{in}(\mathbf{v}_i - \mathbf{v}_n) = -\nabla(p_e + p_i) + n_p m_i \mathbf{g} + \mathbf{j} \times \mathbf{B} \qquad (21)$$

where the inertial terms are neglected as before, $n_p = n_e \approx n_i$ is plasma density. Using $\mathbf{j} = e n_p(\mathbf{v}_i - \mathbf{v}_e)$ yields

$$n_p(m_e \nu_{en} + m_i \nu_{in})(\mathbf{v}_i - \mathbf{v}_n) = -\nabla(p_e + p_i) + n_p m_i \mathbf{g} + \frac{m_e}{e} \nu_{en} \mathbf{j} + \mathbf{j} \times \mathbf{B} \qquad (22)$$

257

or, alternatively,

$$n_p(m_e\nu_{en} + m_i\nu_{in})(\mathbf{v}_e - \mathbf{v}_n) = -\nabla(p_e + p_i) + n_p m_i \mathbf{g} - \frac{m_i}{e}\nu_{in}\mathbf{j} + \mathbf{j} \times \mathbf{B} \tag{23}$$

On the one hand, the scalar parallel component equation of Equation (22) provides

$$n_p(m_e\nu_{en} + m_i\nu_{in})(v_{i\parallel} - v_{n\parallel}) = -\nabla_\parallel(p_e + p_i) + n_p m_i g_\parallel + \frac{m_e}{e}\nu_{en}j_\parallel \tag{24}$$

By defining an ambipolar diffusion coefficient D_a expressed by

$$D_a = \frac{2k_B T_i}{m_e\nu_{en} + m_i\nu_{in}} \tag{25}$$

Equation (24) becomes

$$v_{i\parallel} = v_{n\parallel} - D_a\frac{\nabla_\parallel(p_e + p_i) - n_p m_i g_\parallel}{2n_p k_B T_i} + \frac{m_e\nu_{en}}{en_p(m_e\nu_{en} + m_i\nu_{in})}j_\parallel \tag{26}$$

The dominant electron density perturbation or TEC occurs at F2 altitudes where $\nu_{in} \sim 1/s$, $\nu_{en} \sim 25/s$, and $\nu_{ei} \sim 1500/s$. Thus, $(m_e\nu_{en})/(m_i\nu_{in}) \sim 4 \times 10^{-3} \ll 1$. Considering $j_\parallel/(en_p) = v_{i\parallel} - v_{e\parallel}$. We have

$$n_p v_{i\parallel} = n_p\left(v_{n\parallel} - \frac{m_e\nu_{en}}{m_i\nu_{in}}v_{e\parallel}\right) - D_{ai}\frac{\nabla_\parallel(p_e + p_i) - n_p m_i g_\parallel}{2k_B T_i} \tag{27}$$

in which

$$D_{ai} \approx \frac{2k_B T_i}{m_i\nu_{in}} \tag{28}$$

Thus, from the ion continuity equation

$$\frac{\partial n_i}{\partial t} + \nabla \cdot (n_i\mathbf{v}_{i\perp}) + \nabla \cdot (n_i\mathbf{v}_{i\parallel}) = P_i - n_i L_i \tag{29}$$

in which P_i and L_i are the chemical production and loss rates of ions, respectively, we obtain

$$\frac{\partial n_i}{\partial t} + \nabla \cdot (n_i\mathbf{v}_{i\perp}) + \nabla \cdot \left\{\left[n_i\left(v_{n\parallel} - \frac{m_e\nu_{en}}{m_i\nu_{in}}v_{e\parallel}\right) - D_{ai}\frac{\nabla_\parallel(p_e+p_i)-n_i m_i g_\parallel}{2k_B T_i}\right]\mathbf{b}\right\} = P_i - n_i L_i \tag{30}$$

where $v_{Ti} = \sqrt{2k_B T_i/m_i}$ is ion thermal speed. This result extends the formula given in [71] by involving an extra term: the parallel electron speed $v_{e\parallel}$, which can be large enough to compete with $v_{n\parallel}$ in the presence of factor $(m_e\nu_{en})/(m_i\nu_{in})$.

On the other hand, we take similar steps to the scalar parallel component equation of Equation (23) and obtain

$$n_p(m_e \nu_{en} + m_i \nu_{in})(v_{e\parallel} - v_{n\parallel}) = -\nabla_\parallel(p_e + p_i) + n_p m_i g_\parallel - \frac{m_i}{e}\nu_{in}j_\parallel \tag{31}$$

which becomes

$$n_p v_{e\parallel} = n_p \frac{m_i \nu_{in}}{m_e \nu_{en}}(v_{n\parallel} - v_{i\parallel}) - D_{ae}\frac{\nabla_\parallel(p_e + p_i) - n_p m_i g_\parallel}{2k_B T_i} \tag{32}$$

in which

$$D_{ae} \approx \frac{2k_B T_i}{m_e \nu_{en}} \tag{33}$$

Thus, from the electron continuity equation

$$\frac{\partial n_e}{\partial t} + \nabla \cdot (n_e \mathbf{v}_{e\perp}) + \nabla \cdot (n_e \mathbf{v}_{e\parallel}) = P_e - n_e L_e \tag{34}$$

where P_e and L_e denote the photoionization rate and the chemical loss rate of electrons, respectively, we have

$$\frac{\partial n_e}{\partial t} + \nabla \cdot (n_e \mathbf{v}_e^*) = P_e - n_e L_e \tag{35}$$

in which

$$\mathbf{v}_e^* = \mathbf{v}_e + (v_{e\parallel}^* - v_{e\parallel})\mathbf{b}, \quad \text{where } v_{e\parallel}^* = \frac{m_i \nu_{in}}{m_e \nu_{en}}\left(v_{n\parallel} - v_{i\parallel} - \frac{g_\parallel + g_\parallel^*}{\nu_{in}}\right) \tag{36}$$

Note that $v_{e\parallel}^*$ is a newly introduced retarded speed of electrons caused by the diffusion effect, parallel to the geomagnetic field lines and determined by parallel neutral and ion speeds, and the speed increment of plasma in a time scale of ion-neutral collisions caused by both gravity and pressure gradient, where

$$g_\parallel^* = \frac{\nabla_\parallel(p_e + p_i)}{n_i m_i} = \frac{1}{2}v_{Ti}^2\left[\left(1 + \frac{T_e}{T_i}\right)k_{ne\parallel} + \frac{T_e}{T_i}k_{Te\parallel} + k_{Ti\parallel}\right] \tag{37}$$

is a pseudo-acceleration driven by plasma pressure $(p_e + p_i)$, and $k_{ne\parallel}$, $k_{Te\parallel}$, and, $k_{Ti\parallel}$ are the three inhomogeneous numbers, respectively, in plasma density, electron and ion temperatures in the parallel direction to the geomagnetic field. The parameters in Equations (36) and (37) are defined as follows:

$$\left.\begin{array}{l} v_{n\parallel} = \mathbf{v}_n \cdot \mathbf{b} = V\frac{B_y}{B}, \quad v_{i\parallel} = \mathbf{v}_i \cdot \mathbf{b} = v_{iy}\frac{B_y}{B} + v_{iz}\frac{B_z}{B}, \quad g_\parallel = \mathbf{g} \cdot \mathbf{b} = -g\frac{B_z}{B} \\ k_{ne\parallel} = \frac{\nabla n_e}{n_e} \cdot \mathbf{b} = k_{ne}\frac{B_z}{B}, \quad k_{Te\parallel} = \frac{\nabla T_e}{T_e} \cdot \mathbf{b} = k_{Te}\frac{B_z}{B}, \quad k_{Ti\parallel} = \frac{\nabla T_i}{T_i} \cdot \mathbf{b} = k_{Ti}\frac{B_z}{B} \end{array}\right\} \tag{38}$$

in which $k_{ne} = (dn_e/dz)/n_e$, $k_{Te} = (dT_e/dz)/T_e$, and $k_{Ti} = (dT_i/dz)/T_i$ are the inhomogeneity numbers to represent the vertical gradients in electron density, electron and ion temperatures, respectively.

Equation (35) updates Hooke's continuity equation of electrons which are adopted in literature to deal with ionospheric responses to propagating gravity waves (e.g., [12]). This updated equation includes an additional term originated from a competition between the retarded speed due to diffusion ($v_{e\parallel}^*$) and the field-aligned speed of electrons ($v_{e\parallel}$).

4. Ionospheric Dynamo Electric Field and Electron and Ion Speeds

The ionospheric plasma is quasi-neutral with $n_e \approx n_i = n_0$, however, with a space charge density, $n_{sc} = n_i - n_e$, dominantly as a result of the ionospheric dynamo process. Other mechanisms, such as electron precipitation, may also contribute to n_{sc} in polar regions. This space charge density is always several orders lower than either n_i or n_e, but drives a non-negligible ionospheric electric field **E**. We rely on Maxwell's equations to obtain its solution.

For gravity waves in a time scale from tens of minutes to a few hours, we exclude the daily changes in **E** and **B**. As a result, the displacement current from the time-dependent changes in **E**, $\epsilon_0 \partial\mathbf{E}/\partial t$, and the induced electric field from the time-dependent changes in **B**, $\partial\mathbf{B}/\partial t$, are dropped out. Maxwell's electrodynamic equations reduce to the following [89]:

$$\nabla \cdot \mathbf{E} = \frac{en_{sc}}{\epsilon_0}, \quad \nabla \times \mathbf{E} = 0, \quad \nabla \cdot \mathbf{B} = 0, \quad \nabla \cdot \mathbf{j} = -\frac{\partial(en_{sc})}{\partial t} \tag{39}$$

the second equation of which shows that **E** is derivable from a potential function φ through $\mathbf{E} = -\nabla\varphi$, while the first and the last equations provide an ionospheric time scale τ of

$$\tau \sim \frac{en_{sc}}{\nabla \cdot \mathbf{j}} = \epsilon_0 \frac{\nabla \cdot \mathbf{E}}{\nabla \cdot \mathbf{j}} \sim \frac{\epsilon_0}{\sigma} < 1 \ (\mu s) \tag{40}$$

during which **E** is established due to the appearance of n_{sc} anywhere in ionosphere so as to cancel any divergence of **j** and keep

$$\nabla \cdot \mathbf{j} = 0 \tag{41}$$

in ionosphere. In Equation (40) $\mathbf{j} = \sigma\mathbf{E}$ with a uniform and isotropic σ is assumed to simplify the estimation (Section 2.3 in [91]).

From the lower right panel of Figure 1, we know that $\sigma_{\parallel} \gg 1$. This results in an infinitesimal electric field along magnetic field lines, leading to $\nabla \cdot \mathbf{E}_{\parallel} \approx 0$. Thus, together with $\mathbf{E} = -\nabla \varphi$, Equations (19) and (41) provide that

$$\nabla^2 \varphi = \mathbf{B} \cdot (\nabla \times \mathbf{v}_n) + \frac{\sigma_H}{\sigma_P} \mathbf{b} \cdot [\nabla \times (\mathbf{v}_n \times \mathbf{B})] \tag{42}$$

where $\mathbf{b}$ is defined by a dipole model in terms of a given latitude ϕ (geographic) as follows [11,97,98]; note that [11] defined different orientations of the horizontal coordinates from the present paper):

$$\mathbf{b} = \frac{\mathbf{B}}{B} = \left\{ 0, \frac{B_y}{B}, \frac{B_z}{B} \right\} = \left\{ 0, \frac{\cos\phi}{\sqrt{1 + 3\sin^2\phi}}, -\frac{2\sin\phi}{\sqrt{1 + 3\sin^2\phi}} \right\} \tag{43}$$

and $B = B_{eq}\sqrt{1 + 3\sin^2\phi}$ where B_{eq} is the magnetic field strength at the equator. Neglecting the small horizontal electric field components compared to the vertical one [namely, $\partial\varphi/\partial(x,y) \ll \partial\varphi/\partial z$], Equation (42) reduces to the following:

$$\frac{\partial^2 \varphi}{\partial z^2} = B_y \left[\frac{dv_{nx}}{dz} - \frac{\sigma_H}{\sigma_P} \left(\frac{B_z}{B} \frac{dv_{ny}}{dz} - \frac{B_y}{B} \frac{dv_{nz}}{dz} \right) \right] \tag{44}$$

which is an 1D Poisson equation with the solution of

$$\frac{dE_z}{dz} = B_y \left(\frac{B_z}{B} \frac{\sigma_H}{\sigma_P} \frac{dv_{ny}}{dz} - \frac{dv_{nx}}{dz} \right) \approx -B_y \frac{dv_{nx}}{dz} \tag{45}$$

where $\sigma_H \ll \sigma_P$ is applied in the last step based on the lower right panel of Figure 1. Equation (45) gives

$$E_z = B_y [v_{nx}(z_0) - v_{nx}(z)] \tag{46}$$

where z_0 is a reference altitude at which $E_z = 0$ under the condition of $\sigma_H/\sigma_P \ll 1$. It is also the altitude which divides F-dynamo and E-dynamo regions in ionosphere. By checking the conductivity profiles at lower altitudes, we find this condition holds deep to ionospheric E region (100–150 km altitudes). For example, at the F2 bottom (220 km altitude) the ratio is 57; it reduces to 23 at 130 km height, and down to 1 at 105 km height. Above z_0 the zonal wind solely determines the magnitude of the ionospheric electric field, while below it both the zonal and meridional winds contribute to the electric field.

With E_z at hand by solving Equation (45) numerically or Equation (46) analytically, the components of electron and ion velocities are derived from Equations (16) and (17), respectively, by using following notations:

$$\left.\begin{array}{l}
\mathbf{v}_n \cdot \hat{\mathbf{e}}_x = v_{nx}, \ \mathbf{v}_n \cdot \hat{\mathbf{e}}_y = v_{ny}, \ \mathbf{v}_n \cdot \hat{\mathbf{e}}_z = 0, \ \mathbf{v}_n \cdot \mathbf{b} = v_{n\parallel} = v_{ny}\frac{B_y}{B} \\[4pt]
(\mathbf{v}_n \times \mathbf{b}) \cdot \hat{\mathbf{e}}_x = v_{ny}\frac{B_z}{B_0}, \ (\mathbf{v}_n \times \mathbf{b}) \cdot \hat{\mathbf{e}}_y = -v_{nx}\frac{B_z}{B_0}, \ (\mathbf{v}_n \times \mathbf{b}) \cdot \hat{\mathbf{e}}_z = -v_{nx}\frac{B_y}{B_0} \\[4pt]
\mathbf{E} \cdot \hat{\mathbf{e}}_x = \mathbf{E} \cdot \hat{\mathbf{e}}_y = 0, \ \mathbf{E} \cdot \hat{\mathbf{e}}_z = E_z; \ \mathbf{b} \cdot \hat{\mathbf{e}}_x = 0, \ \mathbf{b} \cdot \hat{\mathbf{e}}_y = \frac{B_y}{B}, \ \mathbf{b} \cdot \hat{\mathbf{e}}_z = \frac{B_z}{B} \\[4pt]
(\mathbf{E} \times \mathbf{b}) \cdot \hat{\mathbf{e}}_x = -E_z\frac{B_y}{B_0}, \ (\mathbf{E} \times \mathbf{b}) \cdot \hat{\mathbf{e}}_y = 0, \ (\mathbf{E} \times \mathbf{b}) \cdot \hat{\mathbf{e}}_z = 0 \\[4pt]
\mathbf{E}_\parallel \cdot \hat{\mathbf{e}}_x = 0, \ \mathbf{E}_\parallel \cdot \hat{\mathbf{e}}_y = E_z\frac{B_y B_z}{B^2}, \ \mathbf{E}_\parallel \cdot \hat{\mathbf{e}}_z = E_z\frac{B_z^2}{B^2}
\end{array}\right\} \quad (47)$$

We obtain

$$\left.\begin{array}{l}
v_{ex} = \dfrac{v_{en}^2}{v_{en}^2+\Omega_e^2}v_{nx} - \dfrac{v_{en}\Omega_e}{v_{en}^2+\Omega_e^2}\dfrac{B_z}{B}v_{ny} - \dfrac{\Omega_e^2}{v_{en}^2+\Omega_e^2}\dfrac{B_y}{B}\dfrac{E_z}{B} \\[10pt]
v_{ey} = \dfrac{v_{en}\Omega_e}{v_{en}^2+\Omega_e^2}\dfrac{B_z}{B_0}v_{nx} + \dfrac{v_{en}^2+\Omega_e^2\frac{B_y^2}{B^2}}{v_{en}^2+\Omega_e^2}v_{ny} - \dfrac{\Omega_e}{v_{en}}\dfrac{\Omega_e^2}{v_{en}^2+\Omega_e^2}\dfrac{B_y B_z}{B^2}\dfrac{E_z}{B} \\[10pt]
v_{ez} = \dfrac{v_{en}\Omega_e}{v_{en}^2+\Omega_e^2}\dfrac{B_y}{B_0}v_{nx} + \dfrac{\Omega_e^2}{v_{en}^2+\Omega_e^2}\dfrac{B_y B_z}{B^2}v_{ny} - \dfrac{\Omega_e}{v_{en}}\dfrac{v_{en}^2+\Omega_e^2\frac{B_z^2}{B^2}}{v_{en}^2+\Omega_e^2}\dfrac{E_z}{B}
\end{array}\right\} \quad (48)$$

$$\left.\begin{array}{l}
v_{ix} = \dfrac{v_{in}^2}{v_{in}^2+\Omega_i^2}v_{nx} + \dfrac{v_{in}\Omega_i}{v_{in}^2+\Omega_i^2}\dfrac{B_z}{B_0}v_{ny} - \dfrac{\Omega_i^2}{v_{in}^2+\Omega_i^2}\dfrac{B_y}{B_0}\dfrac{E_z}{B} \\[10pt]
v_{iy} = -\dfrac{v_{in}\Omega_i}{v_{in}^2+\Omega_i^2}\dfrac{B_z}{B_0}v_{nx} + \dfrac{v_{in}^2+\Omega_i^2\frac{B_y^2}{B^2}}{v_{in}^2+\Omega_i^2}v_{ny} + \dfrac{\Omega_i}{v_{in}}\dfrac{\Omega_i^2}{v_{in}^2+\Omega_i^2}\dfrac{B_y B_z}{B_0^2}\dfrac{E_z}{B} \\[10pt]
v_{iz} = -\dfrac{v_{in}\Omega_i}{v_{in}^2+\Omega_i^2}\dfrac{B_y}{B_0}v_{nx} + \dfrac{\Omega_i^2}{v_{in}^2+\Omega_i^2}\dfrac{B_y B_z}{B^2}v_{ny} + \dfrac{\Omega_i}{v_{in}}\dfrac{v_{in}^2+\Omega_i^2\frac{B_z^2}{B^2}}{v_{in}^2+\Omega_i^2}\dfrac{E_z}{B}
\end{array}\right\} \quad (49)$$

Using Equations (48) and (49), Figure 4 displays the vertical profiles of v_{ex} and v_{ix}, v_{ey} and v_{iy}, and v_{ez} and v_{iz} in the upper left, upper right, and the lower panels, respectively. Above 200 km altitude, the speeds satisfy $v_{ey} = v_{iy}$, and $v_{ez} = v_{iz}$, however, $v_{ex} \neq v_{ix}$. This indicates that the current has only the zonal component $j_x = en_e(v_{ix} - v_{ex})$ the maximal value of which appears at 225 km altitude where $v_{ix} = 3.25 \times 10^{-3}$ m/s, $v_{ex} = -5.30 \times 10^{-6}$ m/s. With $n_e = 7.2 \times 10^{11}$ m^{-3}, we obtain $j_x \approx 0.4$ nA/m^2, 2–3 orders higher than the "fair weather" current at the surface of the Earth. Note that the speeds are dominantly determined by the zonal and meridional neutral winds which are assumed horizontally stratified, they are independent of locations consequently.

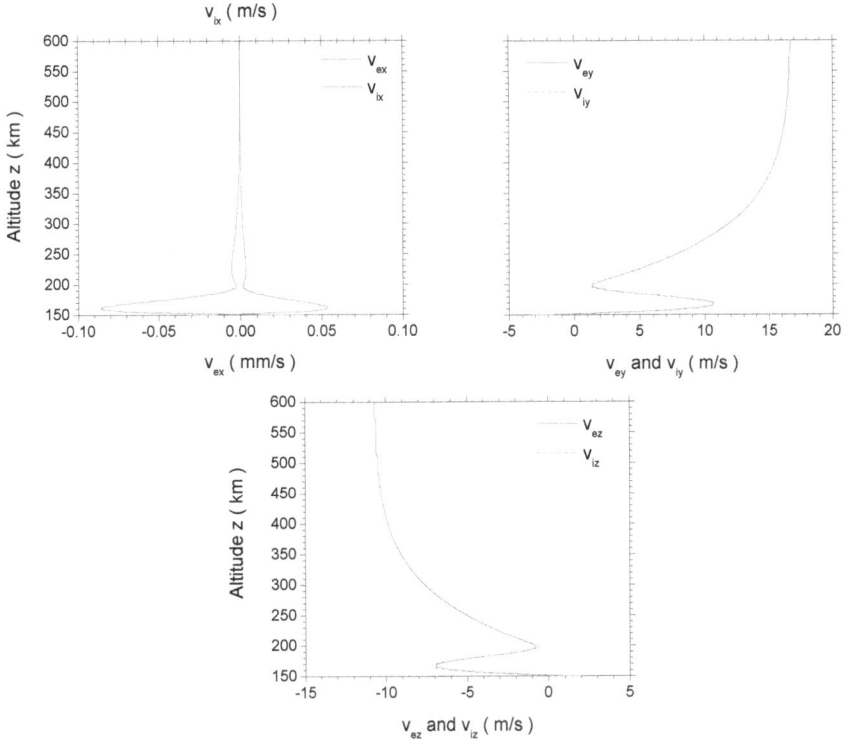

Figure 4. Vertical profiles of ionospheric electron and ion speeds, v_{ex} and v_{ix} (upper left panel), v_{ey} and v_{iy} (upper right panel), v_{ez} and v_{iz} (lower panel), in the presence of ionospheric dynamo electric field. Equations (48) and (49) are used.

By selecting $z_0 = 140$ km as the reference altitude, where $E_z = 0$, to calculate electric field strength as expressed by Equation (46), the LHS panel of Figure 5 depicts the vertical profile of the electric field E_z in the ionospheric F layer. The field strength is in an order of mV/m, enhancing to 2 mV/m at $\sim$200 km altitude, then decreasing exponentially versus altitude upward. Above 500 km altitude, it finally stabilizes at about -0.78 mV/m. In the plasma rest frame, this field induces an $\mathbf{E} \times \mathbf{B}$ drift of $(-25\sim+50)$ m/s, the same order as the neutral wind speeds. It is therefore necessary to take into account the effect of the dynamo electric field in solving relevant ionospheric problems. Note that this field is contributed by the neutral zonal wind, and thus uniform in the horizontal plane due to the assumed stratified atmospheric model.

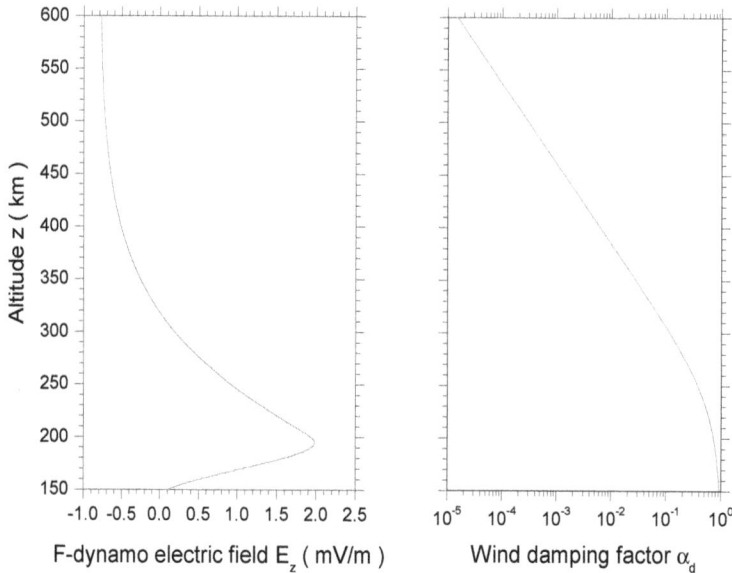

Figure 5. LHS: Vertical profile of F-dynamo electric field E_z with $z_0 = 140$ km at which $E_z = 0$ (from Equation (46)); RHS: Vertical profile of atmospheric wind perturbation damping factor, α_d, used in n'_e and total electron content (TEC) calculations (from Figure 5a of [11]).

5. Electron Density Perturbations Driven by Tsunami-Excited Gravity Waves

5.1. Magnitude of n'_e/n_{e0}

Adopting the standard liearization to Equation (35) yields the perturbed equation of electron density as follows:

$$
\frac{n'_e}{n_{e0}} = \frac{\frac{1}{n_{e0}}P'_e - L'_e - (k_{ne} + k_{ve})v'^*_{ez} - i\mathbf{k} \cdot \mathbf{v}'^*_e}{[L_{e0} + k^*_{ve}v^*_{ez} - i(\omega - \mathbf{k} \cdot \mathbf{v}^*_e)]} \tag{50}
$$

where $k_{ve} = (dv_{ez}/dz)/v_{ez}$, $k^*_{ve} = (dv^*_{ez}/dz)/v^*_{ez}$, and the primed quantities are the departures from respective equilibrium values with subscript "$_0$" in the presence of gravity waves. As follows the terminology and the modeling used to obtain the electron density perturbation is similar to that of Hooke [25].

Above 140 km altitude, the most important reactions are given by

(1) $O + h\nu \rightarrow O^+ + e^-$;
(2) $O^+ + XY \rightarrow XY^+ + O$, with a loss rate of $\beta/[XY] \sim 10^{-12}$ cm^3/s; and,
(3) $XY^+ + e^- \rightarrow X + Y$, with a loss rate of $\alpha \sim 10^{-7}$ cm^3/s.

In the above, (X,Y) denotes (N,O) or (O,O), [XY] is the number density of XY, and, XY^+ is molecular ion. Let O^+ and XY^+ have number densities $[O^+]$ and $[XY^+]$, respectively. In the F-region above 250 km altitude, the empirical IRI-2012 model exposes that $[XY^+]$ decreases from 12.3% in concentration to zero in 50 km upward and $[O^+]$ increases from 79.1% in concentration at 250 km to 98.9% at 300 km altitude till 600 km where it reduces to 79.3%. As noted previously in Section 3.1, these IRI results refer to the arbitrary location selected as an example at UT 7:30 pm and geographic position of $29°12'N\ 81°2'W$ on 19 April 2015. For case studies of data-fit modeling in following papers, the concerned location and the universal time will be updated specifically for the exact percentage of the particle concentrations.

Besides, the total rate of photoionization in the atmosphere has a maximum of about 10^6 m^{-3} per second in the Fl-layer (about 150–220 km in altitude and only during daylight hours; cf. [99]). The rate of photoionization decreases approximately exponentially with increasing altitude. The produced principal ion is O^+ ([100]). Thus, $[O^+]$ dominates the chemical reactions related to electron loss.

Moreover, the neutral atmospheric scale height is $\sim$100 km near the F2-peak altitude around 300 km, several scale heights above the maximal photoionization region in the F1-layer. At this altitude, the perturbed photoionization rate P_e' is contributed by constituent O as described by Equation (11) in [25], satisfying

$$P_e' \simeq \frac{n'}{n_0(z)} P_0 \tag{51}$$

in which $P_0 \simeq \nabla \cdot \{[O^+]\mathbf{u}_0(O^+)\}$ ([101]), and $n_0(z)$ is the equilibrium number density of O at height z and n' is the perturbed density.

Finally, we estimate the total electron loss rate L_e as follows. Related reactions and rates k_i (cm^3/s) are given in Table A1 of [13]:

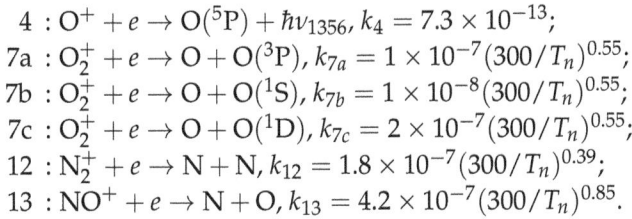

$4 : O^+ + e \rightarrow O(^5P) + \hbar\nu_{1356}, k_4 = 7.3 \times 10^{-13};$
$7a : O_2^+ + e \rightarrow O + O(^3P), k_{7a} = 1 \times 10^{-7}(300/T_n)^{0.55};$
$7b : O_2^+ + e \rightarrow O + O(^1S), k_{7b} = 1 \times 10^{-8}(300/T_n)^{0.55};$
$7c : O_2^+ + e \rightarrow O + O(^1D), k_{7c} = 2 \times 10^{-7}(300/T_n)^{0.55};$
$12 : N_2^+ + e \rightarrow N + N, k_{12} = 1.8 \times 10^{-7}(300/T_n)^{0.39};$
$13 : NO^+ + e \rightarrow N + O, k_{13} = 4.2 \times 10^{-7}(300/T_n)^{0.85}.$

The above reactions offer $L_e = k_4[O^+] + k_7[O_2^+] + k_{12}[N_2^+] + k_{13}[NO^+] \approx k_4[O^+]$ (in which $[O_2^+] = [N_2^+] = [NO^+] \approx 0$ as given by the IRI-2012 model). Considering the charge neutrality condition, we obtain $n_e' \approx [O^+]'$, and

$$P_0 \simeq \nabla \cdot (n_{e0}\mathbf{v}_i) = \frac{d(n_{e0}v_{iz})}{dz}, \quad L_{e0} = k_4[O^+]_0 \approx k_4 n_{e0}, \quad L_e' = k_4[O^+]' \approx k_4 n_e' \tag{52}$$

in which $\mathbf{v}_i = \{v_{ix}, v_{iy}, v_{iz}\}$ is the mean-field ion velocity the components of which are given in Equation (49).

Applying $\Omega_e \gg \nu_{en}$, $\Omega_i \gg \nu_{in}$, and Equation (46) to Equations (48) and (49) yields

$$\left. \begin{array}{l} v_{ex} = -\left(\frac{B_y}{B}\right)^2 [v_{nx}(z_0) - v_{nx}(z)], \quad v_{ey} = -\frac{\Omega_e}{\nu_{en}} \frac{B_y^2 B_z}{B^3} [v_{nx}(z_0) - v_{nx}(z)], \quad v_{ez} = \frac{B_z}{B_y} v_{ey} \\[2mm] v_{ix} = v_{ex}, \quad v_{iy} = \frac{\Omega_i}{\nu_{in}} \frac{B_y^2 B_z}{B^3} [v_{nx}(z_0) - v_{nx}(z)], \quad v_{iz} = \frac{B_z}{B_y} v_{iy} \end{array} \right\} \tag{53}$$

Clearly, $v_{ex} \ll v_{ey}$ or v_{ez}, and

$$v_{ex}^* = v_{ex}, \quad v_{ey}^* = v_{ey} + (v_{e\|}^* - v_{e\|})\frac{B_y}{B}, \quad v_{ez}^* = v_{ez} + (v_{e\|}^* - v_{e\|})\frac{B_z}{B} \tag{54}$$

which gives

$$v_{ex}^{\prime*} = v_{ex}^{\prime}, \quad v_{ey}^{\prime*} = v_{ey}^{\prime} + (v_{e\|}^* - v_{e\|})^{\prime}\frac{B_y}{B}, \quad v_{ez}^{\prime*} = v_{ez}^{\prime} + (v_{e\|}^* - v_{e\|})^{\prime}\frac{B_z}{B} \tag{55}$$

and $v_{ex}^* \ll v_{ey}^*$ or v_{ez}^*, $v_{ex}^{\prime*} \ll v_{ey}^{\prime*}$ or $v_{ez}^{\prime*}$. Considering k_x and k_y are in the same order in magnitudes and k_x (or k_y) is smaller than k_z by a factor of 5–10, thus, $k_z v_{ez}^*$ and $k_z v_{ez}^{\prime*}$ dominate $\mathbf{k} \cdot \mathbf{v}_e^*$ and $\mathbf{k} \cdot \mathbf{v}_e^{\prime*}$, respectively, Equation (50) reduces to the following:

$$\frac{n_e^{\prime}}{n_{e0}} = \frac{\epsilon_1 \frac{n^{\prime}}{n_0} - \alpha_d \epsilon_2 \frac{v_{nx}^{\prime}(z_0) - v_{nx}^{\prime}(z)}{v_{nx}(z_0) - v_{nx}(z)}}{\epsilon_3 \frac{v_L - i v_\omega}{v_{nx}(z_0) - v_{nx}(z)} + \epsilon_4} \tag{56}$$

in which $v_\omega = \omega / (k_{ne} + k_{ve})$, $v_L = 2 k_4 n_{e0} / (k_{ne} + k_{ve})$, and

$$\epsilon_1 = \frac{\nu_{en}\Omega_i}{\Omega_e \nu_{in}}, \quad \epsilon_2 = \frac{k_{ne} + k_{ve} + i k_z}{k_{ne} + k_{ve}}, \quad \epsilon_3 = \frac{\nu_{en}}{\Omega_e} \frac{B^3}{B_y B_z^2}, \quad \epsilon_4 = \frac{k_{ve} + i k_z}{k_{ne} + k_{ve}} \tag{57}$$

Parameter α_d in Equation (56) is the damping factor of the atmospheric wind perturbation, defined as the ratio of the magnitudes of the wind perturbations under dissipative and non-dissipative conditions, respectively. It is calculated from Figure 5a of [11], and plotted in the RHS panel of Figure 5. Below 150 km altitude, α_d is nearly 1; above the altitude, it reduces exponentially versus height. In the core F2 region (250–450 km) it drops 2 orders of magnitude from 0.1 to 0.001, as shown in the RHS panel of Figure 5. This reflects that the atmospheric response is unrealistically large for wave propagation under nondissipative conditions [11] where the neutral wind can be perturbed up to a few hundreds m/s, as presented in the lower three panels of Figure 2, consistent with the results provided in [8]. Interestingly, if we choose a higher horizontal wavenumber, $k_h \sim 2\pi/50$ 1/km, rather than the present lower one, $k_h \sim 2\pi/400$ 1/km, α_d shifts to 1. Thus, tsunami-excited waves of higher energy may possess stronger potential to resist any dissipations in their upward propagations. More discussions of this topic is beyond the scope of this paper and left to be touched in details in another sister paper.

If the dynamo electric field is neglected, Equations (48) and (49) reduce to the following two sets of equations, respectively:

$$
\left.
\begin{aligned}
v_{ex} &= \frac{v_{en}^2}{v_{en}^2+\Omega_e^2}v_{nx} - \frac{v_{en}\Omega_e}{v_{en}^2+\Omega_e^2}\frac{B_z}{B}v_{ny} \\
v_{ey} &= \frac{v_{en}\Omega_e}{v_{en}^2+\Omega_e^2}\frac{B_z}{B_0}v_{nx} + \frac{v_{en}^2+\Omega_e^2\frac{B_y^2}{B^2}}{v_{en}^2+\Omega_e^2}v_{ny} \\
v_{ez} &= \frac{v_{en}\Omega_e}{v_{en}^2+\Omega_e^2}\frac{B_y}{B_0}v_{nx} + \frac{\Omega_e^2}{v_{en}^2+\Omega_e^2}\frac{B_y B_z}{B^2}v_{ny}
\end{aligned}
\right\}
\tag{58}
$$

$$
\left.
\begin{aligned}
v_{ix} &= \frac{v_{in}^2}{v_{in}^2+\Omega_i^2}v_{nx} + \frac{v_{in}\Omega_i}{v_{in}^2+\Omega_i^2}\frac{B_z}{B_0}v_{ny} \\
v_{iy} &= -\frac{v_{in}\Omega_i}{v_{in}^2+\Omega_i^2}\frac{B_z}{B_0}v_{nx} + \frac{v_{in}^2+\Omega_i^2\frac{B_y^2}{B^2}}{v_{in}^2+\Omega_i^2}v_{ny} \\
v_{iz} &= -\frac{v_{in}\Omega_i}{v_{in}^2+\Omega_i^2}\frac{B_y}{B_0}v_{nx} + \frac{\Omega_i^2}{v_{in}^2+\Omega_i^2}\frac{B_y B_z}{B^2}v_{ny}
\end{aligned}
\right\}
\tag{59}
$$

which gives

$$
v_{ex} = -\frac{v_{en}}{\Omega_e}\frac{B_z}{B}v_{ny}, \quad v_{ix} = -\frac{\Omega_e v_{in}}{\Omega_i v_{en}}v_{ex}, \quad v_{ey} = v_{iy} = \frac{B_y^2}{B^2}v_{ny}, \quad v_{ez} = v_{iz} = \frac{B_y B_z}{B^2}v_{ny}
\tag{60}
$$

under the constraints of $\Omega_e \gg v_{en}$ and $\Omega_i \gg v_{in}$. Clearly, $v_{ex} \ll v_{ey}$ or v_{ez}. We then have

$$
v_{ex}^* = v_{ex}, \quad v_{ey}^* = v_{ey} + (v_{e\parallel}^* - v_{e\parallel})\frac{B_y}{B}, \quad v_{ez}^* = v_{ez} + (v_{e\parallel}^* - v_{e\parallel})\frac{B_z}{B}
\tag{61}
$$

and

$$
v_{ex}'^* = v_{ex}', \quad v_{ey}'^* = v_{ey}' + (v_{e\parallel}^* - v_{e\parallel})'\frac{B_y}{B}, \quad v_{ez}'^* = v_{ez}' + (v_{e\parallel}^* - v_{e\parallel})'\frac{B_z}{B}
\tag{62}
$$

We obtain that $v_{ex}^* \ll v_{ey}^*$ or v_{ez}^* and $v_{ex}'^* \ll v_{ey}'^*$ or $v_{ez}'^*$. Adopting the same algebra as those given in the last Subsection produces

$$
\frac{n_e'}{n_{e0}} = \frac{\epsilon_1^*\frac{n'}{n_0} - \alpha_d\epsilon_2^*\frac{v_{ny}'}{v_{ny}}}{\epsilon_3^*\frac{v_L^*-iv_\omega^*}{v_{ny}} + \epsilon_4^*}
\tag{63}
$$

in which $v_\omega^* = \omega/(k_{ne} + k_{vny})$, $v_L^* = 2k_4 n_{e0}/(k_{ne} + k_{vny})$, and

$$
\epsilon_1^* = 1, \quad \epsilon_2^* = \left(1 + i\frac{k_z}{k_{ne} + k_{vny}}\right)\frac{m_i v_{in}}{m_e v_{en}}, \quad \epsilon_3^* = \frac{B^2}{B_y B_z}, \quad \epsilon_4^* = \frac{k_{vny} + ik_z}{k_{ne} + k_{vny}}\frac{m_i v_{in}}{m_e v_{en}}
\tag{64}
$$

It is noteworthy here that, different from Equation (56) where the zonal wind (v_{nx}) and its disturbance dominate the electron density perturbation in the presence of the dynamo electric field, Equation (63) exhibits that it is the meridional wind (v_{ny}) and its disturbance that determines the electron density perturbation in the absence of the dynamo electric field.

5.2. Tsunami-Driven Perturbations

As early as in the 1970s, atmospheric and ionospheric constituents (namely, neutrals, electrons, and ions) were exposed to be featured by wavelike variations in transport properties (namely, density, velocity, and temperature) with respect to spaceborne data from, e.g., AE-C satellite; however, the perturbations demonstrate respective wave characteristics in either amplitudes and/or periods, phases, phase speeds [102]. For example, the electron temperature variations are out of phase with those in the ion density. During the upward propagation of tsunami-driven gravity waves, resonant coupling between the atmospheric wave and ionospheric perturbations happens at some resonant heights, where they both have the same wave characteristics (*i.e.*, wave frequency and wavenumber vector) as each other, leading to detectable perturbations of plasma particles (e.g., [103]). Only at these heights can gravity wave parameters be imposed to the electron equations, Equation (56) and (63), to calculate the magnitudes of the perturbations in electron density and TEC.

Applying the atmospheric perturbations calculated from Equation (4) as inputs to both Equations (56) and (63), we calculate the vertical profiles of electron density and TEC perturbations in the absence (curves in blue) and presence (curves in pink) of the dynamo electric field, E, at middle and low latitudes, 60° N and 29° N, respectively, as displayed in Figure 6. In reference of Hickey *et al.*'s model [11], we choose a typical tsunami source of $L_{tsu}(x) = 2000$ km, $L_{tsu}(y) = 400$ km, and $\alpha_d = 0.01$. In view of rows, the upper two panels illustrate the results at the 60° N location, and the lower two ones manifest those at the 29° N location, closer to the equator. In view of columns, The LHS two panels give the percentages of the perturbations relative to the unperturbed IRI-2012 electron density, and the RHS two ones reveal the corresponding ratios in TEC magnitudes, relevant to a couple of parameters: the unperturbed TEC (*i.e.*, TEC_0) and the perturbed TEC (*i.e.*, TEC') as defined by

$$TEC_0(z) = \int_{150}^{z} n_{e0} dh \text{ (TECU), } TEC'(z) = \int_{150}^{z} n'_e dh \text{ (TECU)} \tag{65}$$

where dh is the element of the increment in the vertical direction, and the height of integration, z, is from 150 km to 600 km altitudes.

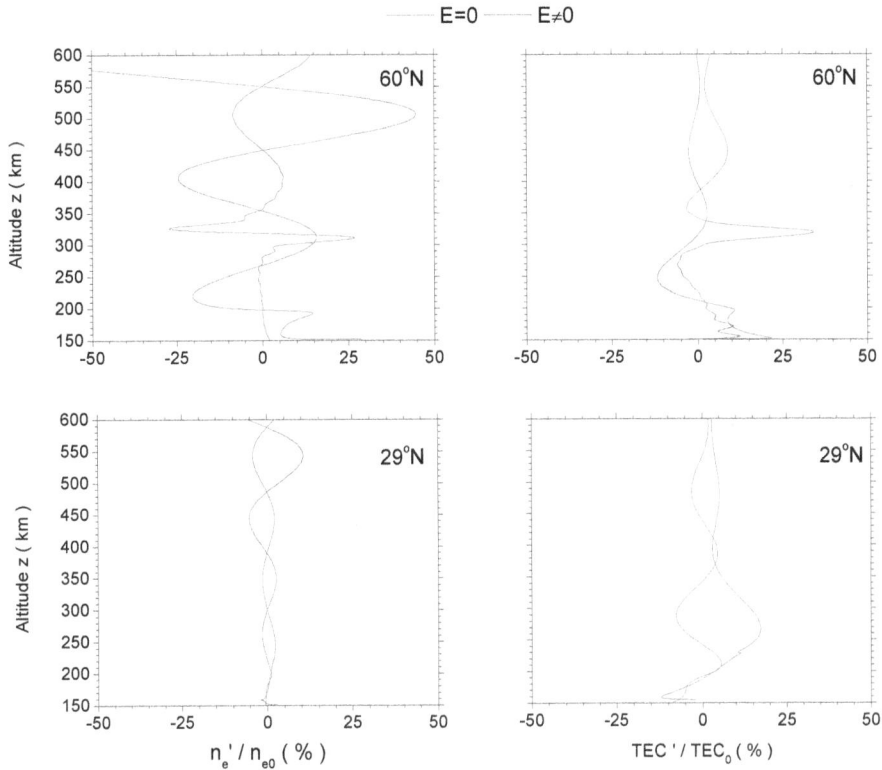

Figure 6. Vertical profiles of electron density and TEC perturbations in the absence (curves in blue) and presence (curves in pink) of the dynamo electric field, E, at middle and low latitudes, $60°$ N and $29°$ N, respectively. A typical tsunami case of $L_{tsu}(x) = 2000$ km, $L_{tsu}(y) = 400$ km, and $\alpha_d = 0.01$, is chosen in reference of Hickey *et al.*'s model [11].

At the $60°$ N location in the upper left panel, the density perturbations express different features in response to the switch of the dynamo electric field, E. If E is off with $E = 0$, n'_e/n_{e0} has a wavelike oscillation with an increased amplitude upward from <25% above 150 km altitude initially to >50% above 500 km altitude. This is the case similar to that described by Hickey *et al.* [11] which provides a percentage of the same magnitude. By contrast, if E is on with $E \neq 0$, n'_e/n_{e0} has a bipolar-pulse waveform within 300 ± 50 km, with a magnitude of >25%; above 350 km, it conveys an anti-phase waveform relative to the previous case in the absence of E, the magnitude of which is <13%, while below 250 km, it fluctuates around zero percentage. Clearly, the presence of E suppresses the density fluctuations substantially at most altitudes, however, offers an abnormal, large-amplitude pulsation around the 300 km altitude.

The corresponding TEC perturbation features are substantiated in the upper right panel. On the one hand, the case of $E = 0$ exposes that the amplitude of the perturbation declines from the highest 11% at ~250 km altitude to less than 1% above 550 km for all altitudes, consistent with the argument that the TEC deviations should be within 10% of the equilibrium TEC [12]; on the other hand, the $E \neq 0$ case demonstrates that there exists an unusual jump up to 34% at ~320 km, in a sharp comparison with the surrounding altitudes where the perturbations are merely less than ~10%. We notice that this result is obtained with the specific tsumani source conditions borrowed from Hickey *et al.*'s model [11]. Under more extreme tsunami conditions, we agree with the postulation that higher TEC perturbations may be yielded, say, up to ~100% [11]. Because E is ubiquitous in ionospheric E and F layers, this altitude-dependent peculiarity in TEC perturbations is of extraordinary importance for us to make use of the GPS-TEC signals detected around this particular height for tsunami analysis.

However, there is a rather significant caveat to the surprising pulse-like variation of large TEC perturbations around 300 km altitudes, occurring at the mid-latitude regions. At the low-latitude location closer to the equator, 29° N, the 300-km peculiarities in both the electron density and TEC perturbations disappear, as exposed in the lower two panels in Figure 6, respectively. The two wave-like perturbations have opposite phases in response to the E-switch, and the former owns an escalated amplitude upward from 0 at 150 km altitude to a little more than 10% at 600 km altitude, whileas and latter holds a reduced one from about 20% lower than 300 km altitude to nearly 0 above 550 km altitude. After a check to the upper left panel of Figure 3, we assume the disappearance of the peculiarities might be explained by the difference in the magnitude of the background electron densities n_{e0} at the two locations: ionospheric plasma looks like a giant filter; the equatorward location has a larger n_{e0} than the polarward one; the enhancement in n_{e0} is large enough to mitigate or filter completely any abrupt electron density perturbations n'_e and therefore TEC'.

5.3. Effects of Atmospheric/Ionospheric Disturbances

According to either Equation (56) or Equation (63), perturbations in electron density (or TEC) are correlated with a few atmospheric/ionospheric disturbances in (1) photoionization gain and chemical loss; (2) plasma velocities; and, (3) dynamo electric field. The effects of these parameters can be estimated conveniently as follows to get more insights into ionospheric plasma irregularities. We take 300 km altitude as an example for the estimations.

(1) Influence of pure photoionization gain and chemical loss.

The effect of the photoionization gain and chemical loss on n_e can be obtained by neglecting all the other terms in Equation (50) and taking $k_4 = 7.3 \times 10^{-19}$ m^3/s;

$n_{e0} \sim 0.88 \times 10^{11}$ /m^3 (IRI-2012); $v_{iz} \sim 0.01$ m/s (Figure 6a in [11]; $k_{ne} \sim 0.005$ 1/km, $n'/n_0 \sim 10$ (Figure 2 in this paper). The percentage of the perturbation is as follows:

$$\left| \frac{n'_e}{n_{e0}} \right| = \frac{k_{ne} v_{iz}}{2 k_4 n_{e0}} \frac{n'}{n_0} \sim 350\% \tag{66}$$

(2) Influence of plasma velocities.

Compared to ionospheric electron/ion velocities, the photoionization gain and chemical loss contribute only a few 1/1000. They are thus neglected in Equation (50) in cases where $\mathbf{v}_e$ and $\mathbf{v}_i$ are included. This gives

$$\left| \frac{n'_e}{n_{e0}} \right| = \left| \frac{\frac{n'}{n_0}(k_{ne} + k_{vi})v_{iz} - (k_{ne} + k_{ve})v'^*_{ez}}{2 k_4 n_{e0} + k^*_{ve} v^*_{ez}} \right| \simeq \left(1 + \frac{k_{ne}}{k^*_{ve}} \right) \left| \frac{v'^*_{ez}}{v^*_{ez}} \right| \sim (48 - 67)\% \tag{67}$$

where $k^*_{ve} \approx k_{ve} = (0.005 - 0.007)$ km^{-1}, $v^*_{ez} \approx v_{ny} \sim 15$ m/s and $v'^*_{ez} \approx v'_{ny} \sim 5$ m/s (Figure 4).

(3) Influence of electric field.

If electric field is involved, the terms containing charge gain and loss and plasma velocities are appreciably small and can be neglected in Equation (50), yielding

$$\left| \frac{n'_e}{n_{e0}} \right| \approx \left| \frac{(k_{ne} + k_{ve})v'^*_{ez}}{k^*_{ve} v^*_{ez}} \right| \simeq \left(1 + \frac{k_{ne}}{k^*_{ve}} \right) \left| \frac{\delta(E/B)}{E/B} \right| \sim 5\% \tag{68}$$

where, applying Equation (46), $k_{ve} \approx d[(E/B)/dz]/(E/B) \approx (dU/dz)/U \sim 0.005$ km^{-1} in which $dU/dz \approx 0.2$ m/s per km, $U \approx 40$ m/s, giving $\delta(E/B)/(E/B) = \delta U/U \approx 0.04$ in which $u \approx 2$ m/s from Figure 2.

Thus, Sections 5.2 and 5.3 expose that, outside the regions of tsunami-driven gravity waves (Section 5.3), in situations where only pure photoionization gain and chemical loss are involved, electron density perturbation, n'_e/n_{e0}, is proportional to the neutral density perturbation, n'/n_0, and can soar to as high as 350% in the F2 peak layer. However, when plasma motions are present in the absence of ionospheric dynamo electric field, $i.e.$, $E = 0$, photoionization and chemical loss become negligible. n'_e/n_{e0} becomes dependent of the meridional wind perturbation, v'_{ny}/v_{ny}, which contributes to a perturbation of 50%–70%. If $E \neq 0$, the dynamo action via the $\mathbf{E} \times \mathbf{B}$ drift suppresses the electron density perturbation to as low as $\sim 5\%$. By contrast, within the wave propagation regions (Section 5.2), the gravity waves bring about a TEC perturbation of around 10% at all altitudes for $E = 0$; and contribute to >30% perturbation in the F2 peak layer, but down to within 10% outside the layer for $E \neq 0$. Because ionospheric E is available everywhere above 150 km

271

altitude, the regions around 300 km altitudes provide us a location to collect GPS-TEC data and extract tsunami information from electron density perturbation signals.

6. Summary and Discussion

In the present study, we concentrate on the theoretical modeling of the ionospheric electron density and TEC perturbations driven by tsunami-excited gravity waves. The purpose of the study lies in suggesting an extended data-fit model which is able to grant data analysis and case study more accurately, so as to design a more reliable algorithm to estimate the tsunami wave front, and subsequently help to confirm and image tsunamis by comparing both the space-borne and ground-based GPS measurements (e.g., [104]) with our modeling results, thus be able to establish a more effective and efficient tsunami warning and alarming system in future work.

For this purpose, this paper extends the procedure described by Hickey *et al.* ([11]) to obtain electron density and TEC perturbations by (1) employing the classical ionospheric electrodynamics to replace *MacLeod*'s ion momentum equation; (2) borrowing Kendall and Pickering's generalized perturbation theory to directly get the electron density perturbation equation; and (3) involving the effect of the ionospheric dynamo electric field on the plasma perturbations. Under nondissipative, windshear, and nonisothermal atmospheric conditions, the study demonstrates that

(1) The magnitude of E is within several mV/m, determined by the crossed product of zonal neutral wind and meridional geomagnetic field;
(2) When $E = 0$ at the mid-latitude location (60° N), the fluctuation in n'_e is dominated by the meridional wind in the F2 region (above 220 km altitude). The percentage of n'_e over n_{e0} has an enhanced amplitude from around 20% at 200–250 km altitudes to larger than 40% at 500 km altitude; by contrast, the amplitude of corresponding TEC perturbation is damped gradually from ~15% to <5% at related altitudes, respectively.
(3) When $E \neq 0$ at the same latitude location, the fluctuation in n'_e is determined by the zonal wind in the same ionospheric region. The percentage of n'_e over n_{e0} drops down to less than 15% at all altitudes, except an appreciable jump to >25% in the F2-peak layer (300–340 km altitudes); within the layer, the related TEC perturbation pulse arrives at 35% while and outside the layer the amplitude of the fluctuation is no more than 10%.
(4) At lower latitudes (say, 29° N), however, the sharp enhancement in the magnitude of the dynamo E-driven TEC perturbation in the F2-peak layer is filtered away by the denser background electron density; in both $E = 0$ and $E \neq 0$ cases, the amplitudes of the fluctuations in n'_e or TEC are roughly the same as each other, but anti-phased.

(5) Although atmospheric/ionospheric fluctuations caused by photoionization gain and chemical loss and plasma velocities are able to enhance the n'_e-amplitude substantially to 350% and 48%~67%, respectively, electric field restrains the divergence significantly to 4% if gravity waves are not involved.

We come to a conclusion that the effect of the ionospheric dynamo electric field cannot be neglected in estimating electron density perturbations driven by tsunami-excited gravity waves. Dynamo E-driven TEC perturbation provides a probe for tsunami monitoring by making use of the GPS-TEC signals outside low-latitude regions. Though only an individual component in the gravity wave spectrum is involved, we hope to make use of the basic knowledge gained in this paper to attack more realistic problems through a series of following incremental steps toward our goal of reconstructing and explaining measured tsunami-related electron density perturbations reported in natural hazards, say, 2004 Sumatra tsunami events (e.g., [15]). Such a problem may be related to the temporal behavior of the waves, in addition to the vertical profiles as discussed in this paper: How long does it take for the tsunami perturbation to reach a height, h, say, 600 km above the sea surface? To solve the problem, we need to rely on the dispersion relation of the gravity waves, $\omega = \omega(k)$. Only after obtaining an explicit expression of the dispersion relation, can we can calculate both the vertical phase speed, $v_{ph} = \omega/k_z$, and the vertical group speed, $v_{gr} = \partial\omega/\partial k_z$. From the two speeds, we can finally illustrate the time lapses, $t_{ph1} = \int_0^h dh/v_{ph}$ and $t_{ph2} = \int_0^h dh/v_{gr}$, for the wave crests' and its energy's travels upward, respectively. A sister paper will introduce new ray-tracing results of tsunami-driven gravity waves propagating upward by developing the classical Hines' isothermal and shear-free model to a more generalized realistic nonisothermal and wind-shearing model.

In addition, we would like to discuss a potential concern which may arise from readers: the ionospheric signature of the tsunami-excited gravity waves has been discussed up to 600 km height under a non-dissipative model. However, in realistic atmosphere, gravity waves can be considerably dissipated by such terms like kinetic viscosity and heat conductivity. Are these results valid for dissipative situations from which GPS-TEC signals are detected?

We explain that this paper is the first one of a series on Tsunami imaging using ionospheric radio occultation data. It did not discuss straightforwardly the dissipative effects. This is because we are dealing with a complicated subject related to wave excitation and propagation in atmosphere and ionosphere where electrodynamics plays a dominant role to drive plasma perturbations, unfortunately neglected before due to understandable reasons. The complexity of the topic requires that we pay attention dominantly to the electric field effect first of all in this first paper, with a purpose to approach to a finally least-error solution through a series of incremental steps, so as to be able to understand the physics and, based on gained knowledge, to develop appropriate algorithms for solving more realistic problems, e.g., using ionospheric radio occultation data to detect tsunami wave fronts while providing increased coverage and data density for the purpose to provide effective and efficient data-fit modeling to GPS signals for constructing a tsunami warning or alerting system. Fortunately, this subject has attracted more attentions in applications and many new results have been reported recently, such as, Yang *et al.*'s detection of the ionospheric disturbances in response to North Korean nuclear tests [105]; Yang *et al.*'s study on the meteor ionospheric impact by means of GPS data [106] and the ionospheric disturbances over Alaska driven by Tohoku-Oki earthquake [107]; and, most recently, Coisson *et al.*'s pioneer work to provide evidence that radio occultation data can be used for tsunami detection [108].

In order to reduce the complexity, this first paper did not discuss explicitly the effects of dissipative terms, such as kinetic viscosity and thermal conductivity. Instead, it applied relevant results implicitly in relevant simulations, while a comprehensive discussion is presented in a sister paper (to be submitted soon), based on a revisit to Vadas and Fritts' work ([109]). According to this sister paper, below 150 km altitude, the dissipation terms has no discernable effects. By contrast, above 200 km altitude, the dissipation considerably damps the atmospheric perturbation. For example, the neutral wind perturbation drops substantially from several hundreds of m/s under nondissipative conditions (as given in Figure 2 of this paper) to merely a few of m/s under dissipative conditions. This result is in consistent with Hickey *et al.* [11]. However, for the ionospheric properties, dissipation can be totally neglected due to the several orders smaller in magnitude in the momentum and energy equations than the Lorentz force and Joule heating impacts, respectively. These results verified Kaladze *et al.*'s gravity wave model in ionosphere [110]. Consequently, the impact of the dissipative terms affect heavily the neutral wind above 200 km altitude, and it exerts little effect on plasma properties at all altitudes. We therefore need to consider these terms only in the calculations related to the magnitude of neutral wind perturbations; in other words, we just need to replace the neutral wind profiles under non-dissipative conditions with those under dissipative conditions to obtain theoretical TEC signatures in realistic dissipative atmosphere.

Fortunately, Hickey *et al.* [11] Figure 5a provided the vertical profiles of the wind magnitudes under both dissipative and non-dissipative conditions, respectively. The ratio of the two magnitudes is further defined as a damping factor, α_d, in this paper. The vertical profile of α_d is plotted in the RHS panel of Figure 5. In calculating the electron density/TEC perturbations in the last Section, all the wind speeds under non-dissipative conditions are substituted by dissipative ones with the introduction of factor α_d. As a result, though starting from a simple, non-dissipative model, which provides readers the simplest picture to gain important insights into data-fit modeling of GPS signals, the electron density/TEC perturbation results presented in this paper respond to realistic atmospheric conditions in the presence of previously neglected ionospheric dynamo electric field: we have in fact tackled a situation for which the dissipative ingredients are also involved to influence GPS signals through the wind damping factor. The results thus offer a reference to produce a tsunami warning or alerting algorithm in realistic situations both qualitatively and quantitatively. The complete picture of the dissipative effects on the tsunami-driven gravity waves will be introduced in a sister paper.

At last, we would like to argue that Figure 5 of Occhipinti *et al.* [23] might be difficult to interpret, since it contains the data of the whole network of stations. In the figure there are saturations (*i.e.*, values exceeding the range shown in the panels where there are points colored with the maximum value of the used color-bars. Considering the amplitude of the signals varies for each phase of TEC oscillations, this kind of figures will usually show a limited range of values that allow identifications of all the oscillations. It therefore reveals the periods of the waves, however, not the full range of observed amplitudes [111].

Acknowledgments: The work is supported by a grant from a NASA-JPL project "Tsunami Imaging Using Ionospheric Radio Occultation Data" collaborated with Embry-Riddle Aeronautical University (ERAU). We thank M. D. Zettergren and J. B. Snively for discussions on the ionospheric electric field model and on the properties of gravity waves, respectively. We show gratitudes to Pierdavide Coisson (IPGP) and other two anonymous referees for comments and suggestions in evaluating this paper, which led the manuscript to a significant improvement; particularly, Coisson made detailed suggestions and contributions to enhance the quality of the paper, including those on the most recent advance in measurements, on the atmospheric and ionospheric modeling, and on the connection between the present theoretical study and GPS signals. Copies of the simulation runs and figures can be obtained by emailing maz@erau.edu.

Author Contributions: J.Z.G.M.: programming, calculation, analysis; M.P.H.: atmospheric theory and modeling; A.K.: atmospheric and ionospheric theory and modeling.

Conflicts of Interest: The authors declare no conflict of interest.

References

1. Cosgrave, J. *Synthesis Report: Expanded Summary. Joint Evaluation of the International Response to the Indian Ocean Tsunami*; Tsunami Evaluation Coalition: London, UK, 2007; pp. 1–41.
2. Hines, C.O. Gravity waves in the atmosphere. *Nature* **1972**, *239*, 73–78.
3. Peltier, W.R.; Hines, C.O. On the possible detection of tsunamis by a monitoring of the ionosphere. *J. Geophys. Res.* **1976**, *81*, 1995–2000.
4. Marshall, J.; Plumb, R.A. *Atmosphere, Ocean and Climate Dynamics*; Academic Press: Waltham, MA, USA, 1989.
5. Huang, C.S.; Sofko, G.J. Numerical simulations of midlatitude ionospheric perturbations produced by gravity waves. *J. Geophys. Res.* **1998**, *103*, 6977–6989.

6. Liu, J.-Y.; Tsai, Y.-B.; Ma, K.-F.; Chen, Y.-I.; Tsai, H.-F.; Lin, C.-H.; Kamogawa, M.; Lee, C.-P. Ionospheric GPS total electron content (TEC) disturbances triggered by the 26 December 2004 Indian Ocean tsunami. *J. Geophys. Res.* **2006**, *111*, A05303, doi:10.1029/2005JA011200.

7. Lognonné, P.; Lambin, J.; Garcia, R.; Crespon, F.; Ducic, V.; Jeansou, E. Ground based GPS tomography of ionospheric post-seismic signal. *Planet. Space Sci.* **2006**, *54*, 528–540.

8. Occhipinti, G.; Kherani, E.A.; Lognonné, P. Geomagnetic dependence of ionospheric disturbances induced by tsunamigenic internal gravity waves. *Geophys. J. Int.* **2008**, *173*, 753–765.

9. Occhipinti, G.; Coisson, P.; Makela, J.J.; Allgeyer, S.; Kherani, A.; Hebert, H.; Lognonn, P. Three-dimensional numerical modeling of tsunami-related internal gravity waves in the Hawaiian atmosphere. *Earth Planets Space* **2011**, *63*, 847–851.

10. Lee, M.C.; Pradipta, R.; Burke, W.J.; Cohen, J.A.; Dorfman, S.E.; Coster, A.J.; Sulzer, M.; Kuo, S.P. Did tsunami-launched gravity waves trigger ionospheric turbulence over Arecibo? *J. Geophys. Res.* **2008**, *113*, A01302, doi:10.1029/2007JA012615.

11. Hickey, M.P.; Schubert, G.; Walterscheid, R.L. Propagation of tsunami-driven gravity waves into the thermosphere and ionosphere. *J. Geophys. Res.* **2009**, *114*, doi:10.1029/2009JA014105.

12. Mai, C.-L.; Kiang, J.-F. Modeling of ionospheric perturbation by 2004 Sumatra tsunami. *Radio Sci.* **2009**, *44*, RS3011, doi:10.1029/2008RS004060.

13. Hickey, M.P.; Schubert, G.; Walterscheid, R.L. Atmospheric airglow fluctuations due to a tsunami-driven gravity wave disturbance. *J. Geophys. Res.* **2010**, *115*, doi:10.1029/2009JA014977.

14. Rolland, L.M.; Occhipinti, G.; Lognonné, P.; Loevenbruck, A. Ionospheric gravity waves detected offshore Hawaii after tsunamis. *Geophys. Res. Lett.* **2010**, *37*, doi:10.1029/2010GL044479.

15. Galvan, D.A.; Komjathy, A.; Hickey, M.P.; Mannucci, A.J. The 2009 Samoa and 2010 Chile tsunamis as observed in the ionosphere using GPS total electron content. *J. Geophys. Res.* **2011**, *116*, A06318, doi:10.1029/2010JA016204.

16. Makela, J.J.; Lognonné, P.; Hébert, H.; Gehrels, T.; Rolland, L.; Allgeyer, S.; Kherani, A.; Occhipinti, G.; Astafyeva, E.; Coïsson, P.; *et al.* Imaging and modeling the ionospheric airglow response over Hawaii to the tsunami generated by the Tohoku earthquake of 11 March 2011. *Geophys. Res. Lett.* **2011**, *38*, L00G02, doi:10.1029/2011GL047860.

17. Rozhnoi, A.; Shalimov, S.; Solovieva, M.; Levin, B.; Hayakawa, M.; Walker, S. Tsunami-induced phase and amplitude perturbations of subionospheric VLF signals. *J. Geophys. Res.* **2012**, *117*, A09313, doi:10.1029/2012JA017761.

18. Garcia, R.F.; Doornbos, E.; Bruinsma, S.; Hebert, H. Atmospheric gravity waves due to the Tohoku-Oki tsunami observed in the thermosphere by GOCE. *J. Geophys. Res. Atmos.* **2014**, *119*, 4498–4506.

19. Artru, J.; Ducic, V.; Kanamori, H.; Lognonné, P.; Murakami, M. Ionospheric detection of gravity waves induced by tsunamis. *Geophys. J. Int.* **2005**, *160*, 840–848.

20. Artru, J.; Lognonne, P.; Occhipinti, G.; Crespon, F.; Garcia, R.; Jeansou, E. Tsunami detection in the ionosphere. *Space Res. Today* **2005**, *163*, 23–27.

21. Occhipinti, G.; Lognonné, P.; Kherani, E.A.; Hébert, H. Three dimensional waveform modeling of ionospheric signature induced by the 2004 Sumatra tsunami. *Geophys. Res. Lett.* **2006**, *33*, L20104, doi:10.1029/2006GL026865.

22. Yiyan, Z.; Yun, W.; Xuejun, Q.; Xunxie, Z. Ionospheric anomalies detected by ground-based GPS before the Mw7.9 Wenchuan earthquake of 12 May 2008, China. *J. Atmos. Sol. -Terr. Phys.* **2009**, *71*, 959–966.

23. Occhipinti, G.; Rolland, L.; Lognonné, P.; Watada, S. From Sumatra 2004 to Tohoku-Oki 2011: The systematic GPS detection of the ionospheric signature induced by tsunamigenic earthquakes. *J. Geophys. Res.* **2013**, *118*, 3626–3636.

24. Rolland, L.M.; Lognonné, P.; Astafyeva, E.; Alam Kherani, E.; Kobayashi, N.; Mann, M.; Munekane, H. The resonant response of the ionosphere imaged after the 2011 off the Pacific coast of Tohoku Earthquake. *Earth Planets Space* **2011**, *63*, 853–857.

25. Hooke, W.H. Ionospheric irregularities produced by internal atmospheric gravity waves. *J. Atmos. Terr. Phys.* **1968**, *30*, 795–823.

26. MacLeod, M.A. Sporadic E theory. I. Collision-geomagnetic equilibrium. *J. Atmos. Sci.* **1966**, *23*, 96–109.

27. Hickey, M.P.; Walterscheid, R.L.; Taylor, M.J.; Ward, W.; Schubert, G.; Zhou, Q.; Garcia, F.; Kelley, M.C.; Shepherd, G.G. Numerical simulations of gravity waves imaged over Arecibo during the 10-day January 1993 campaign. *J. Geophys. Res.* **1997**, *102*, 11475–11489.

28. Hickey, M.P.; Taylor, M.J.; Gardner, C.S.; Gibbons, C.R. Full-wave modeling of small-scale gravity waves using Airborne Lidar and Observations of the Hawaiian Airglow (ALOHA-93) O(^{1}S) images and coincident Na wind/temperature lidar measurements. *J. Geophys. Res.* **1998**, *103*, 6439–6453.

29. Hickey, M.P.; Walterscheid, R.L.; Schubert, G. Gravity wave heating and cooling in Jupiter's thermosphere. *Icarus* **2000**, *148*, 266–281.

30. Hickey, M.P.; Walterscheid, R.L.; Schubert, G. A full-wave model for a binary gas thermosphere: Effects of thermal conductivity and viscosity. *J. Geophys. Res.* **2015**, *120*, 3074–3083.

31. Meng, X.; Komjathy, A.; Verkhoglyadova, O.P.; Yang, Y.-M.; Deng, Y.; Mannucci, A.J. A new physics-based modeling approach for tsunami-ionosphere coupling. *Geophys. Res. Lett.* **2015**, *42*, doi:10.1002/2015GL064610.

32. Friedman, J.P. Propagation of internal gravity waves in a thermally stratified atmosphere. *J. Geophys. Res.* **1966**, *71*, 1033–1054.

33. Volland, H. Full wave calculations of gravity wave propagation through the thermosphere. *J. Geophys. Res.* **1969**, *74*, 1786–1795.

34. Yeh, K.C.; Liu, C.H. Acoustic-gravity waves in the upper atmosphere. *Rev. Geophys. Space Sci.* **1974**, *12*, 193–216.

35. Cole, K.D. Atmospheric excitation and ionization by ions in strong auroral and man-made electric fields. *J. Atmos. Terr. Phys.* **1971**, *33*, 1241–1249.

36. Cole, K.D. Effects of crossed magnetic and spatially dependent electric fields on charged particles motion. *Planet. Space Sci.* **1976**, *24*, 515–518.

37. Temerin, M.; Cerny, K.; Lotko, W.; Mozer, F.S. Observations of double layers and solitary waves in the auroral plasma. *Phys. Rev. Lett.* **1982**, *48*, 1175–1179.

38. Boström, R.; Gustafsson, G.; Holback, B.; Holmgren, G.; Koskinen, H.; Kintner, P. Characteristics of solitary waves and weak double layers in the magnetospheric plasma. *Phys. Rev. Lett.* **1988**, *61*, 82–85.

39. Matsumoto, H.; Kojima, H.; Miyatake, T.; Omura, Y.; Okada, M.; Nagano, I.; Tsutsui, M. Electrostatic solitary waves (ESW) in the magnetotail: BEN wave forms observed by Geotail. *Geophys. Res. Lett.* **1994**, *21*, 2915–2918.

40. Mozer, F.S.; Ergun, R.E.; Temerin, M.; Cattell, C.; Dombeck, J.; Wygant, J. New features of time domain electric-field structures in the auroral acceleration region. *Phys. Rev. Lett.* **1997**, *79*, 1281–1284.

41. Bale, S.D.; Kellogg, P.J.; Larson, D.E.; Lin, R.P.; Goetz, K.; Lepping, R.P. Bipolar electrostatic structures in the shock transition region: Evidence of electron phase space holes. *Geophys. Res. Lett.* **1998**, *25*, 2929–2932.

42. Ergun R.E.; Carlson, C.W.; McFadden, J.P.; Mozer, F.S.; Delory, G.; Peria, W. FAST satellite observations of large-amplitude solitary structures. *Geophys. Res. Lett.* **1998**, *25*, 2041–2044.

43. Franz J.R.; Kintner, P.M.; Seyler, C.E.; Pickett, J.S.; Scudder, J.D. On the perpendicular scale of electron phase-space holes. *Geophys. Res. Lett.* **2000**, *27*, 169–172.

44. McFadden, J.P.; Carlson, C.W.; Ergun, R.E.; Mozer, F.S.; Muschietti, L.; Roth, I. FAST observations of ion solitary waves. *J. Geophys. Res.* **2003**, *108*, 8018, doi:10.1029/2002JA009485.

45. Pickett, J.S.; Chen, L.-J.; Kahler, S.W.; Santolik, O.; Goldstein, M.L.; Lavraud, B.; Decreau, P.M.E.; Kessel, R.; Lucek, E.; Lakhina,G.S.; *et al.* On the generation of solitary waves observed by Cluster in the near-Earth magnetosheath. *Nonlin. Proc. Geophys.* **2005**, *12*, 181–193.

46. Klumpar D.M.; Peterson, W.K.; Shelley, E.G. Direct evidence for two-stage (bimodal) acceleration of ionospheric ions. *J. Geophys. Res.* **1984**, *89*, 10779–10787.

47. Klumpar, D.M. A digest and comprehensive bibliography on transverse auroral ion acceleration. In *Ion Acceleration in the Magnetosphere and Ionosphere*; Chang, T., Hudson, M.K., Jasperse, J.R. Johnson, R.G., Kintner, P.M., Schulz, M., Eds.; AGU: Washington, DC, USA, 1986; pp. 389–398.

48. Chiueh, T.; Diamond, P.H. Two-point theory of current-driven, ion-cyclotron turbulence. *Phys. Fluids* **1986**, *29*, 76–96.

49. Moore, T.E.; Waite, J.H., Jr.; Lockwood, M.; Chappell, C.R. Observations of coherent transverse ion acceleration. In *Ion Acceleration in the Magnetosphere and Ionosphere*; Chang, T., Hudson, M.K., Jasperse, J.R., Johnson, R.G., Kintner, P.M., Schulz, M., Eds.; AGU: Washington, DC, USA, 1986; pp. 50–55.

50. Moore, T.E.; Chandler, M.O.; Pollock, C.J.; Reasoner, D.L.; Arnoldy, R.L.; Austin, B.; Kintner, P.M.; Bonnell, J. Plasma heating and flow in an auroral arc. *J. Geophys. Res.* **1996**, *101*, 5279–5298.

51. Kan, J.R.; Akasofu, S.-I. Electrodynamics of solar wind-magnetosphere-ionosphere interactions. *IEEE Trans. Plasma Sci.* **1989**, *17*, 83–108.

52. Vago, J.L.; Kintner, P.M.; Chesney, S.W.; Arnoldy, R.L.; Lynch, K.A; Moore, T.E. Transverse ion acceleration by localized hybrid waves in the topside auroral ionosphere. *J. Geophys. Res.* **1992**, *97*, 16935–16957.

53. Mottez, F. Instabilities and Formation of Coherent Structures. *Astrophys. Space Sci.* **2001**, *277*, 59–70.

54. Vogelsang, H.; Lühr, H.; Voelker, H.; Woch, J.; Bosinger, T.; Potemra, T.A.; Lindqvist, P.A. An ionospheric travelling convection vortex event observed by ground-based magnetometers and by VIKING. *Geophys. Res. Lett.* **1993**, *20*, 2343–2346.

55. Lund, E.J.; Möbius, E.; Ergun, R.E.; Carlson, C.W. Mass-dependent effects in ion conic production: The role of parallel electric fields. *Geophys. Res. Lett.* **1999**, *26*, 3593–3596.

56. Lund, E.J.; Möbius, E.; Carlson, C.W.; Ergunc, R.E.; Kistlera, L.M.; Kleckerd, B.; Klumpare, D.M.; McFaddenc, J.P.; Popeckia, M.A.; Strangewayf, R.J.; *et al.* Transverse ion acceleration mechanisms in the aurora at solar minimum: Occurrence distributions. *J. Atmo. Sol. -Terr. Phys.* **2000**, *62*, 467–475.

57. Mottez, F.; Chanteur, G.; Roux, A. Filamentation of plasma in the auroral region by an ion-ion instability: A process for the formation of bidimensional potential structures. *J. Geophys. Res.* **1992**, *97*, 10801–10810.

58. Mamun, A.A.; Shukla, P.K.; Stenflo, L. Obliquely propagating electron-acoustic solitary waves. *Phys. Plasmas* **2002**, *9*, 1474–1477.

59. Sauer, K.; Dubinin, E.; McKenzie, J.F. Wave emission by whistler oscillitons: Application to "coherent lion roars". *Geophys. Res. Lett.* **2002**, *29*, 2225, doi:10.1029/2002GL015771.

60. Sauer, K.; Dubinin, E.; McKenzie, J.F. Solitons and oscillitons in multi-ion space plasmas. *Nonlin. Processes Geophys.* **2003**, *10*, 121–130.

61. Janhunen, P.; Olsson, A.; Laakso, H. The occurrence frequency of auroral potential structures and electric fields as a function of altitude using Polar/EFI data. *Ann. Geophys.* **2004**, *22*, 1233–1250.

62. Karlsson, T.; Marklund, G.; Brenning, N.; Axnäs, I. On enhanced aurora and low-altitude parallel electric fields. *Phys. Scr.* **2005**, *72*, 419–422.

63. Cattaert, T.; Verheest, F. Large amplitude parallel propagating electromagnetic oscillitons. *Phys. Plasmas* **2003**, *12*, 012307.

64. Eliasson, B.; Shukla, P.K. Formation and dynamics of coherent structures involving phase-space vortices in plasmas. *Phys. Rep.* **2006**, *42*, 225–290.

65. Sydora, R.D.; Sauer, K.; Silin, I. Coherent whistler waves and oscilliton formation: Kinetic simulations. *Geophys. Res. Lett.* **2007**, *34*, L22105, doi:10.1029/2007GL031839.

66. Lakhina, G.S.; Kakad, A.P.; Singh, S.V.; Verhest, F. Ionand electron-acoustic solitons in two temperature space plasmas. *Phys. Plasmas* **2008**, *15*, 062903, doi:10.1063/1.2930469.

67. Ma, J.Z.G.; St.-Maurice, J.-P. Ion distribution functions in cylindrically symmetric electric fields in the auroral ionosphere: The collision-free case in a uniformly charged configuration. *J. Geophys. Res.* **2008**, *113*, A05312, doi:10.1029/2007JA012815.

68. Ma, J.Z.G.; St.-Maurice, J.-P. Backward mapping solutions of the Boltzmann equation in cylindrically symmetric, uniformly charged auroral ionosphere. *Astrophys. Space Sci.* **2015**, *357*, 104, doi:10.1007/s10509-015-2331-6.

69. Pottelette, R.; Berthomier, M. Nonlinear electron acoustic structures generated on the high-potential side of a double layer. *Nonlin. Processes Geophys.* **2009**, *16*, 373–380.

70. Coïsson, P.; Lognonné, P.; Walwer, D.; Rolland, L.M. First tsunami gravity wave detection in ionospheric radio occultation data. *Earth Space Sci.* **2015**, *2*, 125-133.

71. Kendall, P.C.; Pickering, W.M. Magnetoplasma diffusion at F2-region altitudes. *Planet. Space Sci.* **1967**, *15*, 825–833.

72. Einaudi, F.; Hines, C.O. WKB approximation in application to acoustic-gravity waves. *Can. J. Phys.* **1970**, *48*, 1458–1471.

73. Georges, T.M. HF Doppler studies of traveling ionospheric disturbances. *J. Atmos. Terr. Phys.* **1968**, *30*, 735–746.

74. Lighthill, J. *Waves in Fluids*; Cambridge University Press: Cambridge, UK, 1978.

75. Hébert, H.; Sladen, A.; Schindelé, F. Numerical modeling of the great 2004 Indian Ocean tsunami: Focus on the Mascarene Islands. *Bull. Seismol. Soc. Am.* **2007**, *97*, S208–S222.

76. Hickey, M.P. Atmospheric gravity waves and effects in the upper atmosphere associated with tsunamis. In *The Tsunami Threat—Research and Technology*; Nils-Axel Mårner, Ed.; In Tech: Rijeka, Croatia; Shanghai, China, 2011; pp. 667–690.

77. Harris, I.; Priester, W. Time dependent structure of the upper atmosphere. *J. Atmos. Sci.* **1962**, *19*, 286–301.

78. Pitteway, M.L.V.; Hines, C.O. The viscous damping of atmospheric gravity waves. *Can. J. Phys.* **1963**, *41*, 1935–1948.

79. Volland, H. Full wave calculations of gravity wave propagation through the thermosphere. *J. Geophys. Res.* **1969**, *74*, 1786–1795.

80. Volland, H. The upper atmosphere as a multiple refractive medium for neutral air motions. *J. Atmos. Terr. Phys.* **1969**, *31*, 491–514.

81. Hickey, M.P.; Schubert, G.; Walterscheid, R.L. Acoustic wave heating of the thermosphere. *J. Geophys. Res.* **2001**, *106*, 21543–21548.

82. Walterscheid, R.L.; Hickey, M.P. One-gas models with height-dependent mean molecular weight: Effects on gravity wave propagation. *J. Geophys. Res.* **2001**, *106*, 28831–28839.

83. Schubert, G.; Hickey, M.P.; Walterscheid, R.L. Heating of Jupiter's thermosphere by the dissipation of upward propagating acoustic waves. *ICARUS* **2003**, *163*, 398–413.

84. Schubert, G.; Hickey, M.P.; Walterscheid, R.L. Physical processes in acoustic wave heating of the thermosphere. *J. Geophys. Res.* **2005**, *110*, D07106, doi:10.1029/2004JD005488.

85. Walterscheid, R.L.; Hickey, M.P. Acoustic waves generated by gusty flow over hilly terrain. *J. Geophys. Res.* **2005**, *110*, A10307, doi:10.1029/2005JA011166.

86. Walterscheid, R.L.; Hickey, M.P. Gravity wave propagation in a diffusively separated gas: Effects on the total gas. *J. Geophys. Res.* **2012**, *117*, A05303, doi:10.1029/2011JA017451.

87. Zhou, Q.; Morton, Y.T. Gravity wave propagation in a nonisothermal atmosphere with height varying background wind. *Geophys. Res. Lett.* **2007**, *34*, L23803, doi:10.1029/2007GL031061.

88. Picone, J.M.; Hedin, A.E.; Drob, D.P.; Aikin, A.C. NRLMSISE-00 empirical model of the atmosphere: Statistical comparisons and scientific issues. *J. Geophys. Res.* **2002**, *107*, 1468, doi:10.1029/2002JA009430.

89. Hedin, A.E.; Fleming, E.L.; Manson, A.H.; Schmidlin, F.J.; Avery, S.K.; Clark, R.R.; Franke, S.J.; Fraser, G.J.; Tsuda, T.; Vial, F.; *et al.* Empirical wind model for the upper, middle and lower atmosphere. *J. Atmos. Terr. Phys.* **1996**, *58*, 1421–1447.

90. Schunk, R.W.; Navy, F.A. *Ionospheres: Physics, Plasma Physics, and Chemistry*; Cambridge University Press: Cambridge, UK, 2000.

91. Kelley, M.C. *The Earth's Ionosphere: Plasma Physics and Electrodynamics*, 2nd ed.; Elsevier: New York, NY, USA, 2009.

92. Richmond, A.D.; Thayer, J.P. Ionospheric electrodynamics: A tutorial. In *Magnetospheric Current Systems*; Ohtani, S.-I., Fujii, R., Hesse, M., Lysak, R.L., Eds.; AGU: Washington, DC, USA, 2000; pp. 130–146.

93. Nicolet, M. The collision frequency of electrons in the ionosphere. *J. Atmos. Terr. Phys.* **1953**, *3*, 200–211.

94. Chapman, S. The electric conductivity in the ionosphere: A review. *Nuovo Cimento* **1956**, *4*, 1385–1412.

95. Ruzhin, Y.Y.; Sorokin, V.M.; Yashchenko, A.K. Physical mechanism of ionospheric total electron content perturbations over a seismoactive region. *Geomagn. Aeron.* **2014**, *54*, 337–346.

96. Bilitza, D.; Altadill, D.; Zhang, Y.; Mertens, C.; Truhlik, V.; Richards, P.; McKinnell, L.-A.; Reinisch, B. The International reference ionosphere 2012—A model of international collaboration. *J. Space Weather Space Clim.* **2014**, *4*, 2–12.

97. Volland, H. The upper atmosphere as a multiply refractive medium for neutral air motions. *J. Atmos. Terr. Phys.* **1969**, *31*, 491–514.

98. Hickey, MP.; Cole, K.D. A quantic dispersion equation for internal gravity waves in the thermosphere. *J. Atmos. Terr. Phys.* **1987**, *49*, 889–899.

99. Banks, P.M.; Kockarts, G. *Aeronomy*; Academic Press: New York, NY, USA, 1973.

100. Burke, P.G.; Moiseiwitsch, B.L. *Atomic Process and Applications*; North-Holland Publishing Company: Amsterdam, The Netherlands; New York, NY, USA; Oxford, UK, 1976.

101. Rishbeth, H.; Barron, D.W. Equilibrium electron distributions in the ionospheric F2-layer. *J. Atmos. Terr. Phys.* **1960**, *18*, 234–252.

102. Reber, C.A.; Hedin, A.E.; Pelz, D.T.; Potter, W.E.; Brace, L.H. Phase and amplitude relationships of wave structure observed in the lower thermosphere. *J. Geophys. Res.* **1975**, *80*, 4576–4580.

103. Tu, J.-N. The coupling between atmospheric waves and electron density perturbations. *Ch. J. Space Sci.* **1993**, *13*, 190–195.

104. Komjathy, A.; Galvan, D.A.; Stephens, P.; Butala, M.; Akopian, V.; Wilson, B. Detecting ionospheric TEC perturbations caused by natural hazards using a global network of GPS receivers: The Tohoku case study. *Earth Planets Space* **2012**, *64*, 1287–1294.

105. Yang, Y.-M.; Garrison, J.L.; Lee, S.-C. Ionospheric disturbances observed coincident with the 2006 and 2009 North Korean underground nuclear tests. *Geophys. Res. Lett.* **2012**, *39*, L02103, doi:10.1029/2011GL050428.

106. Yang, Y.-M.; Komjathy, A.; Langley, R.B.; Vergados, P.; Butala, M.D.; Mannucci, A.J. The 2013 Chelyabinsk meteor ionospheric impact studied using GPS measurements. *Radio Sci.* **2014**, *49*, 341–350.

107. Yang, Y.-M.; Meng, X.; Komjathy, A.; Verkholyadova, O.; Langley, R.B.; Tsurutani, B.T. Tohoku-Oki earthquake caused major ionospheric disturbances at 450 km altitude over Alaska. *Radio Sci.* **2014**, *49*, 1206–1213.

108. Coisson, P.; Lognonné, P.; Walwer, D.; Rolland, L.M. First tsunami gravity wave detection in ionospheric radio occultation data. *Earth Space Sci.* **2015**, *2*, 125–133.

109. Vadas, S.L.; Fritts, D.C. Thermospheric responses to gravity waves: Influences of increasing viscosity and thermal diffusivity. *J. Geophys. Res.* **2005**, *110*, D15103.

110. Kaladze, T.D.; Pokhotelov, O.A.; Shah, H.A.; Khana, M.I.; Stenflod, L. Acoustic-gravity waves in the Earth's ionosphere. *J. Atm. Sol. -Terr. Phys.* **2008**, *70*, 1607–1616.

111. Coisson, P. Institut de Physique du Globe de Paris, Sorbonne Paris Cité, Université Paris Diderot, CNRS, Paris, France. Private communication, 2015.

Modulation of Atmospheric Nonisothermality and Wind Shears on the Propagation of Seismic Tsunami-Excited Gravity Waves

John Z. G. Ma

Abstract: We study the modulation of atmospheric nonisothermality and wind shears on the propagation of seismic tsunami-excited gravity waves by virtue of the vertical wavenumber, m (with its imaginary and real parts, m_i and m_r, respectively), within a correlated characteristic range of tsunami wave periods in tens of minutes. A generalized dispersion relation of inertio-acoustic-gravity (IAG) waves is obtained by relaxing constraints on Hines' idealized locally-isothermal, shear-free and rotation-free model to accommodate a realistic atmosphere featured by altitude-dependent nonisothermality (up to 100 K/km) and wind shears (up to 100 m/s per km). The obtained solutions recover all of the known wave modes below the 200-km altitude where dissipative terms are assumed negligible. Results include: (1) nonisothermality and wind shears divide the atmosphere into a sandwich-like structure of five layers within the 200-km altitude in view of the wave growth in amplitudes: Layer I (0–18) km, Layer II (18–87) km, Layer III (87–125) km, Layer IV (125–175) km and Layer V (175–200) km; (2) in Layers I, III and V, the magnitude of m_i is smaller than Hines' imaginary vertical wavenumber (m_{iH}), referring to an attenuated growth in the amplitudes of upward propagating waves; on the contrary, in Layers II and IV, the magnitude of m_i is larger than that of m_{iH}, providing a pumped growth from Hines' model; (3) nonisothermality and wind shears enhance m_r substantially at an $\sim$100-km altitude for a tsunami wave period T_{ts} longer than 30 min. While Hines' model provides that the maximal value of m_r^2 is $\sim$0.05 (1/km^2), this magnitude is doubled by the nonisothermal effect and quadrupled by the joint nonisothermal and wind shear effect. The modulations are weaker at altitudes outside 80–140-km heights; (4) nonisothermality and wind shears expand the definition of the observation-defined "damping factor", β: relative to Hines' classical wave growth with $\beta = 0$, waves are "damped" from Hines' result if $\beta > 0$ and "pumped" if $\beta < 0$. The polarization of β is determined by the angle θ between the wind velocity and wave vector.

Reprinted from *J. Mar. Sci. Eng.* Cite as: Ma, J.Z.G. Modulation of Atmospheric Nonisothermality and Wind Shears on the Propagation of Seismic Tsunami-Excited Gravity Waves. *J. Mar. Sci. Eng.* **2016**, *4*, 4.

1. Introduction

For more than 10 years, LiDAR has recorded both atmospheric nonisothermality (featured with temperature gradients up to 100 °K per km) and large wind shears (e.g., 100 m/s per km) between ~85- and 95-km altitudes [1–5]. Spaceborne data revealed that the criterion of the wind shear-related Richardson number, $R_i \leq 1/4$, is only a necessary, but not sufficient, condition for dynamic instability [6]. Hall *et al.* [7] obtained spatially-averaged R_i data, which appeared to reach one at a 90-km altitude over Svalbard (78° N, 16° E). Importantly, measurements of airglow layer perturbations in $O(^1S)$ (peak emission altitude ~97 km) and OH (peak emission altitude ~87 km) driven by propagating acoustic-gravity waves suggested an exponentially-growing wave amplitudes [8,9]: $A_{O(^1S)} = A_{OH} \exp\left[(1 - \beta)\Delta z/(2H)\right]$, in which $A_{O(^1S)}$ and A_{OH} are the amplitudes at the $O(^1S)$ and OH emission lines, respectively; Δz is the height difference between the OH and $O(^1S)$ emission layers, and β is the so-called "damping factor", which classifies waves with (1) $\beta = 0$: free propagating without damping; (2) $0 < \beta < 1$: weakly damped; (3) $\beta = 1$: saturated without amplitude increase; and (4) $\beta > 1$: over-damped [9,10]. In addition, for the vertical wavelengths of 20–50 km, β is between zero and four, indicating that most waves were damping-dominated.

By contrast, in theoretical studies on acoustic-gravity waves, the earliest work focused on an idealized atmosphere featured with an isothermal temperature, homogeneous horizontal wind speeds, rotation free and dissipation free. For example, Hines [11,12] showed that A increases with height (z) exponentially from the initial values A_0 at $z = 0$: $A = A_0\exp\left[z/(2H)\right]$, where $H = C^2/(\gamma g)$ is the scale height. Here, $C = \sqrt{\gamma k_B T/M}$ is the speed of sound; γ is the ratio of specific heat; g is the gravitational acceleration constant; k_B is Boltzmann's constant; T is the mean-field temperature; and M is the mean molecular mass. The result was then extended to an isothermal, but dissipative atmosphere [13,14]. It was found that growth A becomes attenuated due to the introduction of the imaginary component (m_i) of the vertical wavenumber (m), expressed by a similar formula: $A = A_0 \exp \int_{z_0}^{z} (1/2H - |m_i|) \, dz$, in which m_i increases in altitude. Above some height (e.g., F_2-peak altitude), it is approximately equal to $1/(2H)$, while at higher altitudes, it is larger than $1/(2H)$, leading to a decaying amplitude [15–18]. Based on a "multi-layer" approximation, Hines and Reddy [19] calculated the coefficients of the energy transmission through a stratified atmosphere. They argued that nonideal conditions, like vertically-changing temperature and wind speeds, do not severely attenuate incident waves propagating upward through the mesosphere; however, stronger attenuation can be indeed expected low in the thermosphere. In addition, Hines [20] found that the shear-contributed anisotropic Richardson criterion, $R_i \leq 1/4$, can well portend the onset of isotropic atmospheric turbulence.

However, for symmetric instabilities, it was claimed that the criterion becomes $R_i \leq 1$ ([21]).

From the 1970s, seismic tsunamis began to be recognized as a possible driver to excite atmospheric gravity waves, which subsequently propagate to the upper atmosphere, where the conservation of wave energy causes the wave disturbance amplitudes to be enhanced due to the decrease of atmospheric density with increasing altitudes, based on the isothermal and shear-free model [22,23]. Nevertheless, a realistic atmosphere does own temperature gradients and wind shears. Serious concerns were naturally attracted towards such fundamental questions, like to what extent the nonisothermality and wind shears influence the propagation of acoustic-gravity waves and what the mechanism is for amplitude A to be modulated in wave damping or growing *versus* altitude. Theoretically speaking, while m_i has already been solved either with the linear wave approximation (e.g., [24–29]) or with the numerical "full wave model" approach under the WKB approximation (e.g., [14,18,30–38]), the intrinsic connection between m_i and A, as well as other parameters, like β and R_i, is so complicated in the presence of nonisothermality and wind shears that no appropriate models were proposed to account for the damping and growth of gravity waves.

Merely for the Richardson number, it has different expressions under isothermal and nonisothermal conditions: by definition, it is the ratio between the buoyancy (or Brunt-Väisälä) frequency and the shear S. However, there exist two buoyancy frequencies, ω_b (isothermal) and ω_B (nonisothermal) (e.g., [18,39,40]). Accordingly, there are two Richardson numbers:

$$R_i = \omega_b^2/S^2 \text{ (isothermal), or, } R_I = \omega_B^2/S^2 \text{ (nonisothermal)} \tag{1}$$

in which:

$$\omega_b^2 = (\gamma - 1)\frac{g^2}{C^2}, \ \omega_B^2 = (\gamma - 1)\frac{g^2}{C^2} + \frac{g}{C^2}\frac{dC^2}{dz}, \ S^2 = \left(\frac{dU}{dz}\right)^2 + \left(\frac{dV}{dz}\right)^2 \tag{2}$$

where U and V are the zonal and meridional components of the mean-field horizontal wind with velocity $\mathbf{v}_0 = \{U, V, 0\}$. Similar issues also exist when dealing with the cut-off frequencies of acoustic-gravity waves under different thermodynamic conditions. It deserves mentioning here that the amended R_i-criterion given in [21] (*i.e.*, for $R_i < 0.25$, the K-Hinstabilities dominate; for $0.25 < R_i < 0.95$, the symmetric instabilities dominate; for $R_i > 0.95$, the conventional baroclinic instabilities dominate) is valid even for a stratified shear flow in view of energy balance [41]; and stepping further, a stably stratified turbulence can still survive for $R_i \gg 1$ [42].

How do atmospheric nonisothermality and wind shears influence the damping and growth of seismic tsunami-excited acoustic-gravity waves? We are inspired to turn our attention to this subject in the study of realistic atmospheres surrounding not only the Earth, but also other planets, like Mars. The motivation to tackle this problem is the necessity of an effective physical model to demonstrate the effects of the nonisothermality and wind shears on the modulation of propagating gravity waves driven by hazard events, like tsunamis. We develop the study on the basis of the proper knowledge of: (1) the vertical growth of gravity waves under nonisothermal, wind shear conditions; (2) the relation among the wave period, β, and vertical wavelength; (3) the dependence of β on the zonal and meridional wind shears; and (4) the filtering of waves due to background winds. The region concerned is from sea level to a 200-km altitude within which the atmosphere is non-dissipative (negligible viscosity and heat conductivity), and the ion drag and Coriolis force can be reasonably omitted [14,43–45]. This is also a region that completely covers the lower airglow emission zone below an $\sim$100-km altitude. The structure of the paper is as follows: Section 2 formulates the physical model used to expose the mean-field properties, which are obtained from the empirical neutral atmospheric model, NRLMSISE-00 [46], and the horizontal wind model, HWM93 [47]. A generalized dispersion relation of inertio-acoustic-gravity (IAG) waves under nonisothermal and wind shear conditions is derived. Employing this dispersion relation under different conditions, Section 3 also extends all of the classical wave modes contributed by previous models, including Hines' locally-isothermal, shear-free and rotation-free model [11], Eckart [48] and Eckermann's [49] IAG model and Hines [20] and Hall et al.'s [7] isothermal and wind shear model. In addition, this section also presents the respective influences of nonisothermality, wind shears and the Coriolis parameter on propagating waves. Section 4 offers the conclusion and a discussion.

2. Modeling

A Cartesian frame is suitable to be used for studying the propagation of acoustic-gravity waves in the Earth's spherically-symmetric gravitational field [50]. We choose such a local coordinate system, $\{\hat{\mathbf{e}}_x, \hat{\mathbf{e}}_y, \hat{\mathbf{e}}_z\}$, in which $\hat{\mathbf{e}}_x$ is horizontally due east, $\hat{\mathbf{e}}_y$ due north and $\hat{\mathbf{e}}_z$ vertically upward. The neutral atmosphere is described by a set of hydrodynamic equations based on conservation laws in mass, momentum and energy, as well as the equation of state. Considering that airglow emissions happen at 80–100 km heights (e.g., [51–53]) and that below a 200-km altitude, the atmosphere is non-dissipative, where the viscosity, heat conductivity and the ion drag can all be neglected [14,43–45], we obtain these equations as follows (for the complete set of equations including these terms, see, e.g., [25,45,50,54–57]):

$$\left.\begin{array}{l}\frac{D\rho}{Dt} = -\rho\nabla \cdot \mathbf{v} \\ \rho\frac{D\mathbf{v}}{Dt} = -\nabla p + \rho\mathbf{g} + 2\rho\mathbf{v} \times \mathbf{\Omega} \\ \frac{1}{\gamma}\frac{Dp}{Dt} = -p\nabla \cdot \mathbf{v} \\ p = \rho R_s T\end{array}\right\} \tag{3}$$

in which we still keep the Coriolis term alive so as to be convenient to test our model by a direct comparison with the well-developed inertio-acoustic-gravity (IAG) model [58,59]. The parameters in Equation (3) are defined as follows:

$\mathbf{v}$, ρ, p and T: atmospheric velocity, density, pressure and temperature, respectively;

$D/Dt = \partial/\partial t + \mathbf{v} \cdot \nabla$: substantial derivative over time t;

$\mathbf{g} = \{0,0,-g\}$: gravitational acceleration;

$\mathbf{\Omega} = \{0, \Omega\cos\phi, \Omega\sin\phi\}$: Earth's Coriolis vector where $\Omega = 7.29 \times 10^{-5}$ rad/s and ϕ is latitude;

γ and R_s: adiabatic index and gas constant, respectively.

We adopt standard linearization by neglecting higher-order perturbations. The variables in Equation (3) contain two types of ingredients: the ambient mean-field component to be denoted by subscript "$_0$" and the first-order quantity denoted by subscript "$_1$":

$$\left.\begin{array}{l}\rho = \rho_0 + \rho_1, T = T_0 + T_1, p = p_0 + p_1 \\ \mathbf{v} = \mathbf{v}_0 + \mathbf{v}_1 = \{U, V, 0\} + \{u, v, w\} \\ \left(\frac{\rho_1}{\rho_0}, \frac{p_1}{p_0}, \frac{T_1}{T_0}, \frac{u}{U}, \frac{v}{V}, w\right) \propto e^{i(\mathbf{k}\cdot\mathbf{r}-\omega t)}\end{array}\right\} \tag{4}$$

where U and V are the zonal (eastward) and meridional (northward) components of the mean-field wind velocity (note that the wind is horizontal, and thus, the vertical component W is zero), respectively; u, v, w are the three components of the perturbed velocity, respectively; and $\mathbf{k} = \{k, l, m\}$ (in which $m = m_r + im_i$) is the wave vector, and ω is the wave frequency. Due to the existence of the inhomogeneities in the mean-field properties in a realistic atmosphere, there exist the following input parameters:

$$\left.\begin{array}{l}k_\rho = \frac{1}{H_\rho} = \frac{d(\ln\rho_0)}{dz}, k_p = \frac{1}{H_p} = \frac{d(\ln p_0)}{dz}, k_T = \frac{d(\ln T_0)}{dz}; \\ \omega_v = S = \sqrt{\left(\frac{dU}{dz}\right)^2 + \left(\frac{dV}{dz}\right)^2}\end{array}\right\} \tag{5}$$

in which H_ρ and H_p are the density and pressure scale heights, respectively, k_ρ, k_p and k_T are the density, pressure and temperature scale numbers, respectively, satisfying $k_T = k_p - k_\rho$ from the equation of state. There also exists a simple relation among k_p, g, and C: $k_p = -\gamma g/C^2$. From now on, we use ω_v to replace S in order to expose the spatially-velocity-curl nature of wind shears. Note that the unit of ω_v is m/s per km. In dimensional analysis (a useful tool to check the validity of the algebra of the modeling at the lowest level), this unit has the same physical dimension as that of the wave frequency, rad/s. Thus, the unit of ω_v is "m/s per km", rather than "rad/s".

Note that the linearization introduced above is different from the WKB approximation. The WKB approach assumes linear wavelike solutions in time and 2D horizontal coordinates, but not in the vertical direction only along which the mean-field properties are supposed to vary, while keeping their homogeneities in the horizontal plane (e.g., [24]). A 1D vertical Taylor–Goldstein equation (or, equivalently, a quadratic equation) can thus be derived in the presence of the height-varying temperature and wind shears to describe the vertical propagation of tsunami-excited gravity waves. For details, see, e.g., Equation (4) in [60].

2.1. Mean-Field Properties

The undisturbed mean-field parameters and wind components in the vertical direction up to a 200-km altitude are calculated, as shown in Figure 1, by employing both the empirical neutral atmospheric model, NRLMSISE-00 [46], and the horizontal wind model, HWM93 [47]. The chosen heights cover the airglow layer well within which the peak emissions of $O(^1S)$ and OH are at ~97 km and ~87 km, respectively. We arbitrarily choose a position at $60°$ latitude and $-70°$ longitude for a local apparent solar time of 1600 on the 172th day of a year, with the daily solar $F_{10.7}$ flux index and its 81-day average of 150. The daily geomagnetic index is four.

In the figure, (a) displays the atmospheric mass density ρ_0 (pink) and pressure p_0 (blue), while (b) shows their gradients $d\rho_0/dz$ (pink) and dp_0/dz (blue), respectively. Density ρ_0 decreases all the way up from 1.225 kg/m^3 (or 2.55×10^{25} 1/m^3) at sea level to only 2.69×10^{-10} kg/m^3 (5.6×10^{15}/m^3) at a 200-km altitude. Pressure p_0 has a similar tendency to ρ_0. It reduces from 10^5 Pa at sea level to 7.9×10^{-5} Pa ultimately. Both $d\rho_0/dz$ and dp_0/dz die out *versus* height and are nearly zero above an ~50-km altitude. (c) exposes the density scale height H_ρ (blue) and pressure scale height H_p (pink), while (d) gives the three scale numbers in density, k_ρ (pink), pressure, k_p (black), and temperature, k_T (blue). Both H_ρ and H_p are 8.64 km and 8.22 km, respectively, at sea level, but soar to as high as 32.7 km and 39.3 km, respectively, when approaching a 200-km altitude (note that the two heights are not equal; only under the isothermal condition, $k_T = 0$, can $H_\rho = H_p$ or $k_\rho = k_p$ be valid); accordingly, the altitude profiles of k_ρ and k_p are similar to those of $-H_\rho$ and $-H_p$, respectively; by contrast, k_T experiences adjustments a couple of times from

288

negative to positive and eventually keeps its positive polarization above the 100-km height, which is finally inclined to zero.

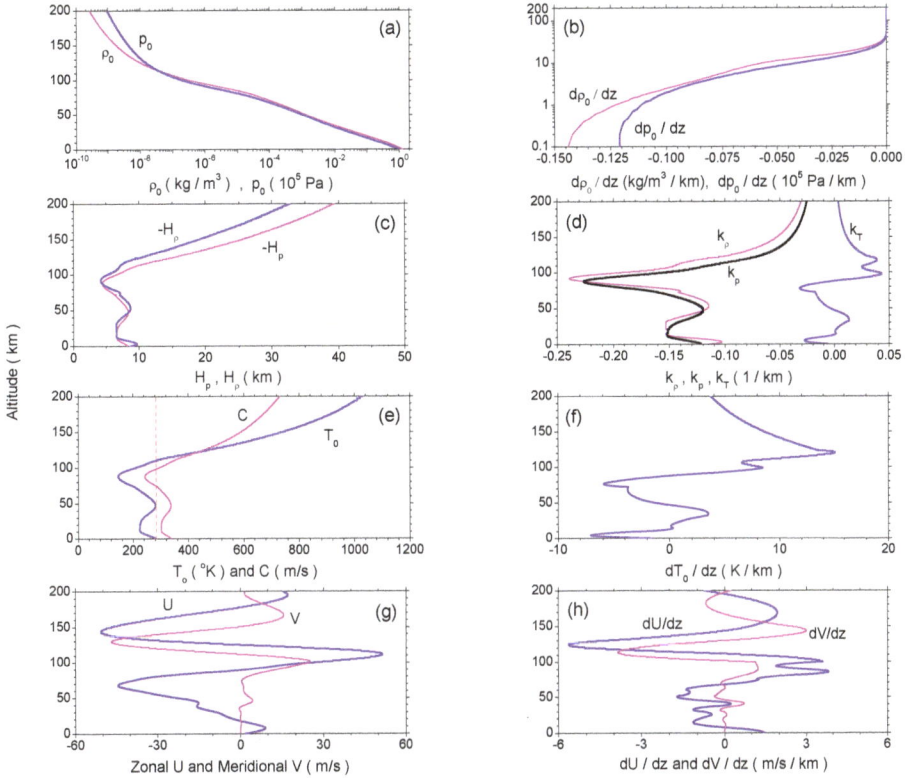

Figure 1. Altitude profiles of mean-field properties. (**a**) Mass density ρ_0 (pink) and pressure p_0 (blue); (**b**) density gradient $d\rho_0/dz$ (pink) and pressure gradient dp_0/dz (blue); (**c**) density scale height H_ρ (blue) and pressure scale height H_p (pink); (**d**) density scale number k_ρ (pink), pressure scale number k_p (black) and temperature scale number k_T (blue); (**e**) temperature T_0 (blue) and sound speed C (pink); (**f**) temperature gradient dT_0/dz; (**g**) zonal (eastward) wind U (blue) and meridional (northward) wind V (pink); (**h**) zonal wind gradient dU/dz (blue) and meridional wind gradient dV/dz (pink); in (e), a dashed line in red is given as a reference to show an ideal atmosphere, which is isothermal at all altitudes.

(e) presents temperature T_0 (blue) and sound speed $C = \sqrt{\gamma R_s T_0}$ (pink) in the LHS one, while (f) illustrates temperature gradient dT_0/dz. Temperature T_0 is 281 °K at sea level. It decreases linearly to 224 °K at 13 km and then returns to 281 °K at 47 km, followed by a reduction again to 146 °K at 88 km. Above this height, the temperature goes up continuously at higher altitudes and reaches an exospheric value of >1000 °K above a 190-km height (at 194 km, it is 1000 °K). As a reference, a dashed line

289

in red is depicted to show an ideal atmosphere that is isothermal at all altitudes; the magnitude of T_0 stays the same as that at sea level. Sound speed C follows the variation of $T_0^{1/2}$. At sea level, it is 337 m/s; at a 200-km altitude, it is 731 m/s. For dT_0/dz, it transits twice from negative to positive below a 100-km altitude, within 10 m/s per km, and monotonously returns to zero above a 120-km height. (g) exhibits the zonal (eastward) wind U (blue) and the meridional (northward) wind V (pink), and (h) displays the zonal wind gradient dU/dz (blue) and the meridional one dV/dz (red). Both of the horizontal wind components oscillate twice dramatically within ± 51 m/s in amplitude, and their gradients, dU/dz and dV/dz, are also featured with obvious undulations. For example, the former jumps from ~ 4 m/s per km---5.5 m/s per km within only a 25 km-thick layer at about a 100-km altitude.

NRLMSISE-00 and HWM93 demonstrated that the horizontal gradients of ρ_0, T_0, p_0, U and V are always at least three orders smaller than those in the vertical direction. We consequently assume, as previous authors did, that the mean-field parameters are uniform and stratified in the horizontal plane, free of any inhomogeneities, i.e., $\partial/\partial x \simeq 0$, $\partial/\partial y \simeq 0$ and $\nabla \cong (\partial/\partial z)\hat{\mathbf{e}}_z$. Besides, we take 350-km and 50-km horizontal wavelengths in our model, based on the data of the relations between horizontal wavelength and wave periods during the SpreadFExcampaign [61].

2.2. Generalized Dispersion Relation

Acoustic-gravity waves originate from small perturbations away from their mean-field properties and propagate in a stratified atmosphere [62]. Employing Equation (4) to linearize Equation (3) yields the following set of perturbed equations:

$$
\left.
\begin{array}{l}
\frac{\partial \rho_1}{\partial t} + \mathbf{v}_0 \cdot \nabla \rho_1 + \mathbf{v}_1 \cdot \nabla \rho_0 + \rho_0 \nabla \cdot \mathbf{v}_1 + \rho_1 \nabla \cdot \mathbf{v}_0 = 0 \\
\frac{\partial \mathbf{v}_1}{\partial t} + \mathbf{v}_1 \cdot \nabla \mathbf{v}_0 + \mathbf{v}_0 \cdot \nabla \mathbf{v}_1 = -\frac{1}{\rho_0}\nabla p_1 + \frac{\rho_1}{\rho_0}\mathbf{g} + 2\mathbf{v}_1 \times \boldsymbol{\Omega} + 2\frac{\rho_1}{\rho_0}\mathbf{v}_0 \times \boldsymbol{\Omega} \\
\frac{\partial p_1}{\partial t} + \mathbf{v}_0 \cdot \nabla p_1 + \mathbf{v}_1 \cdot \nabla p_0 = -\gamma p_0 \nabla \cdot \mathbf{v}_1 - \gamma p_1 \nabla \cdot \mathbf{v}_0 \\
\frac{p_1}{p_0} = \frac{\rho_1}{\rho_0} + \frac{T_1}{T_0}
\end{array}
\right\}
\qquad (6)
$$

which provides the following dispersion equation:

$$
\begin{bmatrix}
\omega & k & l & m - ik_\rho & 0 \\
0 & \omega & -if & i\frac{dU}{dz} & k \\
0 & if & \omega & i\frac{dV}{dz} & l \\
-ig & 0 & 0 & \omega & m - ik_p \\
0 & k & l & m - i\frac{k_p}{\gamma} & \frac{\omega}{C^2}
\end{bmatrix}
\begin{bmatrix}
\frac{\rho_1}{\rho_0} \\
u \\
v \\
w \\
\frac{p_1}{p_0}
\end{bmatrix}
= 0
\qquad (7)
$$

from which a generalized, complex dispersion relation of inertio-acoustic-gravity (IAG) waves is derived in the presence of nonisothermality and wind shears, if and only if the determinant of the coefficient matrix is zero:

$$\left.\begin{aligned}
&\omega_*^4 - \left(C^2K^2 + f^2 + gk_T\right)\omega_*^2 + (C^2m^2 + gk_T)f^2 - \\
&- (\gamma - 1)gk_h\omega_*\omega_{\mathbf{v}}\cos\theta + C^2k_h^2\omega_B^2 = \\
&= iC^2m\left[\frac{\gamma g}{C^2}\left(\omega_*^2 - f^2\right) - k_h\omega_*\omega_{\mathbf{v}}\cos\theta\right]
\end{aligned}\right\} \quad (8)$$

in which $f = 2\Omega\sin\phi$ is the Coriolis parameter (where ϕ is the latitude); θ is the angle between horizontal wave vector $\mathbf{k}_h$ and mean-field wind velocity $\mathbf{v}_0$, defined by:

$$\cos\theta = \frac{\mathbf{k}_h \cdot \mathbf{v}_0}{k_h\sqrt{U^2 + V^2}} \quad (9)$$

and,

$$\omega_* = \omega - \mathbf{k} \cdot \mathbf{v}_0, \quad K^2 = k_h^2 + m^2, \quad k_h^2 = k^2 + l^2 \quad (10)$$

For simplicity, we omit "$_*$" attached to ω in following texts.

Because m is complex, use $(m_r + im_i)$ instead of m in Equation (8). This yields the final expression of the dispersion relation:

$$\left.\begin{aligned}
&m_i = -k_g\left[1 - \frac{\omega^2}{2(\omega^2 - f^2)}\frac{\omega_{\mathbf{v}}}{k_g V_{ph}}\cos\theta\right]; \\
&\omega^4 - \left(C^2k_h^2 + f^2 - \frac{2-\gamma}{2}g\frac{\omega_{\mathbf{v}}}{V_{pH}}\cos\theta\right)\omega^2 - \left(C^2m_r^2 + \omega_A^2\right)(\omega^2 - f^2) + \\
&+ C^2k_h^2\omega_B^2\left(1 - \frac{0.25}{R_I}\frac{\omega^2}{\omega^2 - f^2}\cos^2\theta\right) = 0 \\
&\text{or, alternatively,} \\
&m_r^2 = \frac{\omega^2 - \omega_A^2}{C^2} + k_h^2\left[\frac{\omega_B^2 - \omega^2}{\omega^2 - f^2} - \frac{1}{2}\frac{\omega_{\mathbf{v}}^2\omega^2}{(\omega^2 - f^2)^2}\left(\frac{2-\gamma}{\gamma}\frac{\omega^2 - f^2}{k_h^2 V_p V_{ph}} + \frac{1}{2}\cos\theta\right)\cos\theta\right]
\end{aligned}\right\} \quad (11)$$

in which:

$$\left.\begin{aligned}
&k_g = \frac{\gamma g}{2C^2}, \quad k_{gT}^2 = \frac{gk_T}{C^2}, \quad k_G^2 = k_g^2 + k_{gT}^2; \quad R_I = \frac{\omega_B^2}{\omega_{\mathbf{v}}^2} = R_i + \frac{gk_T}{\omega_{\mathbf{v}}^2}, \quad R_i = \frac{\omega_b^2}{\omega_{\mathbf{v}}^2}; \\
&V_{ph} = \frac{\omega}{k_h}, \quad V_p = \frac{\omega}{k_p}; \quad \omega_a^2 = C^2k_g^2 = \frac{\gamma^2}{4(\gamma-1)}\omega_b^2, \quad \omega_A^2 = C^2k_G^2 = \omega_a^2 + C^2k_{gT}^2
\end{aligned}\right\} \quad (12)$$

where V_{ph} is the horizontal phase speed and R_I is the updated expression of R_i in a nonisothermal atmosphere.

Figure 2 illustrates the vertical profiles of these parameters for a tsunami period of 33.3 min and horizontal wavelengths of $(k, l) = (400, 2000)$ km. (a) reveals that the isothermal Richardson number, R_i (as represented by its inverse, $1/R_i$ in blue), is mostly larger than the nonisothermal one, R_I (as represented by its inverse, $1/R_I$ in pink), below the 85-km altitude, while it is smaller above the 85-km altitude. This is due to the mostly negative k_T below the height and the positive k_T above the height. The maximal value of $1/R_i$ is 0.197, much less than four, indicating that the velocity shear is far incapable of overcoming the tendency of a stratified fluid to remain stratified, and thus, instabilities are sufficiently suppressed (e.g., [20,63]).

In the lower four panels, (b) shows the curves of k_{gT}^2 (black), k_g^2 (pink) and k_G^2 (blue). Take a reference from the k_T-curve in Figure 1. Due to the double polarities of k_T *versus* altitude, the value of k_{gT}^2 can be either positive or negative, depending on the changes of k_T. The values of k_g^2 and k_G^2 are always positive. However, the influence of k_{gT}^2 on k_G^2 cannot be neglected, though the two lines of k_g^2 and k_G^2 appear to be twins: between 20 and 50 km and above 90 km, $k_G^2 > k_g^2$; while in other regions, $k_G^2 < k_g^2$. This feature is important due to the fact that k_g^2 and k_G^2 are directly correlated with the two acoustic cut-off frequencies, ω_a and ω_A, under isothermal and non-isothermal conditions, respectively. Have a glance at (c). Here, two pairs of curves are presented: the above-mentioned ω_a (dash pink) and ω_A (thin pink); and, the two gravity-wave cut-off frequencies, ω_b (dash blue) and ω_B (thin blue), under isothermal and non-isothermal conditions, respectively. At all altitudes, ω_a is always larger than ω_b; and below an ~180-km altitude, ω_A is always larger than ω_B. That is to say, the buoyancy frequency can never be larger than the cut-off frequency in either the isothermal case or the non-isothermal one up to an ~180-km altitude. Nevertheless, this result does not exclude at some altitudes, when we compare the difference of the isothermal and nonisothermal cases, $\omega_a < \omega_B$ (say, above a 90-km altitude) or $\omega_A < \omega_b$ (e.g., 60–80 km). This warns us to be cautious about the different isothermal conditions when using the two sets of frequencies in applications. Some authors confused them by using the nonisothermal ω_B as the buoyancy frequency, but the isothermal ω_a as the cut-off frequency. Frequencies under the two conditions should not be mixed up, especially in wave analysis and data-fit modeling.

In accordance with these four frequencies, (e) depicts the four different periods: τ_B (thin blue) and τ_b (dash blue); τ_A (thin pink) and τ_a (dash pink). The shortest cut-off period occurs at an ~95-km altitude in the nonisothermal case, only $\tau_A = 3.3$ min. The longest period occurs at a 200-km altitude, $\tau_b = 12$ min. Finally, (d) gives the profiles of both $\cos\theta$ and ω_v. Obviously, $\cos\theta$ is not constant *versus* height, but oscillates twice up to a 200-km altitude. The wind shear ω_v is always larger than the Coriolis frequency Ω. It peaks at a 123-km altitude, 6.09 m/s per km, ~84 Ω.

Compared to Hines' idealized atmospheric model with a local isothermality (*i.e.*, the vertical temperature gradient is assumed zero) and a uniform horizontal wind field (*i.e.*, the vertical wind sheared effect is neglected), the NRLMSISE-00 and HWM93 empirical models provide us a more realistic model, which shows that the atmosphere is neither locally isothermal (*i.e.*, the vertical temperature gradient is nonzero), nor uniform (*i.e.*, the wind shear exists in the vertical direction).

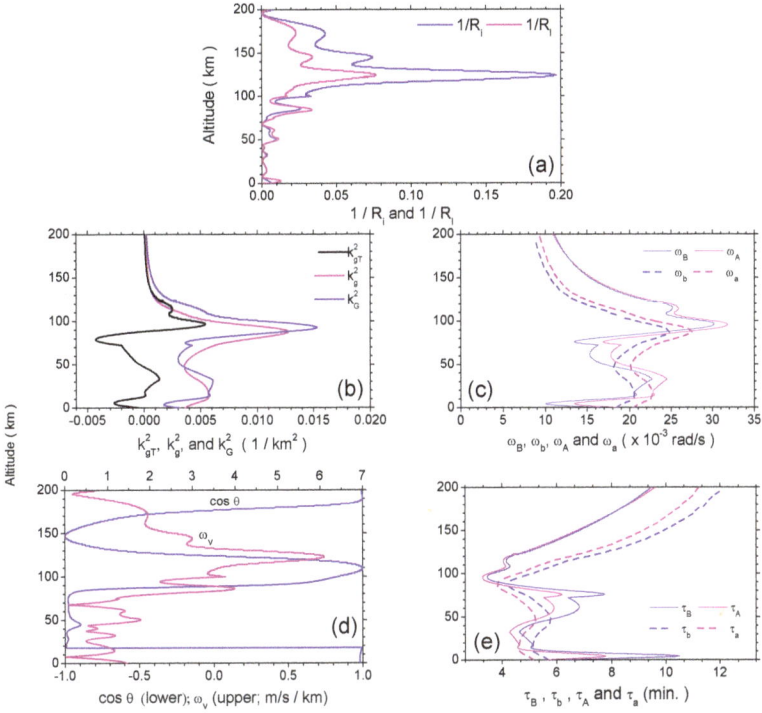

Figure 2. Vertical profiles of input parameters in Equation (11). (**a**) Richardson number R_i (blue) and R_I (pink); (**b**) k_{gT}^2 (black), k_g^2 (pink) and k_G^2 (blue); (**c**) buoyancy frequencies ω_B (thin blue) and ω_b (dash blue) and cut-off frequencies ω_A (thin pink) and ω_a (dash pink); (**d**) $\cos\theta$ (blue) and $\omega_\mathbf{v}$ (pink); (**e**) the four periods, τ_B (thin blue), τ_b (dash blue), τ_A (thin pink) and τ_a (dash pink), corresponding to the four frequencies in the upper right panel.

3. Results

Equation (11) provides a generalized dispersion relation of realistic atmosphere below a 200-km altitude, where the atmosphere is inviscid, nonisothermal and wind sheared. As mentioned previously, the ion drag, viscosity, heat conductivity and Coriolis effect can be reasonably neglected within this region, as already discussed in detail in early work (e.g., [14,43–45]). To test our model by the full IAG formalism for an isothermal and windless atmosphere (e.g., [59]), we include the Coriolis term. It is interesting to note that: (1) the non-isothermal effect, as represented by the the vertical derivative of the log of temperature k_T, never influences the vertical growth rate, m_i; (2) if the horizontal wave vector is perpendicular to the wind velocity, *i.e.*, $\mathbf{k}_h \perp \mathbf{v}_0$ (or $\theta = 90\circ$), the wind shear effect disappears; (3) only in the presence of wind shears can horizontal phase speed V_{ph} come into play. It influences both the vertical wavenumber m_r and the vertical growth rate m_i, inferring that the wave growth is not only dependent on the scale height, but the wave frequency ω, as well.

Equation (11) recovers all of the previous classical wave modes under locally-isothermal and shear-free conditions, *i.e.*, vertical gradients in both wind velocity and temperature are not considered. As follows, we obtain these modes directly from Equation (11) and extend the isothermal results to non-isothermal ones by relaxing these constraints. Then, we pay attention to the influence of the nonisothermality and wind shears on the propagation of gravity waves from sea level to a 200-km altitude, and we present the exact analytical expression of β.

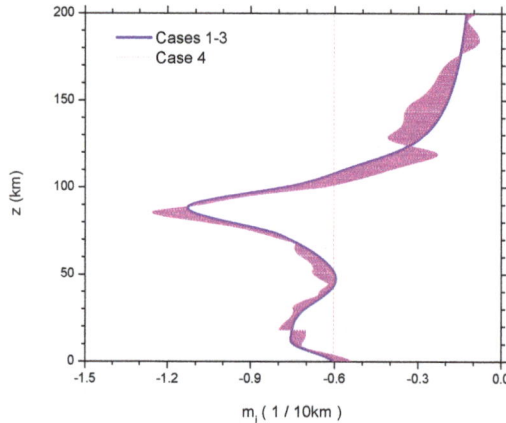

Figure 3. Imaginary vertical wavenumber, m_i (1/10 km), of different tsunami-excited wave modes propagating in an atmosphere. Case 1: Hines' locally-isothermal and shear-free model; Case 2: the extended Hines' model under nonisothermal conditions; Case 3: inertio-acoustic-gravity waves under nonisothermal conditions; Case 4: acoustic-gravity waves under nonisothermal and wind shear conditions. In Cases 1–3, m_i-curves are superimposed upon each other (in blue); in Case 4, the m_i-band fluctuates upon those of Cases 1–3 (in pink). As a reference, a red straight line is shown in the figure to represent the result of m_i for an ideal atmosphere, which is isothermal at all altitudes in response to the constant T_0 in Figure 1.

3.1. Case 1: Hines' Locally-Isothermal and Shear-Free Model

In this basic situation, the atmosphere was assumed locally-isothermal ($k_T = 0$) and shear-free ($dU/dz = dV/dz = 0$) in the absence of the Coriolis term (*i.e.*, rotation-free with $f = 0$). Under these conditions, Equation (11) reduces to the following:

$$m_i = -k_g = m_{iH}, \quad m_r^2 = \frac{\omega^2 - \omega_a^2}{C^2} + k_h^2 \left(\frac{\omega_b^2}{\omega^2} - 1 \right) \tag{13}$$

which is the exact dispersion relation of Hines' classical acoustic-gravity waves [11], where m_{iH} denotes Hines' imaginary wave number. Note that in this locally-isothermal case, the acoustic cut-off frequency and the buoyancy frequency are ω_a and ω_b, respectively. When the horizontal wavenumber k_h has an opposite sign, the solutions of both m_i and m_r^2 in Equation (13) do not change, respectively, as demonstrated by, e.g., Equation (10.29) of [64]. The profiles of m_i and m_r^2 are illustrated in Figures 3 and 4, respectively, together with the additional three cases to be introduced below in the following subsections.

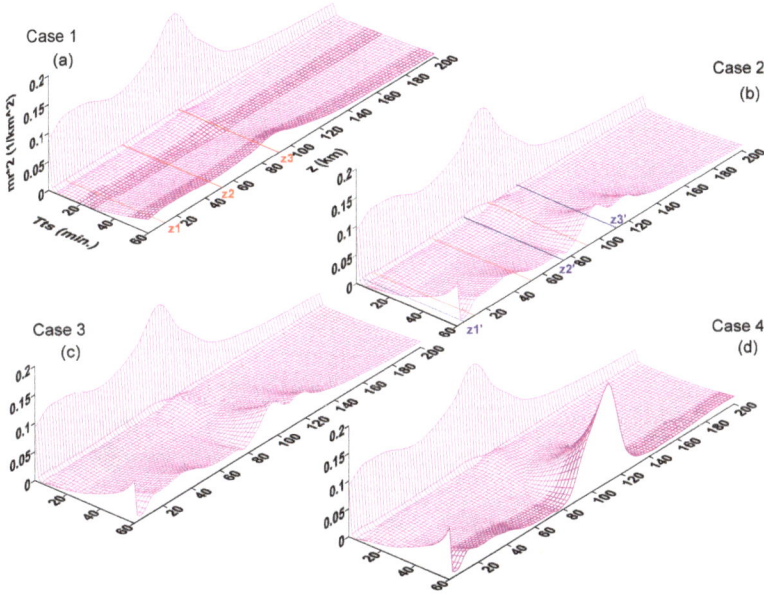

Figure 4. Squared real vertical wavenumber, m_r^2 ($1/km^2$), of different tsunami-excited wave modes propagating in an atmosphere. Case 1: (**a**) Hines' locally-isothermal and shear-free model; Case 2: (**b**) the extended Hines' model under nonisothermal conditions; Case 3: (**c**) inertio-acoustic-gravity waves under nonisothermal conditions; Case 4: (**d**) acoustic-gravity waves under non-isothermal and wind shear conditions. Note that there exists a "quasi-straight line" of $m_r^2 = 0$ in every panel throughout all altitudes at $\sim$4 min in the tsunami period. This line separates the acoustic waveband of <4 min in wave periods from the gravity waveband of >4 min in wave periods. In (a), there are three straight lines in red, which are located at $z_1 \sim 12$ km, $z_2 \sim 50$ km and $z_3 \sim 90$ km, respectively, to separate the space into three regions; in (b), there are three additional straight lines in blue, which are located at $z_1' \sim 4.5$ km, $z_2' \sim 75$ km and $z_3' \sim 110$ km, respectively.

Equation (13) says that the imaginary vertical wavenumber, m_i, does not rely on tsunami wave frequency (ω; or period T_{ts}). The blue curve in Figure 3 displays the vertical profile of m_i (in units of $1/10$ km). Note that this curve is superimposed upon those of Cases 2 and 3. As a reference, a red straight line is shown in the figure to represent the result of m_i for an ideal atmosphere, which is isothermal at all altitudes. It is a constant; the magnitude is that obtained by using the atmospheric temperature at sea level. A direct impression lies in the fact that, relative to the reference line, the profile of Hines' m_i changes in the same way as that of atmospheric temperature T_0. Check the mean-field temperature in Figure 1. Clearly, it is T_0 that dominates the vertical profile of m_i.

By contrast, the features of m_r do rely on wave periods. (a) in Figure 4 exposes the squared real vertical wavenumber, m_r^2 (in units of $1/$km^2), of Hines' mode. Note that there exists a "quasi-straight line" $m_r^2 = 0$ in the panel throughout all altitudes (z) at $\sim$4 min in the tsunami period (T_{ts}). This line separates the acoustic waveband of <4 min from the gravity waveband of >4 min. This tells us that, for tsunami-excited gravity waves with a typical phase speed (V_{ph}) of 200 m/s, a period of $T_{ts} =$(4–60) min corresponds to a horizontal wavelength of $\lambda_h =$(48–720) km.

It deserves to stress here that the "quasi-straight line" shown in the panel to separate the acoustic and gravity wave bands is not a "constant line", as a matter of fact, over the whole range of altitudes. This is exposed in (e) of Figure 2, where the feature of the cut-off frequencies varying with altitude is displayed to tell us that a wave with a period less than ω_b under isothermal conditions (or ω_B under nonisothermal conditions) would not propagate vertically as it becomes evanescent. However, in the timescale up to 60 min used in Figure 4, several minutes of the cut-off periods are so contracted in the panels as to appear as an expression of "quasi-straight lines", although they are actually "curves". In addition, measured tsunami-excited waves are characterized by wave periods that are longer than the cut-off periods and, thus, in the regime of gravity waves only. Consequently, to deal with the tsunami-excited gravity waves in this paper, we concentrate on the gravity wave branch in Figure 4, and so on, in the rest of the text. The narrow acoustic wave band in the figure is presented to provide a direct comparison of the m_r^2-features between the two different wave regimes, rather than to help to show the transition between the two regimes (a different topic beyond the scope of the present work). Note that between the cut-off frequencies, ω_b and ω_a, under isothermal conditions (or ω_B and ω_A under nonisothermal conditions), there might exist evanescent waves that do not propagate vertically, but are allowed to propagate horizontally.

A wave becomes evanescent if $m_r^2 \to 0$ (or infinite wavelength λ_z). After enlarging the panel in the figure, we see that this condition applies approximately for regions of $T_{ts} \sim$ (4–20) min and $z > 150$ km. Thus, Hines' model allows tsunami-excited gravity waves to be alive for $T_{ts} > 20$ min and $z < 150$ km. By contrast, in the acoustic wave regime, $m_r^2 > 0$ is always valid, and waves never disappear, except in the adjacent region close to $T_{ts} = 4$ min. For waves propagating up to $z \sim 150$ km, they can be either partially reflected back from the wind jet into the lower atmosphere (e.g., [65] and the references therein) or dissipated away via terms, like ion drag, kinetic viscosity and/or heat conductivity (e.g., [66] and the references therein).

Furthermore, there are three red straight lines in (a) at $z_1 \sim 12$ km, $z_2 \sim 50$ km and $z_3 \sim 90$ km, respectively, to separate the space into three regions. At z_1, z_2 and z_3, the contours on the plane with fixed T_{ts} are featured with either crests or troughs in m_r^2. The maximal value of $m_r^2 = 0.03889$ (1/km^2) occurs at an 88-km altitude for $T_{ts} = 60$ min. This gives $\lambda_z = 31.4$ km. For $T_{ts} = 33.3$ min, $0.00151 < m_r^2 < 0.00492$ (1/km^2), giving $90 < \lambda_z < 157$ km.

3.2. Case 2: Extended Hines' Model under Non-Isothermal Condition

Hines' local isothermal model excludes the influence of temperature gradient in altitudes on the propagation of acoustic-gravity waves, i.e., $k_T = 0$. When this constraint is relaxed to $k_T \neq 0$, and keeping other conditions unchanged, Equation (11) offers a nonisothermal model:

$$m_i = -k_g = m_{iH} \quad m_r^2 = \frac{\omega^2 - \omega_A^2}{C^2} + k_h^2 \left(\frac{\omega_B^2}{\omega^2} - 1 \right) \tag{14}$$

where the acoustic cut-off frequency, ω_A, and the buoyancy frequency, ω_B, are updated from ω_a and ω_b, respectively, in Equations (13) by taking into account the k_T-effect. Apparently, Equations (13) and (14) have the same appearance. However, the former represents the most basic model under the locally-isothermal condition; while the latter describes a more realistic atmosphere where the temperature gradient brings about impacts on the dispersion relation. Note that m_i stays unchanged, still following Hines' locally-isothermal result, as shown in Figure 3, where the vertical profile of m_i (in blue) follows exactly that in Case 1. This indicates that the temperature inhomogeneity does not affect the vertical growth rate of wave amplitudes.

(b) in Figure 4 illustrates the m_r^2 contours of this non-isothermal model. Generally speaking, the development of m_r^2 is roughly the same as that in (a). For example, the three characteristic heights, z_1, z_2 and z_3, are still alive to characterize the features of Hines' locally-isothermal model. However, there exist a couple of obvious differences: (1) there exist three additional heights, $z_1' \sim 4.5$ km, $z_2' \sim 75$ km and $z_3' \sim 110$ km, as given by the straight lines in blue, respectively; at these altitudes, the contours are influentially disturbed in view of wave frequency; (2) there exists a shift of all of the contours from longer wave periods (or lower wave frequencies) to shorter ones (or higher wave frequencies) at all altitudes. This shift reduces the evanescent regions of $T_{ts} < 20$ min in (a). For instance, starting from $T_{ts} = 4$ min, the magnitude of m_r^2 increases at all altitudes. This indicates that waves can propagate upward to higher altitudes in this case than in Case 1.

3.3. Case 3: Inertio-Acoustic-Gravity Waves under Nonisothermal Condition

In the presence of the rotational Coriolis effect ($f \neq 0$), while the atmosphere stays locally-isothermal ($k_T = 0$) and shear-free ($k_U = k_V = 0$), Equation (11) reproduces the inertio-acoustic-gravity (IAG) modes [58]:

$$\omega^4 - \left(C^2 K_*^2 + \omega_a^2 + f^2\right) \omega^2 + C^2 k_h^2 \omega_b^2 + \left(C^2 m_r^2 + \omega_a^2\right) f^2 = 0 \qquad (15)$$

which reproduces Equation (14) of the IAG formulation in [59]. Related formulae were also given in [48,49].

Now, remove the isothermal limit by allowing $k_T \neq 0$. Equation (11) produces:

$$m_i = -k_g = m_{iH}; \quad m_r^2 = \frac{\omega^2 - \omega_A^2}{C^2} + k_h^2 \frac{\omega_B^2 - \omega^2}{\omega^2 - f^2} \qquad (16)$$

from which we see that the temperature gradient term influences both the high-frequency acoustic branch:

$$\omega^2 \sim C^2 K_*^2 + \omega_A^2 + f^2 \qquad (17)$$

where $K_*^2 = k_h^2 + m_r^2$, and the low-frequency gravito-inertial branch (Equation (1a) of Marks and Eckermann 1995),

$$\omega^2 = \frac{k_h^2 \omega_B^2 + \left(m_r^2 + k_G^2\right) f^2}{K_*^2 + k_G^2 + \frac{f^2}{C^2}} \qquad (18)$$

which contains two modes, namely the gravity mode and the inertial mode in nonisothermal situations, as expressed respectively by:

$$\omega^2 = \frac{k_h^2 \omega_B^2}{K_*^2 + k_G^2}, \quad \omega^2 = \frac{\left(m_r^2 + k_G^2\right) f^2}{K_*^2 + k_G^2 + \frac{f^2}{C^2}} \qquad (19)$$

the second formula of which says that at low latitudes ($\phi \sim 0°$), the inertial mode can be neglected.

The expression of m_i in Equation (16) is the same as that in Case 1. Thus, the vertical profile of m_i (in blue) in Figure 3 does not change from that of Case 1 or Case 2. The contours of m_r^2 in Equation (16) are portrayed in (c) of Figure 4. We see that, relative to Case 2, the effect of the rotational Coriolis term, f^2, is not recognizable based on the comparison between the two panels. Let us check the magnitude of the time scale T_f of the Coriolis parameter, f: at $\phi = 60\circ$, $T_f \approx 13.9$ h, 25-times $T_{ts} = 33.3$ min. Thus, f^2 is 625-times smaller than ω^2, and thus, $\omega^2 - f^2 \approx \omega^2$ in Equation (16). This means the inertial term can be reasonably omitted in dealing with nonisothermal inertio-acoustic-gravity (IAG) modes with wave periods of tens of minutes.

3.4. Case 4: Acoustic-Gravity Waves under Nonisothermal and Wind-Shear Conditions

In the presence of wind shears ($dU/dz \neq 0$ and $dV/dz \neq 0$), many authors discussed the measure of the static stability of an isothermal ($k_T = 0$), irrotational ($f = 0$) atmosphere due to the destabilizing effect of the shears by virtue of the dimensionless Richardson number, R_i (e.g., [7,20,39,67,68]). These studies assumed that both the horizontal wavevector and the wind velocity are one-dimensional, say along the x-direction. In this case, $k_h = k$ and $|\mathbf{v_0}| = U$). The criterion was found to be $R_{icr} = 0.25$; below the value, dynamic instabilities and turbulence were expected. By adopting this 1D model, we obtain directly the same R_i-threshold from Equation (11):

$$m_i = -k_g \left(1 - \frac{\omega_v}{2k_g V_{ph}} \right) \tag{20}$$

and:

$$\omega^4 - \left(C^2 K_*^2 + \omega_a^2 - \frac{2-\gamma}{2} g \frac{\omega_v}{V_{pH}} \right) \omega^2 + C^2 k^2 \omega_b^2 \left(1 - \frac{0.25}{R_i} \right) = 0 \tag{21}$$

in which $\cos\theta = 1$ (or $\theta = 0$) is applied (note that this result also fits the situation where $U \gg V$ below an $\sim$85-km altitude, as shown in (g) of Figure 1). Clearly, the shear term is introduced in both m_i and the quartic dispersion equation of ω. Different from the previous three cases where m_i is only a function of altitude, Equation (20) expresses that m_i also depends on wave frequency ω through V_{ph}. Besides, Equation (21) tells us that for $Ri < 0.25$, ω^2 will always be negative, and the atmosphere is convectively unstable, leading to dynamic instabilities and turbulence; on the contrary, if $R_i > 0.25$, the atmosphere may stay stable. If $R_i = 0.25$, acoustic modes may just be maintained with:

$$\omega^2 = C^2 K_*^2 + \omega_a^2 - \frac{2-\gamma}{2} g \frac{\omega_v}{V_{pH}} \tag{22}$$

Nevertheless, in a realistic atmosphere, the isothermal condition is broken. Taking into account $k_T \neq 0$, we obtain an extended dispersion relation from Equation (11) as follows:

$$\omega^4 - \left(C^2 k^2 + \omega_A^2 - \frac{2-\gamma}{2} g \frac{\omega_v}{V_{pH}} \right) \omega^2 + C^2 k_h^2 \omega_B^2 \left(1 - \frac{0.25}{R_I} \right) = 0 \tag{23}$$

Interestingly, this is a result that needs only replacing R_i with R_I, and $\omega_{a,b}$ with $\omega_{A,B}$, respectively, in Equation (21). Notice that m_i still keeps its expression in Equation (20).

In fact, it is not always valid to assume $\theta = 0$ in a realistic atmosphere due to the arbitrary directions of the waves in propagation relative to the mean-field wind velocity. We have to relax this condition in physical modeling. For an arbitrary θ in the absence of the inertial f term, Equation (11) gives:

$$m_i = -k_g \left(1 - \frac{1}{2} \frac{\omega_{\mathrm{v}}}{k_g V_{ph}} \cos\theta \right) \neq m_{i\mathrm{H}} \tag{24}$$

$$m_r^2 = \frac{\omega^2 - \omega_A^2}{C^2} + k_h^2 \left[\frac{\omega_B^2 - \omega^2}{\omega^2} - \frac{1}{2} \frac{\omega_{\mathrm{v}}^2}{\omega^2} \left(\frac{2 - \gamma}{\gamma} \frac{\omega^2}{k_h^2 V_p V_{ph}} + \frac{1}{2} \cos\theta \right) \cos\theta \right] \tag{25}$$

Clearly, θ influences both m_i and m_r.

In Figure 3, the pink band attached to the blue curve of Cases 1–3 demonstrates the vertical profile of m_i expressed by Equation (24). Though the band follows the development of that in the previous three cases, it fluctuates on the LHS or RHS of the blue curve specifically depending on different altitudes. The fluctuations divide the atmosphere into five layers. In the three of them, i.e., below 18 km, 87–125 km and above 175 km, the pink band lies on the RHS of the blue curve with $m_i > m_{i\mathrm{H}}$; by contrast, in the rest of the two layers of 18–87 km and 125–175 km, it is on the LHS with $m_i < m_{i\mathrm{H}}$. A comparison between m_i and $m_{i\mathrm{H}}$ tells us that it is the wave frequency ω that brings about the m_i-variations: we scan ω in simulations with (0–60) min in the wave period, $T_{ts} = 2\pi/\omega$. Due the presence of $V_{ph} = \omega/k_h$ in Equation (24), the change in ω influences m_i at all altitudes. This ω-effect is zero only at the four following altitudes: 18 km, 87 km, 125 km, 175 km; it is maximal at 80–90 km and 110–150 km, where the fluctuations of m_i are up to $m_{i\mathrm{H}} \pm 0.15$.

(d) in Figure 4 demonstrates the effect of wind shears on m_r^2. Relative to Case 2, which is in the absence of the shears, the whole envelop has an obvious elevation upward, especially in the region of $T_{ts} > 20$ min and 80–130-km altitudes; on the contrary, above 140-km altitudes, the m_r^2-contours become flattened to lower magnitudes. See m_r^2 at 90–110-km altitudes in both Case 2 and Case 4. It goes up from < 0.1 $(1/\mathrm{km}^2)$ in Case 2 to 0.2 $(1/\mathrm{km}^2)$ in Case 4. It is predictable that wind shears make it more difficult for gravity waves to transmit to higher altitudes. In other words, the shears play a screening role for gravity waves that only those waves of sufficient energy can propagate upward to higher altitudes. A following paper on ray-tracing studies of gravity wave propagation in the atmosphere will introduce the criterion of the energy level in wave transmission and reflection.

3.5. Case 5: IAG Waves under Nonisothermal and Wind Shear Conditions

By taking into consideration the rotational Coriolis f-term, the acoustic-gravity modes under nonisothermal and wind shear conditions discussed in the last subsection extend to the most generalized inertio-acoustic-gravity modes, as expressed by Equation (11), rewritten as follows:

$$m_i = -k_g \left[1 - \frac{\omega^2}{2(\omega^2 - f^2)} \frac{\omega_v}{k_g V_{ph}} \cos \theta \right] \tag{26}$$

and:

$$m_r^2 = \frac{\omega^2 - \omega_A^2}{C^2} + k_h^2 \left[\frac{\omega_B^2 - \omega^2}{\omega^2 - f^2} - \frac{1}{2} \frac{\omega_v^2 \omega^2}{(\omega^2 - f^2)^2} \left(\frac{2 - \gamma}{\gamma} \frac{\omega^2 - f^2}{k_h^2 V_p V_{ph}} + \frac{1}{2} \cos \theta \right) \cos \theta \right] \tag{27}$$

As mentioned previously, in the regime of gravity waves, the f-effect is infinitesimal and can be reasonably omitted. Thus, Equations (26) and (27) are equivalent to Equations (24) and (25), respectively, for tsunami-excited gravity waves, the periods of which are within 4–60 min. In Figure 5, the 3D m_i-envelop in (a) represents the 2D pink band in Figure 3. Now, it is clear to see that the fluctuations in the pink band originate from the summation of variations in m_i *versus* different T_{ts} values at any specific altitudes. For the m_r^2-envelop in (b), it has no difference from (d) in Figure 4, also because of the negligible f-effect.

Both the pink band in Figure 3 and (a) in Figure 5 exhibit that the atmosphere has five layers concerning polarized m_i-fluctuations below the 200-km altitude: Layer I (0–18) km; Layer II (18–87) km; Layer III (87–125) km; Layer IV (125–175) km; and Layer V (175–200) km. Layers I, III and V own a relation of $|m_i| < |m_{iH}|$, indicating that the growth of propagating waves in realistic atmospheric situations is attenuated from Hines' idealized atmospheric model; on the contrary, Layers II and IV have $|m_i| > |m_{iH}|$, referring to the fact that the amplitude of propagating waves is driven from the lower Hines' model to a higher level. For a clear look at the attenuating or damping characteristics in wave propagation, we write the error, $\mathcal{E}$, caused by the nonisothermality and wind shears defined as follows:

$$\mathcal{E} = \frac{m_i - m_{iH}}{m_{iH}} \times 100\% \tag{28}$$

which turns out to be nothing else but the "damping factor", β, after some simple algebra to connect the theoretical work (e.g., [13,14]) with the modeling of the airglow layers perturbed by waves (e.g., [8,9]). A straightforward manipulation with the m_i-expression in Equation (26) and the k_g-expression in Equation (12) yields:

$$\mathcal{E} = \beta = \frac{\omega^2}{2(\omega^2 - f^2)} \frac{\omega_v \cos\theta}{k_g V_{ph}} \rightarrow H\frac{\omega_v}{V_{ph}} \cos\theta \text{ if } f = 0 \tag{29}$$

in which $H = -H_p = 1/(2k_g)[= C^2/(\gamma g)]$ is used. Figure 6 provides β-contours (or, alternatively, $\mathcal{E}$-contours) *versus* T_{ts} and z. (a) is for the special case of $\theta = 0$, and (b) is for the generalized case of $\theta \neq 0$.

For the special case of $\theta = 0$, *i.e.*, $\mathbf{k}_h \| \mathbf{v}_0$, (a) reveals that β_0 is always positive. As a result, the wave growth is always damped from the growth rate of Hines' classical model. By contrast, in the general case where $\theta \neq 0$, (b) presents the five layers introduced above: in Layers I, III and V, it is always valid that $\beta > 0$, validating the previous argument that the growth of propagating waves in realistic atmospheric situations is attenuated from Hines' idealized atmospheric model; in Layers II and IV, the relation of $\beta < 0$ refers to the amplitude growth of the propagating waves being pumped, rather than damped, from Hines' model. We point out here that the proposed free-propagating state with $\beta = 0$ [8,9] only appears at some specific altitudes, say, 18 km, 87 km, 125 km and 175 km.

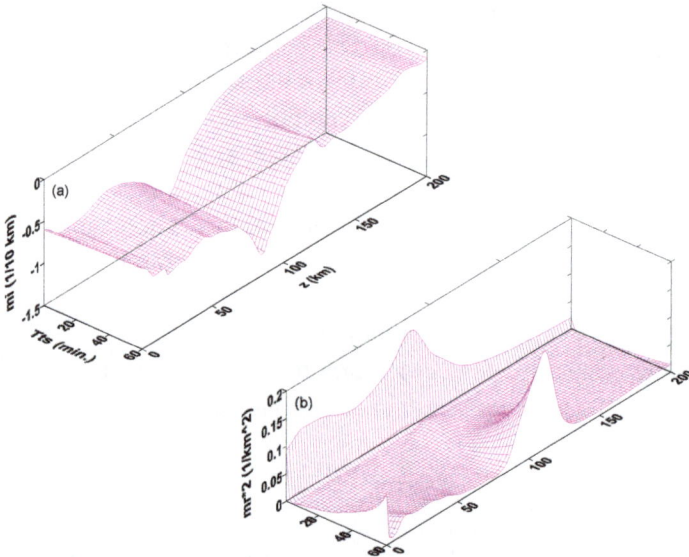

Figure 5. Imaginary and squared real vertical wavenumbers, m_i (1/10 km) and m_r^2 (1/km^2), in Case 5 of IAG waves under non-isothermal and wind shear conditions. (a) m_i-envelop; and (b) m_r^2-envelop. Note that due to the negligible f-effect, (a) gives the pink band in Figure 3; while (b) has no difference from (d) in Figure 4.

Equation (29) shows that the polarities of β are determined by $\cos\theta$, as shown in (d) of Figure 2: in the bottom layer, Layer I, $\cos\theta$ is positive; in Layer II, it is negative; in Layer III, it is positive; in Layer IV, it is negative; and, in the top layer, Layer V, it is positive. In addition, the equation demonstrates that β is inversely proportional to wave frequency ω via V_{ph} and, thus, proportional to the wave period T_{ts}. This feature can be recognized in the two panels of Figure 6: the larger the value of T_{ts}, the higher the magnitude of β_0 or β. Furthermore, the equation discloses that β has a linear relation with the scale height, H (or $-H_p$). See (c) of Figure 1. The magnitude of H oscillates twice till about a 100-km altitude and then increases monotonically upward. This offers a vibrating feature in β_0 or β below the altitude, followed by an enhancement in amplitudes above it, as displayed in Figure 6.

Therefore, the observation-defined "damping factor", β, is found not always to bring about a "damping"(or attenuation) effect in wave amplitude A. This is because:

$$A \sim e^{-\int m_i \mathrm{d}z} = e^{\int (1-\beta)k_g \mathrm{d}z} = e^{\int (1-\beta)\mathrm{d}z/(2H)} \tag{30}$$

where Equations (4), (26) and (29) are used. Clearly, for $\beta = 0$, Equation (30) reduces to Hines' classical result, $A_{\text{Hines}} \sim e^{z/(2H)}$; for $\beta > 0$, A_{Hines} is damped or the wave is attenuated; for $\beta < 0$, A_{Hines} is amplified or the wave is intensified or pumped. This gives results in concordance with the above discussions.

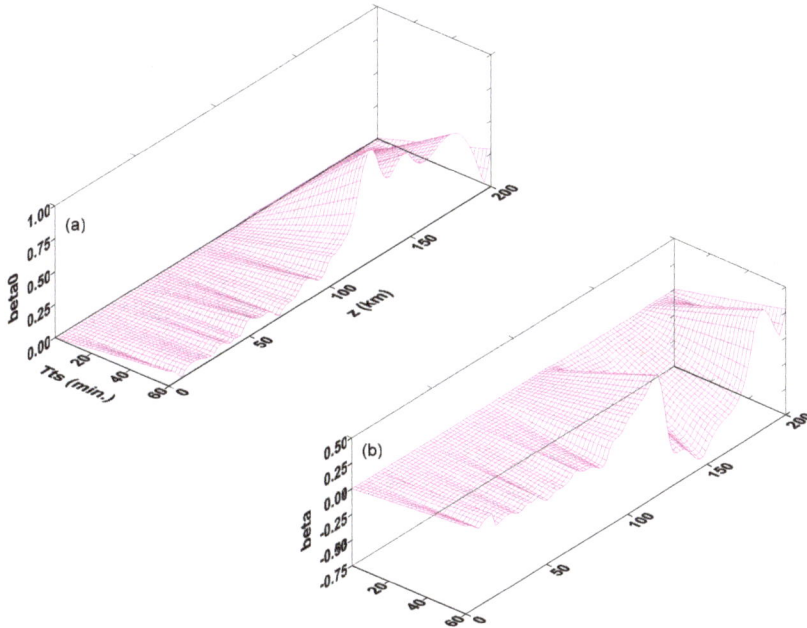

Figure 6. Contours of the "damping factor", β (or, alternatively, the error, $\mathcal{E}$) *versus* T_{ts} and z. (a) special case with $\theta = 0$; and (b) generalized case with $\theta \neq 0$.

3.6. Influence of Phase Speed V_{ph}

Equation (11) reveals that, in the above five cases, phase speed $V_{ph} = \omega/k_h$ influences wave propagation through modulating m_i and m_r^2 simultaneously, except the identical m_i in Cases 1–3. Particularly, in Cases 4 and 5, V_{ph} also has impacts on m_i and m_r^2 through wind shears, as denoted by ω_v. The relationship between V_{ph} and wave propagation thus needs necessary attention.

Observations provided that the characteristic V_{ph} is about 150–160 m/s ([69]). We choose V_{ph} varying from 80 m/s–240 m/s to display how m_i and m_r^2 are influenced by V_{ph} at a characteristic wave period $T_{ts} = 33.3$ min. Due to the negligible f-effect, in addition to the irrelevance of V_{ph} to m_i in Cases 1–3, we just need to present V_{ph}-dependent m_r^2 in Cases 1 and 2 (or 3) and m_i and m_r^2 in Case 4 (or 5). Figure 7 gives the results. (a) is m_r^2 in Case 1; (b) is m_r^2 in Case 2 (or 3); (c) is m_i in Case 4 (or 5); and (d) is m_r^2 in Case 4 (or 5).

The four panels expose the following features:

(1) At specific z, phase speed V_{ph} has an effective range of values, say <200 m/s, within which the dependence of m_i or m_r^2 on V_{ph} is obvious; out of the regime, the influence is negligible. At ~95 km, for example, the upper right panel illustrates that m_r^2 reduces quickly from 0.12 (1/km^2) with $V_{ph} = 80$ m/s; however, m_r^2 tends to be stabilized at 0.005 (1/km^2) for $V_{ph} > 200$ m/s.

(2) At specific V_{ph}, m_i or m_r^2 also changes *versus* z. For instance, at the low V_{ph} end in Case 1, the m_r^2-profile is not constant along z, but has a hump at about 80–100-km altitudes; in Case 4, there are more m_r^2-humps, which nearly fill up all of the altitudes.

(3) The dependence of m_r^2 on V_{ph} is modulated by the atmospheric nonisothermality and wind shears. For instance, in (a) of Hines locally-isothermal and shear-free case, the m_r^2-profile has only one hump with a maximal amplitude of 0.08 (1/km^2); when the isothermal condition is relaxed as shown in (b), the maximal amplitude is enhanced to 0.12 (1/km^2), and more humps and troughs appear to expand to both higher and lower altitudes; stepping further to allow shears present as drawn in (d), higher amplitude fluctuations, peaked at 0.7 (1/km^2), are excited and driven to stretch out toward the higher V_{ph}-region accompanied by increasingly suppressed amplitudes.

(4) Compared to the strong dependence of m_r^2 on V_{ph}, m_i has very weak or little relevance to V_{ph}, as displayed in (c): with the increase of V_{ph}, the m_i-envelop appears constant in the whole range of V_{ph}. This implies that we can ignore the effect of V_{ph} on m_i in dealing with gravity wave growth in space. However, we stress that m_i is heavily dependent on z, as discussed in the last subsections.

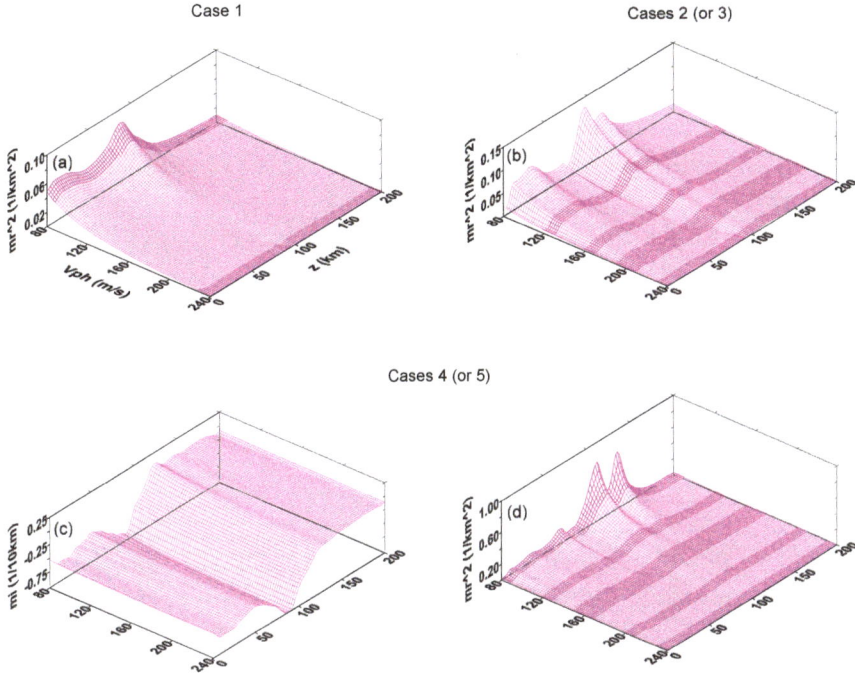

Figure 7. Dependence of m_i and m_r^2 on phase speed V_{ph} at a characteristic wave period of $T_{ts} = 33.3$ min. (**a**) m_r^2 in Case 1; (**b**) m_r^2 in Case 2 (or 3); (**c**) m_i in Case 4 (or 5); and (**d**) m_r^2 in Case 4 (or 5).

4. Summary and Discussion

We generalized Hines' ideal locally-isothermal, shear-free and rotation-free model of gravity waves to accommodate a realistic atmosphere featured with altitude-dependent nonisothermality (up to 100 K/km) and wind shears (up to 100 m/s per km). Although some of the variations in the background state are rather extreme (e.g., the zonal and meridional winds), we first of all applied Equation (4) in the Taylor expansion of all of the physical parameters; then obtained the set of linearized equations, as shown in Equation (7), with vertically-inhomogeneous ones; and finally, manipulated both sides of the equations to obtain Equation (8) or Equation (11), the generalized, complex dispersion relation of inertio-acoustic-gravity (IAG) waves, which recovers all of the known wave modes under different situations below 200-km altitudes where all of the dissipative terms (e.g., viscosity, heat conductivity, ion drag) are neglected.

We studied the modulation of atmospheric nonisothermality and wind shears on the propagation of seismic tsunami-excited gravity waves by virtue of the imaginary and real parts (*i.e.*, m_i and m_r) of the vertical wavenumber, m, within the full band of 4–60 min in tsunami wave periods. In five different situations, we calculated the vertical profiles of m_i and m_r^2: (1) in Hines' classical modes; (2) in the extended Hines' modes in the presence of nonisothermality; (3) in the IAG modes by adding the rotational Coriolis f-effect to the nonisothermal Hines' model; (4) in the generalized AG modes under not only non-isothermal, but also wind shear conditions; and (5) in the generalized IAG wave modes. We also illustrated the influence of phase speed V_{ph} on m_i and m_r^2.

The main results obtained in this paper are summarized and discussed as follows:

It is well known that gravity waves propagate only when their period is longer than the Brunt-Väisälä (BV) period. Below the period, they will become evanescent. For example, in the mesosphere, this period is about 6 min. Those tsunami-excited waves with a >6-minute period will be propagating in the mesosphere. Our first result shows that this BV criterion for wave propagation is a necessary, but not a sufficient, condition. That is to say, even though this condition is satisfied, e.g., a wave period is longer than the BV period, the wave may still be kept evanescent due to $m_r^2 \to 0$, as illustrated with Hines' isothermal model under conditions that the tsunami wave period (T_{ts}) is smaller than 20 min or at altitudes above 150 km. Only beyond these regions, waves can propagate with a nonzero m_r^2. In the presence of nonisothermality, the evanescent regions of $m_r^2 \to 0$ appear to be reduced considerably to free more waves from evanescence to propagation. However, if wind shears are included, an evanescent region emerges again above the 140-km altitude.

Secondly, nonisothermality and wind shears divide the atmosphere into a sandwich-like structure of five layers within the 200-km altitude, in view of the wave growth in amplitudes: Layer I (0–18) km; Layer II (18–87) km; Layer III (87–125) km; Layer IV (125–175) km; and Layer V (175–200) km. In Layers I, III and V, the magnitude of m_i is smaller than that of Hines' result, m_{iH}, referring to an attenuated growth in amplitudes of upward propagating waves in realistic atmospheric situations from Hines' idealized atmosphere; on the contrary, in Layers II and IV, the magnitude of m_i is larger than that of m_{iH}, providing a pumped growth in amplitudes of the waves from Hines' model.

Thirdly, nonisothermality and wind shears enhance m_r substantially at an ∼100-km altitude for a tsunami wave period T_{ts} longer than 30 min. Hines' model gives that the maximal value of m_r^2 is ∼0.05 (1/km²). This magnitude is doubled by the nonisothermal effect and quadrupled by the joint nonisothermal and wind shear effect. The modulations are weaker at altitudes outside 80–140-km heights.

Fourthly, nonisothermality and wind shears expand the meaning of the observation-defined "damping factor", β. It does not merely refer to the "damping" of wave growth anymore. Instead, it is updated with a couple of opposite implications: relative to Hines' classical result in wave growth under $\beta = 0$, waves are damped or attenuated from Hines' isothermal and shear-free result for $\beta > 0$;

nevertheless, waves can also be amplified or pumped from Hines' result for $\beta < 0$. The polarization of β is determined by the angle θ between the wind velocity and wave vector.

Lastly, the nonisothermal and wind shear modulation on the wave propagation is not influenced by the rotational Coriolis effect in the tsunami waveband of up to one hour in wave periods.

This study provided us a better understanding of the nature of tsunami-excited gravity waves under non-Hines' conditions. For example, the involvement of nonisothermality updates Hines' classical formula of the dispersion relation by simply replacing the isothermal parameters, $\omega_{a,b}$ and R_i, with their non-isothermal counterparts, $\omega_{A,B}$ and R_I, respectively. Here, we stress that it is invalid to mix up these pairs in relevant studies (as shown in some modeling and experimental publications), e.g., using ω_b (ω_B) and ω_A (ω_a) at the same time to analyze gravity wave phenomena.

In addition, the angle θ between horizontal wind velocity and the wave vector is an important parameter to reflect the modulation of nonisothermality and wind shears on the propagation of gravity waves. It decides the polarities of β and, thus, has a direct and an effective impact on the damping or intensifying mechanism in wave propagation. The importance of θ was not well recognized in some publications with a single-component horizontal wavevector (e.g., $k_h = k_x$), which was assumed to be parallel to wind velocity, and thus, θ is always zero. This assumption excludes realistic situations where $\theta \neq 0$. In this case, results may be totally different. For example, for $\theta \to 90\circ$ (i.e., the directions of wind and wave tend to be perpendicular to each other), the shear effect tends to zero, as shown by Equation (11).

Furthermore, calculations from the NRLMSISE-00 and HWM93 models provided that either the isothermal $1/R_i$ or the non-isothermal $1/R_I$ are smaller than 0.2 (accordingly, their inverses are larger than five, as shown in the top panel of Figure 2). This suggests that the realistic atmosphere is unable to be teared up easily from its stratified state by any wind shears. Thus, any instabilities below a 200-km altitude appear to be sufficiently suppressed. This argument, though as a result under the nonisothermal situation, reiterates the conclusion obtained in the 1970s under isothermal conditions by, e.g., [20,63].

Nevertheless, we argue that the above result is true only for the empirical atmosphere provided by the NRLMSIS and HWM models. In the realistic atmosphere observed by LiDAR and meteor radar systems, the temperature gradient and wind shear could be much larger than those provided by the models and would bring the atmosphere to a dynamically unstable state due to the action of planetary waves and tides. See the details in the papers by, e.g., *Li et al.* [70,71], on the characteristics of instabilities in the mesopause region and on the observations of gravity wave breakdown into ripples associated with dynamical instabilities, respectively.

Finally, we illustrated clearly the evanescent regions in the five cases that form the boundary layers between the high-frequency acoustic waves (below several minutes in wave periods) and the low-frequency gravity waves (above several to tens of minutes in wave periods). The thickness of the evanescent regions varies in altitude, as exhibited in the figures of Cases 1–5.

The present work offers a detailed model to describe the propagation of tsunami-excited gravity waves in the atmosphere above sea level. It extends the results of Hines' and others' classical work by taking into account the variability of the atmospheric temperature and the wind field. It exposes the influence of the temperature gradient and wind shears on the real and imaginary parts of the vertical wavenumber and presents an explicit expression for the β factor, which is relevant to and generalizes the concept of the damping/amplification of the wave amplitude throughout the atmosphere at least in the range of 0–200 km. While the work focuses mainly on tsunami-generated waves, the results are more general and applicable to gravity waves of any nature and generation source. However, we admit that the present work concerns only those tsunami-generated waves that fall into the regime of the wave properties that would be allowed to propagate vertically after the excitation occurring at sea level. According to the strict work done most recently by *Godin, Zabotin* and *Bullett* on acoustic-gravity waves in the atmosphere generated by infragravity waves in the ocean [72], not every tsunami-generated wave has periodicity in the permitted regime; in particular, these waves are featured with a transition frequency of about 3 mHz (34.9 min in wave periods) below which the infragravity waves continuously radiate their energy into the upper atmosphere in the form of acoustic-gravity waves. Therefore, in applying the results of this paper in relevant data-fit modeling and data analysis, we must be cautious in checking the initial and boundary conditions (not only the tsunami wave periods, but also the zonal and meridional wavelengths, as well as the vertical wave speeds), so as to avoid a wrong employment of the model in coding ray-tracing algorithms to demonstrate wave propagations and in interpreting experimental signals from, e.g., GPS satellites, for a global manifestation of the ocean-generated gravity waves.

In addition, there exists a concern about the application of the present work in the thermosphere, where the composition is a strong function of altitude and, thus, affects the mean molecular weight and all of the thermodynamic quantities related to it. Fortunately, in order to avoid such an infeasibility to thermospheric studies, we have relied on NASA's empirical atmospheric models, NRLMSISE-00 [46] and HWM93 [47], to describe the mean-field atmosphere and the horizontal wind profiles. On the one hand, the MSISE model provides thermospheric temperature and density based on *in situ* data from seven satellites and numerous rocket probes and estimates of temperature and the densities of N_2, O, O_2, He, Ar and H. It (1) uses the low-order spherical harmonics to describe the major variations throughout the atmosphere, including latitude, annual, semiannual and simplified local time and longitude variations; (2) employs a Bates–Walker temperature profile as a function of geopotential height for the upper thermosphere and an inverse polynomial in geopotential height for the lower thermosphere; and (3) expresses the exospheric temperature and other atmospheric quantities as functions of the geographical and solar/magnetic parameters. On the other hand, the HWM model is based on wind data obtained from the AE-Eand DE-2 satellites. It (1) uses a limited set of vector spherical harmonics to describe the zonal and meridional wind components; (2) includes wind data from ground-based incoherent scatter radar and Fabry–Perot optical interferometers, as well as the solar cycle variations and the magnetic activity index (Ap) ones; and (3) describes the transition from predominantly diurnal variations in the upper thermosphere to semidiurnal variations in the lower thermosphere, as well as transitions from summer to winter flow above 140 km and from winter to summer flow below [73]. We therefore consider that the present work will provide a reference in dealing with atmospheric studies, including the thermosphere, where the atmospheric composition varies as a strong function of altitude.

At the end of the paper, we remind readers that the present work temporally neglected the effects of dissipative terms, like viscosity, although they become appreciable above the 150-km altitudes. This is because we are dealing with a very complicated subject related to wave excitation and propagation in the atmosphere, where nonisothermality and wind shears play a dominant role to drive gravity wave propagations, an important subject that, however, needs extensive studies. The complexity of the topic requires that we pay attention first of all to the nonisothermal and wind shearing effects in this paper, with the purpose to approach finally a least-error solution through a series of incremental steps, so as to be able to understand the physics and, based on the gained knowledge, to develop appropriate algorithms for solving more realistic problems, while leaving the studies on the dissipative terms to a following paper. Such a paper was submitted and is under review.

Acknowledgments: Acknowledgments: The work is supported by a grant from a NASA-JPL project "Tsunami Imaging Using Ionospheric Radio Occultation Data" in collaboration with Embry-Riddle Aeronautical University (ERAU). John thanks M. P. Hickey for advice on Hines' gravity wave theory and introduction to his full-wave model (FWM) and code use. John thanks him and J. B. Snively and M. D. Zettergren for their respective partial financial supports.

Conflicts of Interest: Conflicts of Interest: The author declares no conflict of interest.

References

1. Liu, A.Z.; Hocking, W.K.; Franke, S.J.; Thayaparan, T. Comparison of Na LiDAR and meteor radar wind measurements at starfire optical range, NM, USA. *J. Atmos. Sol. Terr. Phys.* **2002**, *64*, 31–40.

2. Fritts, D.C.; Williams, B.P.; She, C.Y.; Vance, J.D.; Rapp, M.; Lübken, F.-J.; Müllemann, A.; Schmidlin, F.J.; Goldberg, R.A. Observations of extreme temperature and wind gradients near the summer mesopause during the MaCWAVE/MIDAS rocket campaign. *Geophys. Res. Lett.* **2004**, *31*, L24S06.

3. Franke, S.J.; Chu, X.; Liu, A.Z.; Hocking, W.K. Comparison of meteor radar and Na Doppler LiDAR measurements of winds in the mesopause region above Maui, HI. *J. Geophys. Res.* **2005**, *110*, D09S02.

4. She, C.Y.; Williams, B.P.; Hoffmann, P.; Latteck, R.; Baumgarten, G.; Vance, J.D.; Fiedler, J.; Acott, P.; Fritts, D.C.; Luebken, F.-J. Observation of anti-correlation between sodium atoms and PMSE/NLC in summer mesopause at ALOMAR, Norway (69N, 12E). *J. Atmos. Sol. Terr. Phys.* **2006**, *68*, 93-101.

5. She, C.Y.; Krueger, D.A.; Akmaev, R.; Schmidt, H.; Talaat, E.; Yee, S. Long-term variability in mesopause region temperatures over Fort Collins, Colorado (41° N, 105° W) based on LiDAR observations from 1990 through 2007, *J. Terr. Sol. Atmos. Phys.* **2009**, *71*, 1558–1564.

6. Yue, J.; She, C.-Y.; Liu, H.-L. Large wind shears and stabilities in the mesopause region observed by Na wind-temperature LiDAR at midlatitude. *J. Geophys. Res.* **2010**, *115*, A10307.

7. Hall, C.M.; Aso, T.; Tsutsumi, M. Atmospheric stability at 90 km, 78° N, 16° E. *Earth Planets Space* **2007**, *59*, 157–164.

8. Liu, A.Z.; Swenson, G.R. A modeling study of O_2 and OH airglow perturbations induced by atmospheric gravity waves. *J. Geophys. Res.* **2003**, *108*, 4151.

9. Vargas, F.; Swenson, G.; Liu, A.; Gobbi, D. $O(^1S)$, OH, and O_2 airglow layer perturbations due to AGWs and their implied effects on the atmosphere. *J. Geophys. Res.* **2007**, *112*, D14102.

10. Takahashi, H.; Onohara, A.; Shiokawa, K.; Vargas, F.; Gobbi, D. Atmospheric wave induced O_2 and OH airglow intensity variations: Effect of vertical wavelength and damping. *Ann. Geophys.* **2011**, *29*, 631–637.

11. Hines, C.O. Internal atmospheric gravity waves at ionospheric heights. *Can. J. Phys.* **1960**, *38*, 1441–1481.

12. Hines, C.O. Atmospheric gravity waves: A new toy for the wave theorist. *Radio Sci. J. Res.* **1965**, *69D*, 375–380.

13. Midgley, J.E.; Liemohn, H.B. Gravity waves in a realistic atmosphere. *J. Geophys. Res.* **1966**, *71*, 3729–3748.

14. Volland, H. The upper atmosphere as a multiple refractive medium for neutral air motions. *J. Atmos. Terr. Phys.* **1969**, *31*, 491–514.

15. Klostermeyer, J. Numerical calculation of gravity wave propagation in a realistic thermosphere. *J. Atmos. Terr. Phys.* **1972**, *34*, 765–774.

16. Klostermeyer, J. Comparison between observed and numerically calculated atmospheric gravity waves in the F-region. *J. Atmos. Terr. Phys.* **1972**, *34*, 1393–1401.

17. Klostermeyer, J. Influence of viscosity, thermal conduction, and ion drag on the propagation of atmospheric gravity waves in the thermosphere. *Z. Geophys.* **1972**, *38*, 881–890.

18. Yeh, K.C.; Liu, C.H. Acoustic-gravity waves in the upper atmosphere. *Rev. Geophys. Space Sci.* **1974**, *12*, 193–216.

19. Hines, C.O.; Reddy, C.A. On the propagation of atmospheric gravity waves through regions of wind shear. *J. Geophys. Res.* **1967**, *72*, 1015–1034.

20. Hines, C.O. Generalization of the Richardson criterion for the onset of atmospheric turbulence. *Q. J. R. Meteorol. Soc.* **1971**, *97*, 429–439.

21. Stone, P.H. On non-geostrophic baroclinic stability. *J. Atmos. Sci.* **1966**, *23*, 390–400.

22. Hines, C.O. Gravity waves in the atmosphere. *Nature* **1972**, *239*, 73–78.

23. Peltier, W.R.; Hines, C.O. On the possible detection of tsunamis by a monitoring of the ionosphere. *J. Geophys. Res.* **1976**, *81*, 1995–2000.

24. Einaudi, F.; Hines, C.O. WKB approximation in application to acoustic-gravity waves. *Can. J. Phys.* **1970**, *48*, 1458–1471.

25. Francis, S.H. Acoustic-gravity modes and large-scale traveling ionospheric disturbances of a realistic, dissipative atmosphere. *J. Geophys. Res.* **1973**, *78*, 2278-2301.

26. Gill, A.E. *Atmosphere-Ocean Dynamics*; International Geophysics Series; Academic Press: Orlando, FL, USA, 1982.

27. Hickey, M.P.; Cole, K.D. A numerical model for gravity wave dissipation in the thermosphere. *J. Atmos. Terr. Phys.* **1988**, *50*, 689–697.

28. Nappo, C.J. *An Introduction to Atmospheric Gravity Waves*; Academic Press: Waltham, MA, USA, 2002.

29. Vadas, S.L. Horizontal and vertical propagation and dissipation of gravity waves in the thermosphere from lower atmospheric and thermospheric sources. *J. Geophys. Res.* **2007**, *112*, A06305.

30. Lindzen, R.S.; Tung, K.-K. Banded convective activity and ducted gravity waves. *Mon. Weather Rev.* **1976**, *104*, 1602–1617.

31. Hickey, M.P.; Walterscheid, R.L.; Taylor, M.J.; Ward, W.; Schubert, G.; Zhou, Q.; Garcia, F.; Kelley, M.C.; Shepherd, G.G. Numerical simulations of gravity waves imaged over Arecibo during the 10-day January 1993 campaign. *J. Geophys. Res.* **1997**, *102*, 11475–11489.

32. Hickey, M.P.; Taylor, M.J.; Gardner, C.S.; Gibbons, C.R. Full-wave modeling of small-scale gravity waves using Airborne Lidar and Observations of the Hawaiian Airglow (ALOHA-93) $O(^1S)$ images and coincident Na wind/temperature LiDAR measurements. *J. Geophys. Res.* **1998**, *103*, 6439–6453.

33. Hickey, M.P.; Walterscheid, R.L.; Schubert, G. Gravity wave heating and cooling in Jupiter's thermosphere. *Icarus* **2000**, *148*, 266–281.

34. Hickey, M.P.; Schubert, G.; Walterscheid, R.L. Acoustic wave heating of the thermosphere. *J. Geophys. Res.* **2001**, *106*, 21543–21548.

35. Liang, J.; Wan, W.; Yuan, H. Ducting of acoustic-gravity waves in a nonisothermal atmosphere around a spherical globe. *J. Geophys. Res.* **1998**, *103*, 11229–11234.

36. Walterscheid, R.L.; Hickey, M.P. One-gas models with height-dependent mean molecular weight: Effects on gravity wave propagation. *J. Geophys. Res.* **2001**, *106*, 28831–28839.

37. Schubert, G.; Hickey, M.P.; Walterscheid, R.L. Heating of Jupiter's thermosphere by the dissipation of upward propagating acoustic waves. *Icarus* **2003**, *163*, 398–413.

38. Schubert, G.; Hickey, M.P.; Walterscheid, R.L. Physical processes in acoustic wave heating of the thermosphere. *J. Geophys. Res.* **2005**, *110*, doi:10.1029/2004JD005488.

39. Kundu, P.K. *Fluid Mechanics*; Academic Press: San Diego, CA, USA, 1990.

40. Sutherland, B.R. *Internal Gravity Waves*; Cambridge University Press: Cambridge, UK, 2010.

41. Miles, J. Richardson's criterion for the stability of stratified shear flow. *Phys. Fluids* **1986**, *29*, 3470–3471.

42. Galperin, B.; Sukoriansky, S.; Anderson, P.S. On the critical Richardson number in stably stratified turbulence. *Atmos. Sci. Lett.* **2007**, *8*, 65–69.

43. Harris, I.; Priester, W. Time dependent structure of the upper atmosphere. *J. Atmos. Sci.* **1962**, *19*, 286–301.

44. Pitteway, M.L.V.; Hines, C.O. The viscous damping of atmospheric gravity waves. *Can. J. Phys.* **1963**, *41*, 1935–1948.

45. Volland, H. Full wave calculations of gravity wave propagation through the thermosphere. *J. Geophys. Res.* **1969**, *74*, 1786–1795.

46. Picone, J.M.; Hedin, A.E.; Drob, D.P.; Aikin, A.C. NRLMSISE-00 empirical model of the atmosphere: Statistical comparisons and scientific issues. *J. Geophys. Res.* **2002**, *107*, 1468.

47. Hedin, A.E.; Fleming, E.L.; Manson, A.H.; Schmidlin, F.J.; Avery, S.K.; Clark, R.R.; Franke, S.J.; Fraser, G.J.; Tsuda, T.; Vial, F.; *et al.* Empirical wind model for the upper, middle and lower atmosphere. *J. Atmos. Terr. Phys.* **1996**, *58*, 1421–1447.

48. Eckart, C. *Hydrodynamics of Oceans and Atmospheres*; Pergamon: New York, NY, USA, 1960.

49. Eckermann, S.D. Influence of wave propagation on the Doppler spreading of atmospheric gravity waves. *J. Atmos. Sci.* **1997**, *54*, 2554–2573.

50. Hickey, M.P.; Cole, K.D. A quartic dispersion equation for internal gravity waves in the thermosphere. *J. Atmos. Terr. Phys.* **1987**, *49*, 889–899.

51. Hickey, M.P.; Schubert, G.; Walterscheid, R.L. Gravity-wave driven fluctuations in the O_2 atmospheric (0–1) nightglow from an extended, dissipative emission region. *J. Geophys. Res.* **1993**, *98*, 13717–13729.

52. Fagundes, P.R.; Takahashi, H.; Sahai, Y.; Gobbi, D. Observations of gravity waves from multispectral mesospheric nightglow emissions observed at $23°$ S. *J. Atm. Sol. Terr. Phys.* **1995**, *57*, 39–40.

53. Azeem, S.M.I.; Sivjee G.G. Multiyear observations of tidal oscillations in OH M(3,1) rotational temperatures at South Pole, Antarctica. *J. Geophys. Res.* **2009**, *114*, A06312.

54. Landau, L.D.; Lifshitz E.M. *Fluid Mechanics*; Pergamon: New York, NY, USA, 1959.

55. Fritts, D.C.; Alexander, M.J. Gravity wave dynamics and effects in the middle atmosphere. *Rev. Geophys.* **2003**, *41*, 1003.

56. Vadas, S.L.; Fritts, D.C. Thermospheric responses to gravity waves: Influences of increasing viscosity and thermal diffusivity. *J. Geophys. Res.* **2005**, *110*, D15103.

57. Liu, X.; Xu, J.; Yue, J.; Vadas, S.L. Numerical modeling study of the momentum deposition of small amplitude gravity waves in the thermosphere. *Ann. Geophys.* **2013**, *31*, 1–14.

58. Kaladze, T.D.; Pokhotelov, O.A.; Stenflo, L.; Shah, H.A.; Jandieri, G.V. Electromagnetic inertio-gravity waves in the ionospheric E-layer. *Phys. Scr.* **2007**, *76*, 343–348.

59. Kaladze, T.D.; Pokhotelov, O.A.; Shah, H.A.; Khan, M.I.; Stenflo, L. Acoustic-gravity waves in the Earth's ionosphere. *J. Atm. Sol. Terr. Phys.* **2008**, *70*, 1607–1616.

60. Ma, J.Z.G; Hickey, M.P.; Komjathy, A. Ionospheric electron density perturbations driven by seismic tsunami-excited gravity waves: Effect of dynamo electric field. *J. Mar. Sci. Eng.* **2015**, *3*, 1194–1226.

61. Taylor, M.J.; Pautet, P.-D.; Medeiros, A.F.; Buriti, R.; Fechine, J.; Fritts, D.C.; Vadas, S.L.; Takahashi, H.; São Sabbas, F.T. Characteristics of mesospheric gravity waves near the magnetic equator, Brazil, during the SpreadFEx campaign. *Ann. Geophys.* **2009**, *27*, 461–472.

62. Gossard, E.; Hooke, W. *Waves in the Atmosphere*; Elsevier: New York, NY, USA, 1975.

63. Turner, J.S. *Buoyancy Effects in Fluids*; Cambridge University Press: Cambridge, UK, 1973.

64. Schunk, R.W.; Navy, A.F. *Ionosphere: Physics, Plasma Physics, and Chemistry*, 2nd ed.; Cambridge University Press: Cambridge, UK, 2009.

65. Broutman, D.; Eckermann, S.D.; Drob, D.P. The Partial Reflection of Tsunami-Generated Gravity Waves. *J. Atmos. Sci.* **2014**, *71*, 3416–3426.

66. Hickey, M.P. Atmospheric gravity waves and effects in the upper atmosphere associated with tsunamis. In *The Tsunami Threat—Research and Technology*; M.ärner, N.-A., Ed.; In Tech: Rijeka, Croatia; Shanghai, China, 2011; pp. 667–690.

67. Weinstock, J. Vertical turbulent diffusion in a stably stratified fluid. *J. Atmos. Sci.* **1978**, *35*, 1022–1027.

68. Roper, R.G.; Brosnahan, J.W. Imaging Doppler interferometry and the measurement of atmospheric turbulence. *Radio Sci.* **1997**, *32*, 1137–1148.

69. Hickey, M.P.; Schubert, G.; Walterscheid, R.L. Propagation of tsunami-driven gravity waves into the thermosphere and ionosphere. *J. Geophys. Res.* **2009**, *114*, A08304.

70. Li, F.; Liu, A.Z.; Swenson G.R. Characteristics of instabilities in the mesopause region over Maui, Hawaii. *J. Geophys. Res.* **2005**, *110*, D09S12.

71. Li, F.; Liu, A.Z.; Swenson G.R.; Hecht, J.H.; Robinson W.A. Observations of gravity wave breakdown into ripples associated with dynamical instabilities. *J. Geophys. Res.* **2005**, *110*, D09S11.

72. Godin, O.A.; Zabotin, N.A.; Bullett, T.W. Acoustic-gravity waves in the atmosphere generated by infragravity waves in the ocean. *Earth Planets Space* **2015**, *67*, 47.

73. European Space Agency. Space Environment Information System 2015. Available online: https://www.spenvis.oma.be/help/background /atmosphere/models.html (accessed on 25 December 2015).

MDPI AG

Klybeckstrasse 64

4057 Basel, Switzerland

Tel. +41 61 683 77 34

Fax +41 61 302 89 18

http://www.mdpi.com/

JMSE Editorial Office

E-mail: jmse@mdpi.com

http://www.mdpi.com/journal/jmse